MATHEMATICS RESEARCH DEVELOPMENTS

WEIGHTED NORM INEQUALITIES FOR INTEGRAL TRANSFORMS WITH PRODUCT KERNALS

MATHEMATICS RESEARCH DEVELOPMENTS

Boundary Properties and Applications of the Differentiated Poisson Integral for Different Domains
Sergo Topuria
2009. ISBN: 978-1-60692-704-5

Quasi-Invariant and Pseudo-Differentiable Measures in Banach Spaces
Sergey Ludkovsky
2009. ISBN: 978-1-60692-734-2

Operator Splittings and their Applications
Istvan Farago and Agnes Havasiy
2009. ISBN: 978-1-60741-776-7

Measure of Non-Compactness for Integral Operators in Weighted Lebesgue Spaces
Alexander Meskhi
2009. ISBN: 978-1-60692-886-8

Mathematics and Mathematical Logic: New Research
Peter Milosav and Irene Ercegovaca (Editors)
2009. ISBN: 978-1-60692-862-2

Role of Nonlinear Dynamics in Endocrine Feedback
Chinmoy K. Bose
2009. ISBN: 978-1-60741-948-8

Geometric Properties and Problems of Thick Knots
Yuanan Diao and Claus Ernst
2009. ISBN: 978-1-60741-070-6

Lie Groups: New Research
Altos B. Canterra (Editors)
2009. ISBN: 978-1-60692-389-4

Emerging Topics on Differential Geometry and Graph Theory
Lucas Bernard and Francois Roux (Editors)
2009. ISBN: 978-1-60741-011-9

Weighted Norm Inequalities for Integral Transforms with Product Kernals
Vakhtang Kokilashvili , Alexander Meskhi and Lars-Erik Persson
2010. ISBN: 978-1-60741-591-6

MATHEMATICS RESEARCH DEVELOPMENTS

WEIGHTED NORM INEQUALITIES FOR INTEGRAL TRANSFORMS WITH PRODUCT KERNALS

VAKHTANG KOKILASHVILI
ALEXANDER MESKHI
AND
LARS-ERIK PERSSON

Nova Science Publishers, Inc.
New York

For permission to use material from this book please contact us:
Telephone 631-231-7269; Fax 631-231-8175
Web Site: http://www.novapublishers.com

LIBRARY OF CONGRESS CATALOGING-IN-PUBLICATION DATA

Kokilashvili, V. M. (Vakhtang Mikhailovich)
Weighted norm inequalities for integral transforms with product kernals / Vakhtang Kokilashvili, Alexander Meskhi, and Lars-Erik Persson.
p. cm.
Includes bibliographical references and index.
ISBN 978-1-60741-591-6 (hardcover)
1. Inequalities (Mathematics) 2. Integral transforms. I. Meskhi, Alexander. II. Persson, Lars Erik, 1944- III. Title.
QA295.K755 2009
515'.723--dc22

2009021063

Published by Nova Science Publishers, Inc. † New York

Contents

Preface

The monograph is devoted to solutions of some challenging open problems in multidimensional fractional calculus and harmonic analysis. This book might be considered as a continuation of the monograph: D. E. Edmunds, V. Kokilashvili and A. Meskhi, *Bounded and compact operators*, Vol. 543, Kluwer Academic Publishers, Dordrecht, Boston, London, 2002, where complete two–weight theory has been developed for a wide class of fractional integrals with single kernel. Similar problems for integral transforms with product kernels is much more complicated and profound.

The authors succeeded in making the first essential steps in the investigation of two–weighted problems of integral transforms with product kernels, namely, for double Riemann-Liouville and Weyl transforms in the most difficult case, when the kernel has an essential (less than one) singularity. The results collected here are scattered throughout various journals and most of them appeared quite recently. However, a considerable part of the monograph contains unpublished material. We believe that a variety and profoundness of multidimensional analysis is reflected in this monograph. It should be emphasized that the authors' results, in fact, are pioneering in the directions mentioned above, and these problems were not studied previously.

The book may be considered as a systematic and detailed analysis of a wide class of integral transforms with product kernels from the point of view of the two–weighted boundedness. It is important that considered product kernels cover that case when the factors of the kernel have essential (less than one) singularities. Nowadays the construction of the two–weight theory for fractional integrals with single kernel and related fractional maximal functions might be considered as completed; however, attempts to derive two–weighted boundedness criteria for multiple integral transforms using the iteration process and corresponding inequalities for integral transforms with single kernel do not lead to the desired results. In fact, the class of weight pairs satisfying the condition guaranteeing the boundedness of a given multiple integral operator from one weighted Lebesgue space into another one for the sake of using one-dimensional two weight inequalities is narrower than the class of weights which completely characterizes the two–weight inequality for this operator. Taking into account this difference, there arises the necessity to elaborate new methods for studying the main problems of the proposed book. The book intends to make a breakthrough in two directions: to cover multidimensional potentials, Hilbert transforms, strong maximal functions and, at the same time, to present two–weight estimates for them which are much more complicated than those of one–weighted ones. On the basis of two–weighted inequalities for potentials and Hilbert transforms with product kernels we derive similar estimates for Fourier multipliers in Triebel–Lizorkin spaces.

One of the main objectives of this book is the Hardy-type transform. The contemporary study of the two–weight theory for one-dimensional Hardy–type transforms was developed in the following book: A. Kufner and L. E. Persson, *Weighted inequalities of Hardy-type.* World Scientific Publishing Co., Inc., River Edge, NJ, 2003.

The given monograph may be regarded as a continuation also of this book with special focus on some new multidimensional results, which complete and support the other parts of our presentation. In particular two–weighted boundedness criteria for multiple Hardy and its limit Pòlya–Knopp transforms are reflected in the case when the two–dimensional weight on one of the sides of the appropriate inequality is of product type, i.e., a product of two weight functions of single variables. In this case we present more simple and transparent criteria than those of E. Sawyer for general weights.

The book contains a complete solution of the trace problem for strong maximal functions. We solve similar problem for potentials with product kernels under some additional natural assumptions for pairs of weights. The results for one–sided potentials with product kernels are applied to establish sufficient conditions on a weight pair governing the two–weight Strichartz estimate for the solution of semilinear wave equation.

A considerable part of this monograph is devoted to two–weighted norm inequalities for the Cesàro and Abel's summability means of multiple Fourier series and Fourier integrals. Both sides of the derived inequalities have different weights and exponents of spaces. These inequalities are established due to two–weighted norm estimates for the multidimensional Steklov average operator, which is also studied in the book. On the base of the above-mentioned results we present norm summability theorems for the two–dimensional Fourier series and an extension of the well-known Bernstein inequality for derivatives of trigonometric polynomials and integral functions of finite degree of several variables in the two weighted setting.

A characteristic feature of the monograph is that most of the statements proved there have the form of criteria. These criteria enable us, for example, to give various explicit examples of pairs of weighted Banach function spaces, governing the boundedness of a wide class of multiple integral transforms.

We provide a list of problems which were open at the time of completion of the book. We hope that this list will stimulate further research.

A few words about the organization of the book are necessary. The enumeration of theorems, lemmas, formulas etc. follows the natural three-digit system, where a consecutive number within the section is used after the chapter and the section. There are three categories of numbering: theorems, lemmas, propositions, definitions and remarks and then come the formulae.

The bibliography contains 177 titles; we have collected the most relevant ones related to the topics presented here and it is by no means complete, our concern was to provide a basic orientation rather than a database.

The book is aimed for a wide audience, ranging from researches in multidimensional fractional calculus and harmonic analysis to experts in Applied Mathematics and postgraduate students. In particular, we hope that this book will serve as a source of inspiration for researchers and students working with inequalities of various kinds of applications (e.g., depending on the fact that it contains many challenging open questions, new information and surprising results). The text of the book is self–contained. This book can be read without looking frequently into other sources, however some knowledge of elements of function spaces, fractional calculus and harmonic analysis is needed; therefore the book can be used for teaching purposes, too.

The main subjects of this book can be useful for applications both within various areas of the mathematical sciences (e.g. Fourier and Harmonic analysis, Fractional Calculus, BVP of PDE in Mathematical Physics, Stochastic Processes, Error Estimates in Numerical Analysis, etc.) as well as directly in some applied sciences (e.g. Signal processing, Image Analysis, Homogenization Theory in material sciences, etc.).

The authors
Tbilisi, Luleå, 2009

Acknowledgment

We wish to express our sincere gratitude to Dr. Tsira Tsanava from A. Razmazde Mathematical Institute for the energy and patience she devoted to the production of the TEX master of the book.

We thank Dr. Maria Johansson at Luleå University of Technology and Dr. Sorina Barza at University of Karlstad for many constructive suggestions and ideas concerning Chapter 1 and Chapter 2, respectively. We thank also Dr. Elena Ushakova at Uppsala University for her careful proofreading of parts of these chapters.

The first two authors are gratefully acknowledged the support of Grant GNSF/ST07/3-169.

Basic Notation

$a \ll b$: there is a positive constant c, which might be depended only on exponents of function spaces, dimensions of Euclidean spaces and parameters of kernels of operators, such that $a \leq cb$;
$a \approx b$: $a \ll b$ and $b \ll a$;
$a := b$ (resp. $a =: b$): b is denoted by a (resp. a is denoted by b);
$\mathbb{R}^n$: n-dimensional Euclidean space;
$\mathbb{R}_+ := [0, \infty)$;
$\mathbb{R}^n_+ := \{x = (x_1, \ldots, x_n) \in \mathbb{R}^n : x_j \geq 0, \ j = 1, \ldots, n\}$;
$\mathbb{R} := \mathbb{R}^1$;
$\mathbb{R}^{1+n}_+ : \mathbb{R}_+ \times \mathbb{R}^n$;
$B(x, r)$: open ball with center x and radius r;
$\bar{B}(x, r)$: closed ball with center x and radius r;
$I(a, r) = (a - r, a + r)$;
$I^- := \big(2a - b, a\big)$, where I is an interval of the form (a, b);
$I^+ := \big(b, 2b - a\big)$, where I is an interval of the form (a, b);
$\mathbb{Z}$: the set of all integers;
$\mathbb{Z}_+$: the set of all non-negative integers;
$\mathbb{Z}_-$: the set of all non-positive integers;
$\mathbb{N}$: the set of all natural numbers;
$w(E) := \int_E w(x)dx$, where w is a weight function;
$u_x(E) := \int_E u(x, y)dy$ and $u_y(E) := \int_E u(x, y)dx$, where u is a two–dimensional weight;
$\rho_{x_1, \ldots, x_{i-1}, x_{i+1}, \ldots, x_n}(E) := \int\limits_E \rho(x_1, \ldots, x_i, \ldots, x_n)dx_i$;
$|E|$: the Lebesgue measure of a set $E \subset \mathbb{R}^n$;
χ_E: the characteristic function of a set E;
$p' := \frac{p}{p-1}$, where p is a constant with $1 < p < \infty$;
L^p: the Lebesgue space with exponent p $(1 \leq p < \infty)$;
L^p_ρ: the weighted Lebesgue space with weight ρ;
L_{loc}: the class of locally integrable functions;
$\|f\|_{L^p_\rho}$ (or $\|f\|_{p,\rho}$): the norm of a function f in the space L^p_ρ;
$\mathbb{T} := [-\pi, \pi]$;
$\square$: end of the proof.

Chapter 1

Hardy and Pólya–Knopp Inequalities

In this Chapter we complete the classical two–dimensional Hardy-type inequality by E. Sawyer (see Theorem 1.1 below) in various ways. In this case three different (independent) conditions are used to characterize the Hardy inequality for the case $1 < p \leq q < \infty$. In Section 1.1 we state and prove various two–weighted criteria in terms of just one condition for the double Hardy transform in the case when the weight on the right–hand side is a product of two weights of single variable. In Section 1.2, as a limiting case (when $p \to \infty$), we get a characterization of the corresponding two–dimensional Pólya-Knopp inequality. We include also a second proof of independent interest. In Section 1.3 we show that n–dimensional Hardy-type inequalities in the case $1 < p \leq q < \infty$ can be characterized by using just one condition both when the right–hand side weight is of product type and also when the left–hand side is. Some similar results for the case $1 < q < p < \infty$ are proved in Section 1.4 but with some additional restrictions on the weights. In Section 1.5 we prove some corresponding n-dimensional limit Pólya-Knopp type inequalities. Section 1.6 is devoted to the criteria guaranteeing the two–weight inequality for the Riemann–Liouville operator with the parameter more than one. In Section 1.7 we present new refined multidimensional Hardy and Pólya-Knopp type inequalities. In Section 1.8 we prove some multidimensional weighted Hardy and Pólya-Knopp type inequalities. In particular, in this section we discuss some refined multidimensional Hardy-type inequalities via superquadracity and Jensen's inequality.

In 1920 G. H. Hardy [66] announced and proved in [67] the following integral inequality (see also [68, Chapter 9])

$$\int_0^\infty \left(\frac{1}{x} \int_0^x f(t)dt \right)^p dx \leq \left(\frac{p}{p-1} \right)^p \int_0^\infty f^p(x)dx, \tag{1}$$

where $p > 1$ and $f \in L^p(0,\infty)$ is a non–negative and measurable function. Moreover, the following related inequality:

$$\int_0^\infty \exp\left(\frac{1}{x}\int_0^x \ln f(t)dt\right)dx \le e\int_0^\infty f(x)dx, \tag{2}$$

may be regarded as a limit inequality of (1) as $p \to \infty$. This limit argument is as follows: if in (1) f is replaced by $f^{\frac{1}{p}}$ and if we let $p \to \infty$, then we obtain (2).

Inequality (1) is the classical Hardy inequality while inequality (2) is sometimes referred to as Knopp's inequality with reference to [82]. However, inequality (2) was certainly known before the appearance of Knopp's result in [82]. Hardy acknowledged this fact (see [67, p. 156]) when he pointed out that it was G. Pólya who made him aware of the elegant limiting argument that directly implied inequality (2). In this regard, inequality (2) will be referred to as Pólya-Knopp's inequality (see also [100], [101] and [102]).

1.1 A Two–dimensional Hardy–type Inequality

The following remarkable result was proved by E. T. Sawyer in [157, Theorem 1].

Theorem 1.1.1. *Let $1 < p \le q < \infty$ and let u and v be weight functions on $\mathbb{R}^2_+$. Then*

$$\left(\int_0^\infty\int_0^\infty\left(\int_0^{x_1}\int_0^{x_2} f(t_1,t_2)dt_1dt_2\right)^q v(x_1,x_2)dx_1dx_2\right)^{\frac{1}{q}}$$
$$\le C\left(\int_0^\infty\int_0^\infty f\left(x_1,x_2\right)^p w(x_1,x_2)dx_1dx_2\right)^{\frac{1}{p}} \tag{1.1.1}$$

holds for all positive and measurable functions f on $\mathbb{R}^2_+$ if and only if

$$\sup_{y_1,y_2>0}\left(\int_{y_1}^\infty\int_{y_2}^\infty v(x_1,x_2)dx_1dx_2\right)^{\frac{1}{q}}$$
$$\times\left(\int_0^{y_1}\int_0^{y_2} w(x_1,x_2)^{1-p'}dx_1dx_2\right)^{\frac{1}{p'}} := A_1 < \infty, \qquad ((1.1.2))$$

$$\sup_{y_1,y_2>0}\frac{\left(\int_0^{y_1}\int_0^{y_2}\left(\int_0^{x_1}\int_0^{x_2} w(t_1,t_2)^{1-p'}dt_1dt_2\right)^q v(x_1,x_2)dx_1dx_2\right)^{\frac{1}{q}}}{\left(\int_0^{y_1}\int_0^{y_2} w(x_1,x_2)^{1-p'}dx_1dx_2\right)^{\frac{1}{p}}} := A_2 < \infty, \tag{1.1.3}$$

and

$$\sup_{y_1,y_2>0} \frac{\left(\int\limits_{y_1}^{\infty}\int\limits_{y_2}^{\infty}\left(\int\limits_{x_1}^{\infty}\int\limits_{x_2}^{\infty} v\left(t_1,t_2\right)dt_1dt_2\right)^{p'} w(x_1,x_2)^{1-p'}dx_1dx_2\right)^{\frac{1}{p'}}}{\left(\int\limits_{y_1}^{\infty}\int\limits_{y_2}^{\infty} v(x_1,x_2)dx_1dx_2\right)^{\frac{1}{q'}}} = A_3 < \infty. \quad (1.1.4)$$

Note that in this two–dimensional case all three conditions (1.1.2)–(1.1.4) are needed in order to characterize the weighted Hardy inequality. However, in our next Theorem we shall point out the fact that in one dimension the situation is simpler. More exactly, in this case the conditions corresponding to (1.1.2)–(1.1.4) are in fact equivalent and each of them can be used to characterize the corresponding one-dimensional Hardy inequality (see also [23] and [163]).

Theorem 1.1.2. *Let $1 < p \le q < \infty$, and let u and v be weight functions on $\mathbb{R}_+$. Then each of the following conditions are necessary and sufficient for the inequality*

$$\left(\int\limits_0^\infty\left(\int\limits_0^x f(t)dt\right)^q v(x)dx\right)^{\frac{1}{q}} \le C\left(\int\limits_0^\infty f^p(x)w(x)dx\right)^{\frac{1}{p}} \quad (1.1.5)$$

to hold for all positive and measurable functions on $\mathbb{R}_+$.

a) *The Muckenhoupt condition,*

$$A_M := \sup_{x>0}\left(\int\limits_x^\infty v(t)dt\right)^{\frac{1}{q}}\left(\int\limits_0^x w(t)^{1-p'}dt\right)^{\frac{1}{p'}} < \infty. \quad (1.1.6)$$

Moreover, the best constant C in (1.1.5) can be estimated as follows:

$$A_M \le C \le \left(1+\frac{q}{p'}\right)^{\frac{1}{q}}\left(1+\frac{p'}{q}\right)^{\frac{1}{p'}} A_M.$$

b) *The condition of L. E Persson and V. D. Stepanov,*

$$A_{PS} := \sup_{x>0} W(x)^{-\frac{1}{p}}\left(\int\limits_0^x v(t)W(t)^q dt\right)^{\frac{1}{q}} < \infty, \quad W(x) := \int\limits_0^x w(t)^{1-p'}dt. \quad (1.1.7)$$

Moreover, the best constant C in (1.1.5) satisfies the following estimates:

$$A_{PS} \le C \le p' A_{PS}.$$

c) *The condition (c.f.* (1.1.4)),

$$A^*_{PS} := \sup_{x>0} \left(\int_x^\infty \left(\int_t^\infty v(s)ds \right)^{p'} w^{1-p'}(t)dt \right)^{\frac{1}{p'}} \left(\int_x^\infty v(t)dt \right)^{-\frac{1}{q'}} < \infty. \qquad (1.1.8)$$

Moreover, the best constant C in (1.1.5) *can be estimated as follows:*

$$A^*_{PS} \le C \le q A^*_{PS}. \qquad (1.1.9)$$

Remark 1.1.1. A direct proof of Theorem 1.1.2 can be found in [174]. For the original proofs of a) and b) see [123] and [142], respectively, and (1.1.8) is just the dual condition of (1.1.7). A much more general result where (1.1.6)–(1.1.8) is replaced by 4 different scales of conditions is proved in [60]. For more results, historical remarks and references in the one dimensional case see the recent book [100] and historical article [101].

Moreover, it was recently discovered by A. Wedestig in her Ph.D. thesis [174] (see also [175]) that if the weight on the right–hand side is of product type, then, in fact, (1.1.1) can be characterized by just one condition (or, more generally, just one of infinite possible conditions). In particular, the following statement holds:

Theorem 1.1.3. *Let $1 < p \le q < \infty$, $s_1, s_2 \in (1, p)$ and let u be a weight functions on $\mathbb{R}^2_+$ and let w_1 and w_2 be weight functions on $\mathbb{R}_+$. Then the inequality*

$$\left(\int_0^\infty \int_0^\infty \left(\int_0^{x_1} \int_0^{x_2} f(t_1, t_2) dt_1 dt_2 \right)^q v(x_1, x_2) dx_1 dx_2 \right)^{\frac{1}{q}}$$

$$\le C \left(\int_0^\infty \int_0^\infty f^p(x_1, x_2) w_1(x_1) w_2(x_2)\, dx_1 dx_2 \right)^{\frac{1}{p}} \qquad (1.1.10)$$

holds for all measurable functions $f \ge 0$ if

$$A_W(s_1, s_2) := \sup_{t_1, t_2 > 0} W_1(t_1)^{\frac{s_1-1}{p}} W_2(t_2)^{\frac{s_2-1}{p}}$$

$$\times \left(\int_{t_1}^\infty \int_{t_2}^\infty v(x_1,x_2) W_1(x_1)^{q\left(\frac{p-s_1}{p}\right)} W_2(x_2)^{q\left(\frac{p-s_2}{p}\right)} dx_1 dx_2 \right)^{\frac{1}{q}} < \infty, \qquad (1.1.11)$$

where $W_1(t_1) := \int_0^{t_1} w_1(x_1)^{1-p'} dx_1$ and $W_2(t_2) := \int_0^{t_2} W_2(x_2)^{1-p'} dx_2$.

Moreover, if C is the best possible constant in (1.1.10), *then*

$$\sup_{1<s_1,s_2<p}\left(\frac{\left(\frac{p}{p-s_1}\right)^p}{\left(\frac{p}{p-s_1}\right)^p+\frac{1}{s_1-1}}\right)^{\frac{1}{p}}\left(\frac{\left(\frac{p}{p-s_2}\right)^p}{\left(\frac{p}{p-s_2}\right)^p+\frac{1}{s_2-1}}\right)^{\frac{1}{p}} A_W(s_1,s_2)\le C$$

$$\le \sup_{1<s_1,s_2<p} A_W(s_1,s_2)\left(\frac{p-1}{p-s_1}\right)^{\frac{1}{p'}}\left(\frac{p-1}{p-s_2}\right)^{\frac{1}{p'}}. \tag{1.1.12}$$

Proof. Let $f^p(x_1,x_2)w_1(x)w_2(x_2)=g(x_1,x_2)$ in (1.1.10). Then (1.1.10) is equivalent to the inequality

$$\left(\int_0^\infty\int_0^\infty\left(\int_0^{x_1}\int_0^{x_1} g(t_1,t_2)^{\frac{1}{p}}w_1(t_1)^{-\frac{1}{p}}w_2(t_2)^{-\frac{1}{p}}dt_1dt_2\right)^q v(x_1,x_2)dx_1dx_2\right)^{\frac{1}{q}}$$

$$\le C\left(\int_0^\infty\int_0^\infty g(x_1,x_2)dx_1dx_2\right)^{\frac{1}{p}}. \tag{1.1.13}$$

Assume that (1.1.11) holds. By applying Hölder's inequality, the fact that $\frac{d}{dt_1}W_1(t_1)=w_1(t_1)^{1-p'}=w_1(t_1)^{-\frac{p'}{p}}$, $\frac{d}{dt_2}W_2(t_2)=w_2(t_2)^{1-p'}=w_2(t_2)^{-\frac{p'}{p}}$ and Minkowski's inequality we have

$$\left(\int_0^\infty\int_0^\infty\left(\int_0^{x_1}\int_0^{x_2} g(t_1,t_2)^{\frac{1}{p}}w_1(t_1)^{-\frac{1}{p}}w_2(t_2)^{-\frac{1}{p}}dt_1dt_2\right)^q v(x_1,x_2)dx_1dx_2\right)^{\frac{1}{q}}$$

$$= \left(\int_0^\infty\int_0^\infty\left(\int_0^{x_1}\int_0^{x_2} g(t_1,t_2)^{\frac{1}{p}}W_1(t_1)^{\frac{s_1-1}{p}}W_2(t_2)^{\frac{s_2-1}{p}}\right.\right.$$

$$\left.\left.\times W_1(t_1)^{-\frac{s_1-1}{p}}w_1(t_1)^{-\frac{1}{p}}W_2(t_2)^{-\frac{s_2-1}{p}}w_2(t_2)^{-\frac{1}{p}}dt_1dt_2\right)^q v(x_1,x_2)dx_1dx_2\right)^{\frac{1}{q}}$$

$$\le \left(\int_0^\infty\int_0^\infty\left(\int_0^{x_1}\int_0^{x2} g(t_1,t_2)W_1(t_1)^{s_1-1}W_2(t_2)^{s_2-1}dt_1dt_2\right)^{\frac{q}{p}}\right.$$

$$\times\left(\int_0^{x_1} W_1(t_1)^{-\frac{(s_1-1)p'}{p}}w_1(t_1)^{-\frac{p'}{p}}dt_1\right)^{\frac{q}{p'}}$$

$$\left.\times\left(\int_0^{x_2} W_2(t_2)^{-\frac{(s_2-1)p'}{p}}w_2(t_2)^{-\frac{p'}{p}}dt_2\right)^{\frac{q}{p'}} v(x_1,x_2)dx_1dx_2\right)^{\frac{1}{q}}$$

$$= \left(\frac{p-1}{p-s_1}\right)^{\frac{1}{p'}} \left(\frac{p-1}{p-s_2}\right)^{\frac{1}{p'}} \left(\int_0^\infty \int_0^\infty \left(\int_0^{x_1} \int_0^{x_2} g(t_1,t_2) W_1(t_1)^{s_1-1} \right.\right.$$

$$\left.\left. \times W_2(t_2)^{s_2-1} dt_1 dt_2 \right)^{\frac{q}{p}} W_1(x_1)^{q\left(\frac{p-s_1}{p}\right)} W_2(x_2)^{q\left(\frac{p-s_2}{p}\right)} v(x_1,x_2) dx_1 dx_2 \right)^{\frac{1}{q}}$$

$$\leq \left(\frac{p-1}{p-s_1}\right)^{\frac{1}{p'}} \left(\frac{p-1}{p-s_2}\right)^{\frac{1}{p'}} \left(\int_0^\infty \int_0^\infty g(t_1,t_2) W_1(t_1)^{s_1-1} W_2(t_2)^{s_2-1} \right.$$

$$\left. \times \left(\int_{t_1}^\infty \int_{t_2}^\infty W_1(x_1)^{q\left(\frac{p-s_1}{p}\right)} W_2(x_2)^{q\left(\frac{p-s_2}{p}\right)} v(x_1,x_2) dx_1 dx_2 \right)^{\frac{p}{q}} dt_1 dt_2 \right)^{\frac{1}{p}}$$

$$\leq \left(\frac{p-1}{p-s_1}\right)^{\frac{1}{p'}} \left(\frac{p-1}{p-s_2}\right)^{\frac{1}{p'}} A_W(s_1,s_2) \left(\int_0^\infty \int_0^\infty g(t_1,t_2) dt_1 dt_2 \right)^{\frac{1}{p}}.$$

Hence (1.1.13) and, thus, (1.1.10) holds with a constant satisfying the right–hand side inequality in (1.1.12).

Now we assume that (1.1.10) and, thus, (1.1.13) holds and choose the test function

$$g(x_1,x_2) = \left(\frac{p}{p-s_1}\right)^p \left(\frac{p}{p-s_2}\right)^p$$

$$\times W_1(t_1)^{-s_1} w_1(x_1)^{1-p'} W_2(t_2)^{-s_2} w_2(x_2)^{1-p'} \chi_{(0,t_1)}(x_1) \chi_{(0,t_2)}(x_2)$$

$$+ \left(\frac{p}{p-s_1}\right)^p W_1(t_1)^{-s_1} w_1(x_1)^{1-p'} W_2(x_2)^{-s_2} w_2(x_2)^{1-p'} \chi_{(0,t_1)}(x_1) \chi_{(t_2,\infty)}(x_2)$$

$$+ \left(\frac{p}{p-s_2}\right)^p W_1(x_1)^{-s_1} w_1(x_1)^{1-p'} W_2(t_2)^{-s_2} w_2(x_2)^{1-p'} \chi_{(t_1,\infty)}(x_1) \chi_{(0,t_2)}(x_2)$$

$$+ W_1(x_1)^{-s_1} w_1(x_1)^{1-p'} W_2(x_2)^{-s_2} w_2(x_2)^{1-p'} \chi_{(t_1,\infty)}(x_1) \chi_{(t_2,\infty)}(x_2),$$

where t_1, t_2 are fixed numbers > 0. Then the integral on the right–hand side of (1.1.13) can be estimated as follows:

$$\left(\int_0^\infty \int_0^\infty g(x_1,x_2) dx_1 dx_2 \right)^{\frac{1}{p}}$$

$$= \left(\int_0^{t_1} \left(\frac{p}{p-s_1}\right)^p W_1(t_1)^{-s_1} w_1(x_1)^{1-p'} dx_1 \int_0^{t_2} \left(\frac{p}{p-s_2}\right)^p W_2(t_2)^{-s_2} w_2(x_2)^{1-p'} dx_2 \right.$$

$$+\int_0^{t_1}\left(\frac{p}{p-s_1}\right)^p W_1(t_1)^{-s_1}w_1(x_1)^{1-p'}dx_1\int_{t_2}^{\infty}W_2(x_2)^{-s_{21}}w_2(x_2)^{1-p'}dx_2$$

$$+\int_{t_1}^{\infty}W_1(x_1)^{-s_1}w_1(x_1)^{1-p'}dx_1\int_0^{t_2}\left(\frac{p}{p-s_2}\right)^p W_2(t_2)^{-s_2}w_2(x_2)^{1-p'}dx_2$$

$$+\int_{t_1}^{\infty}W_1(x_1)^{-s_1}w_1(x_1)^{1-p'}dx_1\int_{t_2}^{\infty}W_2(x_2)^{-s_2}w_2(x_2)^{1-p'}dx_2\Bigg)^{\frac{1}{p}}$$

$$\leq\ \left(\left(\frac{p}{p-s_1}\right)^p+\frac{1}{s_1-1}\right)^{\frac{1}{p}}\left(\left(\frac{p}{p-s_2}\right)^p+\frac{1}{s_2-1}\right)^{\frac{1}{p}}W_1(t_1)^{\frac{1-s_1}{p}}W_2(t_2)^{\frac{1-s_2}{p}}.$$

Moreover, the left–hand side of (1.1.13) is greater than

$$\Bigg(\int_{t_1}^{\infty}\int_{t_2}^{\infty}\Bigg[\Bigg(\int_0^{t_1}\frac{p}{p-s_1}W_1(t_1)^{-\frac{s_1}{p}}w_1(y_1)^{1-p'}dy_1\Bigg)$$

$$\times\Bigg(\int_0^{t_2}\frac{p}{p-s_2}W_2(t_2)^{-\frac{s_2}{p}}w_2(y_2)^{1-p'}dy_2\Bigg)$$

$$+\Bigg(\int_0^{t_1}\frac{p}{p-s_1}W_1(t_1)^{-\frac{s_1}{p}}w_1(y_1)^{1-p'}dy_1\Bigg)\Bigg(\int_{t_2}^{x_2}W_2(y_1)^{-\frac{s_2}{p}}w_2(y_2)^{1-p'}dy_2\Bigg)$$

$$+\Bigg(\int_{t_1}^{x_1}W_1(y_1)^{-\frac{s_1}{p}}w_1(y_1)^{1-p'}dy_1\Bigg)\Bigg(\int_0^{t_2}\frac{p}{p-s_2}W_2(t_2)^{-\frac{s_2}{p}}w_2(y_2)^{1-p'}dy_2\Bigg)$$

$$+\Bigg(\int_{t_1}^{x_1}W_1(y_1)^{-\frac{s_1}{p}}w_1(y_1)^{1-p'}dy_1\Bigg)\Bigg(\int_{t_2}^{x_2}W_2(y_1)^{-\frac{s_2}{p}}w_2(y_2)^{1-p'}dy_2\Bigg)\Bigg]^q$$

$$\times v(x_1,x_2)dx_1dx_2\Bigg)^{\frac{1}{q}}$$

$$=\cdots=\frac{p}{p-s_1}\frac{p}{p-s_2}\left(\int_{t_1}^{\infty}\int_{t_2}^{\infty}v(x_1,x_2)W_1(x_1)^{q\left(\frac{p-s_1}{p}\right)}W_2(x_2)^{q\left(\frac{p-s_2}{p}\right)}dx_1dx_2\right)^{\frac{1}{q}}$$

[by using (1.1.13)]

$$= \frac{p}{p-s_1}\frac{p}{p-s_2}\Big(\int_{t_1}^{\infty}\int_{t_2}^{\infty} v(x_1,x_2)W_1(x_1)^{q\left(\frac{p-s_1}{p}\right)}W_2(x_2)^{q\left(\frac{p-s_2}{p}\right)}dx_1dx_2\Big)^{\frac{1}{q}}$$

$$\leq C\Big(\Big(\frac{p}{p-s_1}\Big)^p+\frac{1}{s_1-1}\Big)^{\frac{1}{p}}\Big(\Big(\frac{p}{p-s_2}\Big)^p+\frac{1}{s_2-1}\Big)^{\frac{1}{p}}W_1(t_1)^{\frac{1-s_1}{p}}W_2(t_2)^{\frac{1-s_2}{p}},$$

i.e., that

$$\left(\frac{\left(\frac{p}{p-s_1}\right)^p}{\left(\frac{p}{p-s_1}\right)^p+\frac{1}{s_1-1}}\right)^{\frac{1}{p}}\left(\frac{\left(\frac{p}{p-s_2}\right)^p}{\left(\frac{p}{p-s_2}\right)^p+\frac{1}{s_2-1}}\right)^{\frac{1}{p}}W_1(t_1)^{\frac{s_1-1}{p}}W_2(t_2)^{\frac{s_2-1}{p}}$$

$$\times\left(\int_{t_1}^{\infty}\int_{t_2}^{\infty} v(x_1,x_2)W_1(x_1)^{q\left(\frac{p-s_1}{p}\right)}W_2(x_2)^{q\left(\frac{p-s_2}{p}\right)}dx_1dx_2\right)^{\frac{1}{q}}\leq C.$$

We conclude that (1.1.11) and the left–hand side of the estimate of (1.1.12) hold. The proof is complete. □

Now we are going to show that if the weight function w is of product type, i.e., it has the form $w(x,y)=w_1(x)w_2(y)$, then inequality (1.1.1) holds if and only if the Muckenhoupt–type condition (1.1.2) is satisfied.

To formulate the next result we recall some well-known weighted criteria for the discrete Hardy inequality (see, e.g., [20], [48], [100]). We give the proof for completeness.

Theorem 1.1.4. *Let $1<p\leq q<\infty$ and let $\{a_n\}$, $\{b_n\}$ be positive sequences. The inequality*

$$\Big(\sum_{n=-\infty}^{\infty}\Big|\sum_{k=-\infty}^{n} g_k\Big|^q a_n^q\Big)^{1/q}\leq c\Big(\sum_{n=-\infty}^{\infty}|g_n|^p b_n^p\Big)^{1/p} \tag{1.1.14}$$

with the positive constant c independent of $\{g_k\}$, holds if and only if

$$B:=\sup_{n\in\mathbb{Z}}\Big(\sum_{k=n}^{\infty}a_k^q\Big)^{1/q}\Big(\sum_{k=-\infty}^{n}b_k^{-p'}\Big)^{1/p'}<\infty.$$

Moreover, if c is the best constant in (1.1.14), then

$$B\leq c\leq Bq^{\frac{1}{q}}\Big(\frac{q}{q-1}\Big)^{(p-1)/p}.$$

Proof. *Sufficiency.* Let $\alpha_n := \big(\sum_{k=-\infty}^{n} b_k^{-p'}\big)^{\frac{1}{qp'}}$. Due to Hölder's inequality we have

$$\Big(\sum_{n=-\infty}^{\infty} a_n^q \Big|\sum_{k=-\infty}^{n} f_k\Big|^q\Big)^{p/q}$$

$$\leq \Big(\sum_{n=-\infty}^{\infty} a_n^q \Big(\sum_{k=-\infty}^{n} |f_k\alpha_k b_k|^p\Big)^{q/p}\Big(\sum_{k=-\infty}^{n} (\alpha_k b_k)^{-p'}\Big)^{\frac{q}{p'}}\Big)^{p/q} := \Big(\sum_{n=-\infty}^{\infty} A_n\Big)^{\frac{p}{q}}.$$

By Minkowsky's inequality ($\frac{q}{p} \geq 1$) we obtain

$$\Big(\sum_{n=-\infty}^{\infty} A_n\Big)^{p/q} = \Big(\sum_{n=-\infty}^{\infty}\Big(\sum_{k=-\infty}^{n} |f_k\alpha_k b_k|^p a_n^p \Big(\sum_{k=-\infty}^{n} (\alpha_k b_k)^{-p'}\Big)^{\frac{p}{p'}}\Big)^{\frac{q}{p}}\Big)^{\frac{p}{q}}$$

$$\leq \sum_{k=-\infty}^{\infty} |f_k\alpha_k b_k|^p \Big(\sum_{n=k}^{\infty} a_n^q \Big(\sum_{k=-\infty}^{n} (\alpha_k b_k)^{-p'}\Big)^{\frac{q}{p'}}\Big)^{\frac{p}{q}} := \sum_k D_k.$$

For the intrinsic sum we have

$$\sum_{n=k}^{\infty} a_n^q \Big(\sum_{k=-\infty}^{n} (\alpha_k b_k)^{-p'}\Big)^{\frac{q}{p'}} = \sum_{n=k}^{\infty} a_n^q \Big(\sum_{k=-\infty}^{n}\Big(\sum_{i=-\infty}^{k} b_k^{-p'}\Big)^{-\frac{1}{q}} b_k^{-p'}\Big)^{\frac{q}{p'}}.$$

Moreover, the next easily verifiable inequality

$$\sum_{k=-\infty}^{n} b_k^{-p'}\Big(\sum_{i=0}^{k} b_i^{-p'}\Big)^{-\frac{1}{q}} \leq q'\Big(\sum_{k=-\infty}^{n} b_k^{-p'}\Big)^{1/q'}$$

gives

$$\sum_{n=k}^{\infty} a_n^q \Big(\sum_{i=-\infty}^{n} (\alpha_i b_i)^{-p'}\Big)^{\frac{q}{p'}} = (q')^{\frac{q}{p'}} \sum_{n=k}^{\infty} a_n^q \Big(\sum_{i=-\infty}^{n} b_i^{-p'}\Big)^{\frac{q}{q'p'}}.$$

Further, the latter sum does not exceed

$$\lambda := B^{\frac{q}{q'}} (q')^{\frac{q}{p'}} \sum_{n=k}^{\infty} a_n^q \Big(\sum_{i=n}^{\infty} a_i^q\Big)^{-\frac{1}{q'}}.$$

The inequality

$$\sum_{n=k}^{\infty} a_n^q \Big(\sum_{i=n}^{\infty} a_i^q\Big)^{-\frac{1}{q'}} \leq q\Big(\sum_{i=k}^{\infty} a_i^q\Big)^{1/q},$$

can also be easily verified. Therefore we obtain

$$\lambda \leq qB^{\frac{q}{q'}} (q')^{\frac{q}{p'}} \Big(\sum_{i=k}^{\infty} a_i^q\Big)^{1/q} \leq qB \cdot B^{\frac{q}{q'}} (q')^{\frac{q}{p'}} \Big(\sum_{i=0}^{k} b_i^{-p'}\Big)^{-\frac{1}{p'}} = qB^q (q')^{\frac{q}{p'}} \alpha_k^{-q}.$$

Summarazing the estimates derived above we have that

$$\sum_{k\in Z} D_k \le q^{\frac{p}{q}} B^p (q')^{\frac{p}{p'}} \sum_{k=-\infty}^{\infty} |f_k \alpha_k b_k|^p \alpha_k^{-p} = q^{\frac{p}{q}} B^p (q')^{\frac{p}{p'}} \sum_{k=-\infty}^{\infty} |f_k|^p b_k^p.$$

From this we conclude that (1.1.14) holds with the claimed right–hand side estimate of the constant c.

In order to prove necessity we take the sequence

$$g_k = \begin{cases} b_k^{-p'}, & k \le n, \\ 0, & k > n. \end{cases}$$

Then we have

$$\Big(\sum_{i\in\mathbb{Z}} \Big(\sum_{k=-\infty}^{i} g_k\Big)^q a_i^q\Big)^{1/q} \ge \Big(\sum_{i=n}^{\infty} a_i^q\Big)^{1/q} \Big(\sum_{k=-\infty}^{n} b_k^{-p'}\Big).$$

On the other hand,

$$\Big(\sum_{k=-\infty}^{\infty} |g_k|^p b_k^p\Big)^{1/p} = \Big(\sum_{k=-\infty}^{n} b_k^{-p'}\Big)^{1/p}.$$

Finally, we find that

$$B < \infty$$

and that the left–hand side estimate of the constant c holds. □

Analogously, it follows the next statement.

Theorem 1.1.5. *Let $1 < p \le q < \infty$ and let m be an integer. Suppose that $\{a_n\}_{n=-\infty}^{m}$, $\{b_n\}_{n=-\infty}^{m}$ are positive sequences. Then the two–weight inequality*

$$\Big(\sum_{n=-\infty}^{m} \Big|\sum_{k=-\infty}^{n} g_k\Big|^q a_n^q\Big)^{1/q} \le c\Big(\sum_{n=-\infty}^{m} |g_n|^p b_n^p\Big)^{1/p} \tag{1.1.15}$$

holds if and only if

$$B_m := \sup_{-\infty<n\le m} \Big(\sum_{k=n}^{m} a_k^q\Big)^{1/q} \Big(\sum_{k=-\infty}^{n} b_k^{-p'}\Big)^{1/p'} < \infty.$$

Moreover, if c is the best constant in (1.1.15), then

$$B_m \le c \le B_m q^{\frac{1}{q}} \Big(\frac{q}{q-1}\Big)^{(p-1)/p}.$$

Let ρ be a measurable almost everywhere positive function on $\mathbb{R}^n_+$. We say that $f \in L^p_\rho(\mathbb{R}^n_+)$, $1 \le p < \infty$, if

$$\|f\|_{L^p_\rho(\mathbb{R}^n_+)} := \Big(\int\limits_{\mathbb{R}+} \cdots \int\limits_{\mathbb{R}+} |f(x_1,\dots,x_n)|^p \rho(x_1,\dots,x_n)dx_1\cdots dx_n\Big)^{1/p} < \infty.$$

The next statement due to A. Meskhi [120] (see also [121]) shows that if a two–dimensional weight w is a product of one-dimensional weights, then the double Hardy operator

$$H_2 f(x,y) = \int\limits_0^x \int\limits_0^y f(t,\tau)dtd\tau, \quad x,y \in \mathbb{R}_+,$$

is bounded from $L^p_w(\mathbb{R}^2_+)$ to $L^q_v(\mathbb{R}^2_+)$ if and only if the Muckenhoupt–type condition (1.1.2) holds.

Theorem 1.1.6. *Let* $1 < p \le q < \infty$ *and let* $w(x,y) = w_1(x)w_2(y)$. *Then inequality* (1.1.1) *holds if and only if condition* (1.1.2) *is fulfilled.*

Proof. First suppose that $S := \int_0^\infty w_2^{1-p'}(y)dy = \infty$. Let $\{x_k\}_{k=-\infty}^{+\infty}$ be a sequence of positive numbers for which the equality

$$2^k = \int\limits_0^{x_k} w_2^{1-p'}(y)dy \tag{1.1.16}$$

holds for all $k \in \mathbb{Z}$. It is clear that $\{x_k\}$ is increasing and $\mathbb{R}_+ = \cup_{k\in\mathbb{Z}}[x_k, x_{k+1})$. Moreover, it is easy to verify that

$$2^k = \int\limits_{x_k}^{x_{k+1}} w_2^{1-p'}(y)dy.$$

Let $f \ge 0$. We have that

$$\|f\|^q_{L^q_v(\mathbb{R}^2_+)} = \iint\limits_{\mathbb{R}^2_+} v(x,y)\big(H_2 f\big)^q(x,y)dxdy$$

$$\le \sum_{k\in\mathbb{Z}} \int\limits_0^\infty \int\limits_{x_k}^{x_{k+1}} v(x,y)\Big(\int\limits_0^x \int\limits_0^y f(t,\tau)dtd\tau\Big)^q dxdy$$

$$\le \sum_{k\in\mathbb{Z}} \int\limits_0^\infty \Big(\int\limits_{x_k}^{x_{k+1}} v(x,y)dy\Big)\Big(\int\limits_0^x \Big(\int\limits_0^{x_{k+1}} f(t,\tau)d\tau\Big)dt\Big)^q dx$$

$$= \sum_{k\in\mathbb{Z}} \int_0^\infty V_k(x)\Big(\int_0^x F_k(t)dt\Big)^q dx,$$

where

$$V_k(x) := \int_{x_k}^{x_{k+1}} v(x,y)dy; \;\; F_k(t) := \int_0^{x_{k+1}} f(t,\tau)d\tau.$$

It is obvious that

$$A_1^q \geq \sup_{\substack{a>0\\ j\in\mathbb{Z}}} \Big(\int_a^\infty \int_{x_j}^{x_{j+1}} v(x,y)dxdy\Big)\Big(\int_0^a\int_0^{x_j} w^{1-p'}(x,y)dxdy\Big)^{q/p'}.$$

Further,

$$\|f\|^q_{L^q_v(\mathbb{R}^n_+)} \leq cA_1^q \sum_{j\in\mathbb{Z}}\Big[\int_0^\infty w_1(x)\Big(\int_0^{x_j} w_2^{1-p'}(y)dy\Big)^{1-p}(F_k(x))^p dx\Big]^{q/p}$$

$$\leq cA_1^q\Big[\int_0^\infty w_1(x)\sum_{j\in Z}\Big(\int_0^{x_j} w_2^{1-p'}(y)dy\Big)^{1-p}$$

$$\times\Big(\sum_{k=-\infty}^{j}\int_{x_k}^{x_{k+1}} f(x,\tau)d\tau\Big)^p dx\Big]^{q/p}.$$

On the other hand, (1.1.16) yields

$$\sum_{k=n}^{+\infty}\Big(\int_0^{x_k} w_2^{1-p'}(y)dy\Big)^{1-p}\Big(\sum_{k=-\infty}^{n}\int_{x_k}^{x_{k+1}} w_2^{1-p'}(y)dy\Big)^{p-1}$$

$$= \sum_{k=n}^{+\infty}\Big(\int_0^{x_k} w_2^{1-p'}(y)dy\Big)^{1-p}\Big(\int_0^{x_{n+1}} w_2^{1-p'}(y)dy\Big)^{p-1} = \Big(\sum_{k=n}^{+\infty} 2^{k(1-p)}\Big)2^{(n+1)(p-1)} \leq c$$

for all $n\in\mathbb{Z}$. Hence by Theorem 1.1.4 and Hölder's inequality we have

$$\|f\|^q_{L^q_v(\mathbb{R}^n_+)} \leq cA_1^q\Big[\int_0^\infty w_1(x)\sum_{j\in\mathbb{Z}}\Big(\int_{x_j}^{x_{j+1}} w_2^{1-p'}(y)dy\Big)^{1-p}\Big(\int_{x_k}^{x_{k+1}} f(x,\tau)d\tau\Big)^p dx\Big]^{q/p}$$

$$\leq cA_1^q\Big[\int_0^\infty w_1(x)\sum_{j\in\mathbb{Z}}\Big(\int_{x_k}^{x_{k+1}} w_2(\tau)f^p(x,\tau)d\tau\Big)dx\Big]^{q/p} = cA_1^q\|f\|^q_{L^p(R^2_+)}.$$

If $S < \infty$, then without loss of generality we can assume that $S = 1$. In this case we choose the sequence $\{x_k\}_{k=-\infty}^{0}$ for which (1.1.16) holds for all $k \in \mathbb{Z}_-$. Arguing as in the case $S = \infty$ and using Theorem 1.1.5 instead of Theorem 1.1.4, we finally obtain the desired result. □

Remark 1.1.1. Note that Theorem 1.1.6 may be regarded as a natural endpoint result related to Theorem 1.1.3, namely if we were permitted to put $s_1 = s_2 = 1/p$ in Theorem 1.1.3. Theorem 1.1.6 holds also in an n-dimensional setting. See our Theorem 1.3.1, where another technique of proof (as well as the direct proof) is used.

1.2 The Two–dimensional Pólya–Knopp Type Inequality

The main result in this section is just the following natural limit result of the two–dimensional two–weighted Hardy inequality.

Theorem 1.2.1 *Let $0 < p \leq q < \infty$ and let u and v be strictly positive and measurable functions on $\mathbb{R}^2_+$. Then*

$$\left(\int_0^\infty\int_0^\infty \left[\exp\left(\frac{1}{x_1x_2}\int_0^{x_1}\int_0^{x_2}\log f(y_1,y_2)dy_1dy_2\right)\right]^q v(x_1,x_2)dx_1dx_2\right)^{\frac{1}{q}}$$

$$\leq C\left(\int_0^\infty\int_0^\infty f^p(x_1,x_2)w(x_1,x_2)dx_1dx_2\right)^{\frac{1}{p}} \tag{1.2.1}$$

for all positive and measurable functions on $[0,\infty]\times[0,\infty]$ if and only if

$$D_W(s_1,s_2) := \sup_{\substack{y_1>0\\ y_2>0}} y_1^{\frac{s_1-1}{p}} y_2^{\frac{s_2-1}{p}} \left(\int_{y_1}^\infty\int_{y_2}^\infty x_1^{-\frac{s_1q}{p}} x_2^{-\frac{s_2q}{p}} u(x_1,x_2)dx_1dx_2\right)^{\frac{1}{q}} < \infty, \tag{1.2.2}$$

where $s_1, s_2 > 1$ and

$$u(x_1,x_2) = \left[\exp\left(\frac{1}{x_1x_2}\int_0^{x_1}\int_0^{x_2}\log\frac{1}{w(t_1,t_2)}dt_1dt_2\right)\right]^{\frac{q}{p}} v(x_1,x_2) \tag{1.2.3}$$

and the best possible constant C in (1.2.1) can be estimated in the following way:

$$\sup_{s_1,s_2>1}\left(\frac{e^{s_1}(s_1-1)}{e^{s_1}(s_1-1)+1}\right)^{\frac{1}{p}}\left(\frac{e^{s_2}(s_2-1)}{e^{s_2}(s_2-1)+1}\right)^{\frac{1}{p}} D_W(s_1,s_2) \leq C \tag{1.2.4}$$

$$\leq \inf_{s_1,s_2>1} e^{\frac{s_1+s_2-2}{p}} D_W(s_1,s_2).$$

Remark 1.2.1. As mentioned in our introduction it is easy to see that the Pólya-Knopp inequality

$$\int_0^\infty \exp\left(\frac{1}{x}\int_0^x \ln f(t)dt\right)dx \leq e\left(\int_0^\infty f(x)dx\right)$$

may be regarded as a limit case (as $p \to \infty$) of the original Hardy inequality

$$\int_0^\infty \left(\frac{1}{x}\int_0^x f(t)dt\right)^p dx \leq \left(\frac{p}{p-1}\right)^p \int_0^\infty f^p(x)dx.$$

This inequality is sometimes referred to as Knopp's inequality but it was obvious known to Pólya before (see [100]) and the references given there, so nowadays it is usually referred to as the Pólya-Knopp inequality. Therefore it is natural to regard (1.2.1) as a two-dimensional Pólya-Knopp type inequality. Moreover, it is known that a version of Theorem 1.2.1 holds also on $(0,b_1)\times(0,b_2)$ for any $0<b_1,b_2\leq\infty$ (see [174]).

Remark 1.2.2. For the case $p=q=1$ a similar result was recently proved by H. P. Heinig, R. Kerman and M. Krbec [69] but without the estimates of the operator norm (= the best constant C) in (1.2.1) pointed out in (1.2.4) here. Theorem 1.2.1 may also be regarded as a two-dimensional result of B. Opic and P. Gurka [139] but our (limit argument) proof is completely different and shows the connection to the corresponding scale of Hardy–type inequalities.

Proof. If we in inequality (1.2.1) replace $f^p(x_1,x_2)w(x_1,x_2)$ with $f^p(x_1,x_2)$ and let $u(x_1,x_2)$ be defined as in (1.2.3), then (1.2.1) is equivalent to

$$\left(\int_0^\infty\int_0^\infty \left[\exp\left(\frac{1}{x_1x_2}\int_0^{x_1}\int_0^{x_2} \log f(y_1,y_2)dy_1dy_2\right)\right]^q u(x_1,x_2)dx_1dx_2\right)^{\frac{1}{q}}$$

$$\leq C\left(\int_0^\infty\int_0^\infty f^p(x_1,x_2)dx_1dx_2\right)^{\frac{1}{p}}.$$

Further, by using Theorem 1.1.3 with the special weights $v(x_1,x_2)=u(x_1,x_2)x_1^{-q}x_2^{-q}$ and $w_1(x_1)=w_2(x_2)=1$ we have that

$$\left(\int_0^\infty\int_0^\infty \left(\frac{1}{x_1x_2}\int_0^{x_1}\int_0^{x_2} f(t_1,t_2)dt_1dt_2\right)^q u(x_1,x_2)dx_1dx_2\right)^{\frac{1}{q}}$$

$$\leq C\left(\int_0^\infty\int_0^\infty f^p(x_1,x_2)dx_1dx_2\right)^{\frac{1}{p}} \tag{1.2.5}$$

holds for all $f \geq 0$ if and only if

$$A_W(s_1,s_2) = \sup_{t_1,t_2>0} t_1^{\frac{s_1-1}{p}} t_2^{\frac{s_2-1}{p}} \left(\int_{t_1}^\infty\int_{t_2}^\infty u(x_1,x_2)x_1^{-s_1\frac{q}{p}} x_2^{-s_2\frac{q}{p}} dx_1dx_2\right)^{\frac{1}{q}} < \infty. \tag{1.2.6}$$

We note that $A_W(s_1,s_2)$ coincides with the constant $D_W(s_1,s_2) = D_W(s_1,s_2,q,p)$ defined by (1.2.2) and (1.2.3). Moreover, if C is the best possible constant in (1.2.5), then

$$\inf_{1<s_1,s_2<p}\left(\frac{\left(\frac{p}{p-s_1}\right)^p}{\left(\frac{p}{p-s_1}\right)^p+\frac{1}{s_1-1}}\right)^{\frac{1}{p}}\left(\frac{\left(\frac{p}{p-s_2}\right)^p}{\left(\frac{p}{p-s_2}\right)^p+\frac{1}{s_2-1}}\right)^{\frac{1}{p}} D_W(s_1,s_2) \leq C$$

$$\leq \inf_{1<s_1,s_2<p} D_W(s_1,s_2)\left(\frac{p-1}{p-s_1}\right)^{\frac{1}{p'}}\left(\frac{p-1}{p-s_2}\right)^{\frac{1}{p'}}. \tag{1.2.7}$$

Now, if we replace f in (1.2.5) with f^α, $0<\alpha<p$ and after that replace p with $\frac{p}{\alpha}$ and q with $\frac{q}{\alpha}$ in (1.2.5)–(1.2.7), then we find that, for $1<s_1,s_2<\frac{p}{\alpha}$,

$$\left(\int_0^\infty\int_0^\infty\left(\frac{1}{x_1x_2}\int_0^{x_1}\int_0^{x_2} f^\alpha(t_1,t_2)dt_1dt_2\right)^q u(x_1,x_2)dx_1dx_2\right)^{\frac{1}{q}}$$

$$\leq C_\alpha\left(\int_0^\infty\int_0^\infty f^p(x_1,x_2)dx_1dx_2\right)^{\frac{1}{p}} \tag{1.2.8}$$

holds for all $f\geq 0$ if and only if $D_W(s_1,s_2,\frac{q}{\alpha},\frac{p}{\alpha}) = D_W^\alpha(s_1,s_2,q,p) < \infty$. Moreover, if C_α is the best possible constant in (1.2.8), then

$$\sup_{1<s_1,s_2<p}\left(\frac{\left(\frac{p}{p-\alpha s_1}\right)^{\frac{p}{\alpha}}}{\left(\frac{p}{p-\alpha s_1}\right)^p+\frac{1}{s_1-1}}\right)^{\frac{1}{p}}\left(\frac{\left(\frac{p}{p-\alpha s_2}\right)^{\frac{p}{\alpha}}}{\left(\frac{p}{p-\alpha s_2}\right)^p+\frac{1}{s_2-1}}\right)^{\frac{1}{p}} D_W^\alpha(s_1,s_2,q,p) \leq C_\alpha$$

$$\leq \inf_{1<s_1,s_2<p} D_W^\alpha(s_1,s_2,q,p)\left(\frac{p-\alpha}{p-\alpha s_1}\right)^{\frac{p-\alpha}{\alpha p}}\left(\frac{p-\alpha}{p-\alpha s_2}\right)^{\frac{p-\alpha}{\alpha p}}. \tag{1.2.9}$$

We also note that

$$\left(\frac{1}{x_1x_2}\int_0^{x_1}\int_0^{x_2} f^\alpha(t_1,t_2)dt_1dt_2\right)^{\frac{1}{\alpha}} \downarrow \exp\frac{1}{x_1x_2}\int_0^{x_1}\int_0^{x_2}\ln f(t_1,t_2)dt_1dt_2,$$

as $\alpha \to 0_+$.

We conclude that (1.2.2) holds exactly when $\limsup_{\alpha\to 0_+} C_\alpha < \infty$ and this holds, according to (1.2.9), exactly when (1.2.6) holds. Moreover, when $\alpha \to 0_+$ (1.2.9) implies that (1.2.4) holds. For the lower estimate we apply the test function defined by

$$f(x_1,x_2) = t_1^{-1}t_2^{-1}\chi_{(0,t_1)}(x_1)\,\chi_{(0,t_2)}(x_2) + t_1^{-1}\chi_{(0,t_1)}(x_1)\,\frac{e^{-s_2}t_2^{s_2-1}}{x_2^{s_2}}\chi_{(t_2,\infty)}(x_2)$$

$$+\frac{e^{-s_1}t_1^{s_1-1}}{x_1^{s_1}}\chi_{(t_1,\infty)}(x_1)\,t_2^{-1}\chi_{(0,t_2)}(x_2) + \frac{e^{-(s_1+s_2)}t_1^{s_1-1}t_2^{s_2-1}}{x_1^{s_1}x_2^{s_2}}\chi_{(t_1,\infty)}(x_1)\,\chi_{(t_2,\infty)}(x_2).$$

The details are given in our alternative proof below. The proof is complete. □

Remark 1.2.3. This proof shows that the Pólya–Knopp inequality characterized in Theorem 1.2.4 may be regarded as a natural limiting inequality of the (Sawyer type) Hardy inequality characterized in Theorem 1.1.3.

We will finish this section by presenting an alternative proof of Theorem 1.2.1 which is independent of Theorem 1.1.3 but heavily depending of the following well–known two dimensional version of the Minkowski integral inequality:

Lemma 1.2.1. *Let $r > 1$, $-\infty \le a_1 < b_1 \le \infty$, $-\infty \le a_2 < b_2 \le \infty$ and let Φ and Ψ be positive measurable functions on $[a_1, b_1] \times [a_2, b_2]$. Then*

$$\int_{a_1}^{b_1}\int_{a_2}^{b_2} \Phi(x_1,x_2)\left(\int_{a_1}^{x_1}\int_{a_2}^{x_2}\Psi(y_1,y_2)dy_1dy_2\right)^r dx_1dx_2$$

$$\le \int_{a_1}^{b_1}\int_{a_1}^{b_2} \Psi(y_1,y_2)\left(\int_{y_1}^{b_1}\int_{y_1}^{b_2}\Phi(x_1,x_2)dx_1dx_2\right)^{\frac{1}{r}} dy_1dy_2. \tag{1.2.10}$$

For an elementary proof of Lemma 1.2.5 see, e.g., [174, p. 41]

Alternative proof of Theorem 1.2.1. Let $g(x_1,x_2) = f^p(x_1,x_2)w(x_1,x_2)$ in (1.2.1). Then we see that (1.2.1) is equivalent to the inequality

$$\left(\int_0^{b_1}\int_0^{b_2}\left[\exp\left(\frac{1}{x_1x_2}\int_0^{x_1}\int_0^{x_2}\log g(y_1,y_2)dy_1dy_2\right)\right]^{\frac{q}{p}}\right.$$

$$\left.\times\left[\exp\left(\frac{1}{x_1x_2}\int_0^{x_1}\int_0^{x_2}\log\frac{1}{v(t_1,t_2)}dt_1dt_2\right)\right]^{\frac{q}{p}} v(x_1,x_2)dx_1dx_2\right)^{\frac{1}{q}}$$

$$\leq C\left(\int_0^{b_1}\int_0^{b_2} g(x_1,x_2)dx_1dx_2\right)^{\frac{1}{p}}. \tag{1.2.11}$$

If $w(x_1,x_2)$ is defined by (1.2.3), then we can equivalently write (1.2.11) as

$$\left(\int_0^{b_1}\int_0^{b_2}\left[\exp\left(\frac{1}{x_1x_2}\int_0^{x_1}\int_0^{x_2}\log g(y_1,y_2)dy_1dy_2\right)\right]^{\frac{q}{p}} u(x_1,x_2)dx_1dx_2\right)^{\frac{1}{q}}$$

$$\leq C\left(\int_0^{b_1}\int_0^{b_2} g(x_1,x_2)dx_1dx_2\right)^{\frac{1}{p}}. \tag{1.2.12}$$

Let $y_1 = x_1t_1$ and $y_2 = x_2t_2$ and (1.2.12) becomes

$$\left(\int_0^{b_1}\int_0^{b_2}\left[\exp\left(\int_0^{1}\int_0^{1}\log g(x_1t_1,x_2t_2)dt_1dt_2\right)\right]^{\frac{q}{p}} u(x_1,x_2)dx_1dx_2\right)^{\frac{1}{q}}$$

$$\leq C\left(\int_0^{b_1}\int_0^{b_2} g(x_1,x_2)dx_1dx_2\right)^{\frac{1}{p}}. \tag{1.2.13}$$

By using that

$$\left(\exp\int_0^1\int_0^1 \log t_1^{s_1-1}t_2^{s_2-1}dt_1dt_2\right)^{\frac{q}{p}} = e^{-(s_1+s_2-2)\frac{q}{p}}$$

and Jensen's inequality, the left–hand side of (1.2.13) becomes

$$e^{\frac{s_1+s_2-2}{p}}\left(\int_0^{b_1}\int_0^{b_2}\left[\exp\int_0^1\int_0^1 \log\left(t_1^{s_1-1}t_2^{s_2-1}g(x_1t_1,x_2t_2)\right)dt_1dt_2\right]^{\frac{q}{p}} u(x_1,x_2)dx_1dx_2\right)^{\frac{1}{q}}$$

$$\leq e^{\frac{s_1+s_2-2}{p}}\left(\int_0^{b_1}\int_0^{b_2}\left[\int_0^1\int_0^1 t_1^{s_1-1}t_2^{s_2-1}g(x_1t_1,x_2t_2)dt_1dt_2\right]^{\frac{q}{p}} u(x_1,x_2)dx_1dx_2\right)^{\frac{1}{q}}$$

$$= e^{\frac{s_1+s_2-2}{p}}\left(\int_0^{b_1}\int_0^{b_2}\left[\int_0^{x_1}\int_0^{x_2} y_1^{s_1-1}y_2^{s_2-1}g(y_1,y_2)dy_1dy_2\right]^{\frac{q}{p}} \frac{u(x_1,x_2)}{x_1^{s_1\frac{q}{p}}x_2^{s_2\frac{q}{p}}}dx_1dx_2\right)^{\frac{1}{q}}.$$

Therefore, by using also Lemma 1.2.1 with $r=\frac{q}{p}$ for $p<q$ and Fubini's theorem for $p=q$, we find that the left–hand side in (1.2.13) can be estimated as follows:

$$\leq e^{\frac{s_1+s_2-2}{p}}\left(\int_0^{b_1}\int_0^{b_2} y_1^{s_1-1}y_2^{s_2-1}g(y_1,y_2)\left(\int_{y_1}^{b_1}\int_{y_2}^{b_2} x_1^{-\frac{s_1 q}{p}}x_2^{-\frac{s_2 q}{p}}u(x_1,x_2)dx_1dx_2\right)^{\frac{p}{q}}dy_1dy_2\right)^{\frac{1}{p}}$$

$$\leq e^{\frac{s_1+s_2-2}{p}}D_W(s_1,s_2)\left(\int_0^{b_1}\int_0^{b_2} g(y_1,y_2)dy_1dy_2\right)^{\frac{1}{p}}.$$

Hence, (1.2.13) and, thus, (1.2.1) holds with a constant C satisfying the right–hand side estimate in (1.2.4).

Now, assume that (1.2.1) and, thus, (1.2.12) holds. For fixed t_1 and t_2, $0<t_1<b_1$, $0<t_2<b_2$, we choose the test function

$$\begin{aligned} g(x_1,x_2) = g_0(x_1,x_2) &= t_1^{-1}t_2^{-1}\chi_{(0,t_1)}(x_1)\,\chi_{(0,t_2)}(x_2) \\ &+t_1^{-1}\chi_{(0,t_1)}(x_1)\,\frac{e^{-s_2}t_2^{s_2-1}}{x_2^{s_2}}\chi_{(t_2,\infty)}(x_2) \\ &+\frac{e^{-s_1}t_1^{s_1-1}}{x_1^{s_1}}\chi_{(t_1,\infty)}(x_1)\,t_2^{-1}\chi_{(0,t_2)}(x_2) \\ &+\frac{e^{-(s_1+s_2)}t_1^{s_1-1}t_2^{s_2-1}}{x_1^{s_1}x_2^{s_2}}\chi_{(t_1,\infty)}(x_1)\,\chi_{(t_2,\infty)}(x_2). \end{aligned} \tag{1.2.14}$$

Then for the right side of (1.2.12) it yields that

$$\left(\int_0^{b_1}\int_0^{b_2} g_0(y_1,y_2)\,dy_1dy_2\right)^{\frac{1}{p}}$$

$$=\left(\int_0^{t_1}\int_0^{t_2} t_1^{-1}t_2^{-1}dy_1dy_2+\int_0^{t_1}\int_{t_2}^{b_2} t_1^{-1}\frac{e^{-s_2}t_2^{s_2-1}}{y_2^{s_2}}dy_1dy_2\right.$$

$$\left.+\int_{t_1}^{b_1}\int_0^{t_2} t_2^{-1}\frac{e^{-s_1}t_1^{s_1-1}}{y_1^{s_1}}dy_1dy_2+\int_{t_1}^{b_1}\int_{t_2}^{b_2}\frac{e^{-(s_1+s_2)}t_1^{s_1-1}t_2^{s_1-1}}{y_1^{s_1}y_2^{s_2}}dy_1dy_2\right)^{\frac{1}{p}}$$

$$=\left(1+\frac{e^{-s_2}}{s_2-1}\left(1-\left(\frac{t_2}{b_2}\right)^{s_2-1}\right)+\frac{e^{-s_1}}{s_1-1}\left(1-\left(\frac{t_1}{b_1}\right)^{s_1-1}\right)\right.$$

$$+\frac{e^{-s_1}e^{-s_2}}{(s_1-1)(s_2-1)}\left(1-\left(\frac{t_1}{b_1}\right)^{s_1-1}\right)\left(1-\left(\frac{t_2}{b_2}\right)^{s_2-1}\right)\Bigg)^{\frac{1}{p}}$$

$$\leq\left(1+\frac{e^{-s_2}}{s_2-1}+\frac{e^{-s_1}}{s_1-1}+\frac{e^{-s_1}e^{-s_2}}{(s_1-1)(s_2-1)}\right)^{\frac{1}{p}},$$

i.e.,

$$\left(\int_0^{b_1}\int_0^{b_2} g_0(y_1,y_2)\,dy_1dy_2\right)^{\frac{1}{p}}\leq\left(\frac{e^{s_1}(s_1-1)+1}{e^{s_1}(s_1-1)}\right)^{\frac{1}{p}}\left(\frac{e^{s_2}(s_2-1)+1}{e^{s_2}(s_2-1)}\right)^{\frac{1}{p}}. \quad (1.2.15)$$

Moreover, for the left–hand side in (1.2.12) we have

$$\left(\int_0^{b_1}\int_0^{b_2} u(x_1,x_2)\left[\exp\left(\frac{1}{x_1x_2}\int_0^{x_1}\int_0^{x_2}\log g(y_1,y_2)dy_1dy_2\right)\right]^{\frac{q}{p}}dx_1dx_2\right)^{\frac{1}{q}}$$

$$\geq\left(\int_{t_1}^{b_1}\int_{t_2}^{b_2} u(x_1,x_2)\left[\exp\left(\frac{1}{x_1x_2}\int_0^{x_1}\int_0^{x_2}\log g(y_1,y_2)dy_1dy_2\right)\right]^{\frac{q}{p}}dx_1dx_2\right)^{\frac{1}{q}}. \quad (1.2.16)$$

With the function $g_0(y_1,y_2)$ we get that

$$\exp\left(\frac{1}{x_1x_2}\int_0^{x_1}\int_0^{x_2}\log g_0(y_1,y_2)dy_1dy_2\right)=\exp\left(I_1+I_2+I_3+I_4\right),$$

where

$$I_1=\frac{1}{x_1x_2}\int_0^{t_1}\int_0^{t_2}\log\left(t_1^{-1}t_2^{-1}\right)dy_1dy_2=-\frac{t_1t_2}{x_1x_2}\log t_1-\frac{t_1t_2}{x_1x_2}\log t_2,$$

$$I_2=\frac{1}{x_1x_2}\int_0^{t_1}\int_{t_2}^{x_2}\log\left(t_1^{-1}\frac{e^{-s_2}t_2^{s_2-1}}{y_2^{s_2}}\right)dy_1dy_2$$

$$=-\frac{t_1}{x_1}\log t_1+\frac{t_1t_2}{x_1x_2}\log t_1+(s_2-1)\frac{t_1}{x_1}\log t_2+\frac{t_1t_2}{x_1x_2}\log t_2-s_2\frac{t_1}{x_1}\log x_2,$$

$$I_3=\frac{1}{x_1x_2}\int_{t_1}^{x_1}\int_0^{t_2}\log\left(t_2^{-1}\frac{e^{-s_1}t_1^{s_1-1}}{y_1^{s_1}}\right)dy_1dy_2$$

$$= -\frac{t_2}{x_2}\log t_2 + \frac{t_1 t_2}{x_1 x_2}\log t_2 + (s_1-1)\frac{t_2}{x_2}\log t_1$$
$$+\frac{t_1 t_2}{x_1 x_2}\log t_1 - s_1\frac{t_2}{x_2}\log x_1,$$

and

$$I_4 = (s_1-1)\log t_1 - (s_1-1)\,\frac{t_2}{x_2}\log t_1 + \frac{t_1 t_2}{x_1 x_2}\log t_1$$
$$+ (s_2-1)\log t_2 - (s_2-1)\,\frac{t_1}{x_1}\log t_2 + \frac{t_1 t_2}{x_1 x_2}\log t_2$$
$$-s_1\log x_1 + \frac{t_1}{x_1}\log t_1 + s_1\frac{t_2}{x_2}\log x_1$$
$$+ - s_2\log x_2 + \frac{t_2}{x_2}\log t_2 + s_2\frac{t_1}{x_1}\log x_2.$$

Now we see that

$$I_1+I_2+I_3+I_4 = \log\left(\frac{t_1^{(s_1-1)}t_2^{(s_2-1)}}{x_1^{s_1}x_2^{s_2}}\right)$$

so that, by (1.2.16),

$$\left(\int_{t_1}^{b_1}\int_{t_2}^{b_2} u(x_1,x_2)\left[\exp\left(\frac{1}{x_1x_2}\int_0^{x_1}\int_0^{x_2}\log g_0(y_1,y_2)dy_1dy_2\right)\right]^{\frac{q}{p}} dx_1dx_2\right)^{\frac{1}{q}}$$

$$= \left(\int_{t_1}^{b_1}\int_{t_2}^{b_2} u(x_1,x_2)\left[\frac{t_1^{(s_1-1)}t_2^{(s_2-1)}}{x_1^{s_1}x_2^{s_2}}\right]^{\frac{q}{p}} dx_1dx_2\right)^{\frac{1}{q}}.$$

Hence, by (1.2.12) and (1.2.15),

$$t_1^{\frac{s_1-1}{p}}t_2^{\frac{s_2-1}{p}}\left(\int_{t_1}^{b_1}\int_{t_2}^{b_2} x_1^{-\frac{q}{p}s_1}x_2^{-\frac{q}{p}s_2}u(x_1,x_2)dx_1dx_2\right)^{\frac{1}{q}}$$

$$\le C\left(\frac{e^{s_1}(s_1-1)+1}{e^{s_1}(s_1-1)}\right)^{\frac{1}{p}}\left(\frac{e^{s_2}(s_2-1)+1}{e^{s_2}(s_2-1)}\right)^{\frac{1}{p}},$$

i.e.,

$$\left(\frac{e^{s_1}(s_1-1)}{e^{s_1}(s_1-1)+1}\right)^{\frac{1}{p}}\left(\frac{e^{s_2}(s_2-1)}{e^{s_2}(s_2-1)+1}\right)^{\frac{1}{p}} D_W(s_1,s_2)\le C.$$

We conclude that (1.2.2) and the left–hand side inequality of (1.2.4) hold. The proof is complete. □

Corollary 1.2.1. *Let $0 < p \leq q < \infty$, and $\alpha_1, \alpha_2, \beta_1, \beta_2 \in \mathbb{R}$. Then*

$$\left(\int_0^\infty \int_0^\infty \left[\exp \left(\frac{1}{x_1 x_2} \int_0^{x_1} \int_0^{x_2} \log f(y_1, y_2) dy_1 dy_2 \right) \right]^q x_1^{\alpha_1} x_2^{\alpha_2} dx_1 dx_2 \right)^{\frac{1}{q}}$$
$$\leq C \left(\int_0^\infty \int_0^\infty f^p(x_1, x_2) x_1^{\beta_1} x_2^{\beta_2} dx_1 dx_2 \right)^{\frac{1}{p}} \tag{1.2.17}$$

holds for all positive and measurable functions f on $\mathbb{R}_+^2$ with a finite constant C if and only if

$$\frac{\alpha_1 + 1}{q} = \frac{\beta_1 + 1}{p},$$

and

$$\frac{\alpha_2 + 1}{q} = \frac{\beta_2 + 1}{p},$$

and the best constant C in (1.2.17) *can be estimated as follows:*

$$\sup_{s_1, s_2 > 1} \sup \left(\frac{e^{s_1}(s_1 - 1)}{e^{s_1}(s_1 - 1) + 1} \cdot \frac{e^{s_2}(s_2 - 1)}{e^{s_2}(s_2 - 1) + 1} \right)^{\frac{1}{p}} \left(\frac{1}{s_1 - 1} \cdot \frac{1}{s_2 - 1} \right)^{\frac{1}{q}}$$
$$\times e^{\frac{\beta_1 + \beta_2}{p}} \left(\frac{p}{q} \right)^{\frac{2}{q}} \leq C \leq e^{\frac{\beta_1 + \beta_2}{p} + \frac{2}{q}}. \tag{1.2.18}$$

Proof. Apply Theorem 1.2.1 with the weights $v(x_1, x_2) = x_1^{\alpha_1} x_2^{\alpha_2}$ and $w(x_1, x_2) = x_1^{\beta_1} x_2^{\beta_2}$. The left–hand side estimate of (1.2.18) follows directly and for the right–hand side we can use the optimal values $s_1 = 1 + \frac{p}{q}$ and $s_2 = 1 + \frac{p}{q}$ found by B. Opic and P. Gurka [139] and obtain

$$C \leq \inf_{s_1, s_2 > 1} \left(\frac{1}{s_1 - 1} \cdot \frac{1}{s_2 - 1} \right)^{\frac{1}{q}} e^{\frac{\beta_1 + \beta_2 + s_1 + s_2 - 2}{p}} \left(\frac{p}{q} \right)^{\frac{2}{q}}$$
$$= \inf_{s_1 > 1} \left(\frac{1}{s_1 - 1} \cdot \right)^{\frac{1}{q}} e^{\frac{\beta_1 + s_1 - 1}{p}} \inf_{s_2 > 1} \left(\frac{1}{s_2 - 1} \right)^{\frac{1}{q}} e^{\frac{\beta_2 + s_2 - 1}{p}} \left(\frac{p}{q} \right)^{\frac{2}{q}}$$
$$= e^{\frac{\beta_1 + \beta_2}{p} + \frac{2}{q}}.$$

□

Remark 1.2.4. If $p = q$, then inequality (1.2.17) is sharp with the constant $C = e^{\frac{\beta_1 + \beta_2 + 2}{p}}$, see Theorem 2.2 in [75].

Remark 1.2.5. By using the techniques and results in this section and induction all results can be formulated and proved also in an n-dimensional setting (see [174]). However, in our next Sections we will prove some closely related results by using another technique, which was recently presented in another Ph.D. thesis by E. Ushakova [171].

1.3 The Multidimensional Case: $1 < p \le q < \infty$

In this and the next sections we deal with the inequality

$$\left(\int_{\mathbb{R}^n_+} (H_n f)^q (\mathbf{x}) w(\mathbf{x}) d\mathbf{x} \right)^{\frac{1}{q}} \le C \left(\int_{\mathbb{R}^n_+} f^p(\mathbf{y}) v(\mathbf{y}) d\mathbf{y} \right)^{\frac{1}{p}}, \tag{1.3.1}$$

where

$$(H_n f)(\mathbf{x}) := \int_0^{x_1} \cdots \int_0^{x_n} f(t_1, \dots, t_n) dt_1 \cdots dt_n, \quad \mathbf{x} := (x_1, \dots, x_n) \in \mathbb{R}^n_+$$

and one of the two weight functions v and w is of product type, that is

$$v(\mathbf{y}) = v(y_1, \dots, y_n) = v_1(y_1) \dots v_n(y_n) \tag{1.3.2}$$

or

$$w(\mathbf{x}) = w(x_1, \dots, x_n) = w_1(x_1) \dots w_n(x_n). \tag{1.3.3}$$

Conditions (1.3.2) and (1.3.3) are satisfied, for instance, by a power function of n variables.

In this Section we obtain new necessary and sufficient conditions for the validity of (1.3.1) in the case $1 < p \le q < \infty$ and when (1.3.2) is satisfied. The same problem is considered here with the assumption (1.3.3). Our estimates are n-dimensional analogies of well known criteria for the one-dimensional integral Hardy inequality (see [123], [84], [142], [100] and also and the references given there).

In the next preliminary Lemmas we state some necessary conditions for inequality (1.3.1) to hold in the case $1 < p \le q < \infty$ without any restrictions on the weight functions w and v. These Lemmas are useful in our proofs later on but also of independent interest because they indicate the problem to extend Theorems 1.1.1 and 1.1.3 to the n-dimensional case.

Lemma 1.3.1. *Let $1 < p \le q < \infty$ and assume that inequality (1.3.1) holds for all measurable functions f on $\mathbb{R}^n_+$ with a finite constant C, which is independent on f. Then*

$$\sup_{\substack{t_i > 0 \\ i=1,\dots,n}} V(t_1, \dots, t_n)^{\frac{1}{q}} W(t_1, \dots, t_n)^{\frac{1}{p'}} < \infty, \tag{1.3.4}$$

where

$$V(t_1, \dots, t_n) := V(\mathbf{t}) = \int_{t_1}^{\infty} \dots \int_{t_n}^{\infty} v(\mathbf{x}) d\mathbf{x}$$

and

$$W(t_1, \dots, t_n) := W(\mathbf{t}) = \int_0^{t_1} \dots \int_0^{t_n} w(\mathbf{y})^{1-p'} d\mathbf{y}.$$

Proof. For $\mathbf{t} = (t_1, \dots, t_n)$ such that $t_i > 0$, $i = 1, \dots, n$, we take a test function

$$f_{\mathbf{t}}(\mathbf{y}) := \chi_{[0,t_1]}(y_1) \dots \chi_{[0,t_n]}(y_n) w(\mathbf{y})^{1-p'} \tag{1.3.5}$$

and put it into inequality (1.3.1). Then we have that

$$C \geq \frac{\left(\int_{\mathbb{R}^n_+} \left(\int_0^{x_1} \dots \int_0^{x_n} f_{\mathbf{t}}(\mathbf{y}) d\mathbf{y}\right)^q v(\mathbf{x}) d\mathbf{x}\right)^{\frac{1}{q}}}{\left(\int_{\mathbb{R}^n_+} f_{\mathbf{t}}^p(\mathbf{y}) w(\mathbf{y}) d\mathbf{y}\right)^{\frac{1}{p}}}$$

$$\geq \frac{\left(\int_{t_1}^{\infty} \dots \int_{t_n}^{\infty} v(\mathbf{x}) d\mathbf{x}\right)^{\frac{1}{q}} \left(\int_0^{t_1} \dots \int_0^{t_n} w(\mathbf{y})^{1-p'} d\mathbf{y}\right)}{\left(\int_0^{t_1} \dots \int_0^{t_n} w(\mathbf{y})^{1-p'} d\mathbf{y}\right)^{\frac{1}{p}}} = V(\mathbf{t})^{\frac{1}{q}} W(\mathbf{t})^{\frac{1}{p'}}.$$

Thus, (1.3.4) follows by taking the supremum over all $t_i > 0$, $i = 1, \dots, n$. □

Lemma 1.3.2. *Let $1 < p \leq q < \infty$ and suppose that inequality* (1.3.1) *holds for all measurable functions f on $\mathbb{R}^n_+$ with some finite constant C independent on f. Then*

$$\sup_{\substack{t_i > 0,\\ i=1,\dots,n}} W(t_1, \dots, t_n)^{-\frac{1}{p}} \left(\int_0^{t_1} \dots \int_0^{t_n} v(\mathbf{x}) W(\mathbf{x})^q d\mathbf{x}\right)^{\frac{1}{q}} < \infty. \tag{1.3.6}$$

Proof. This statement follows evidently by substituting into inequality (1.3.1) the function $f_{\mathbf{t}}(\mathbf{y})$ (see (1.3.5)) for $\mathbf{t} = (t_1, \dots, t_n)$ such that $t_i > 0$, $i = 1, \dots, n$. □

Lemma 1.3.3. *Let $1 < p \leq q < \infty$ and assume that inequality* (1.3.1) *holds for all measurable functions f on $\mathbb{R}^n_+$ with some finite constant C independent on f. Then*

$$\sup_{\substack{t_i > 0,\\ i=1,\dots,n}} W(t_1, \dots, t_n)^{-\frac{1}{q'}} \left(\int_{t_1}^{\infty} \dots \int_{t_n}^{\infty} v(\mathbf{x})^{1-p'} W(\mathbf{x})^{p'} d\mathbf{x}\right)^{\frac{1}{p'}} < \infty. \tag{1.3.7}$$

Proof. By duality inequality (1.3.1) is equivalent to the inequality

$$\left(\int_{\mathbb{R}^n_+} (H_n^* g)^{p'}(\mathbf{x}) v^{1-p'}(\mathbf{x}) d\mathbf{x}\right)^{\frac{1}{p'}} \leq C \left(\int_{\mathbb{R}^n_+} g^{q'}(\mathbf{y}) w^{1-q'}(\mathbf{y}) d\mathbf{y}\right)^{\frac{1}{q'}} \tag{1.3.8}$$

with the dual operator H_n^* defined by

$$(H_n^* g)(\mathbf{x}) := \int_{x_1}^{\infty} \dots \int_{x_n}^{\infty} g(\mathbf{y}) d\mathbf{y}, \quad x_1, \dots, x_n > 0. \tag{1.3.9}$$

Now (1.3.7) follows by substituting into inequality (1.3.8) the function

$$g_{\mathbf{t}}(\mathbf{y}) := \chi_{[t_1,\infty)}(y_1)\dots\chi_{[t_n,\infty)}(y_n)v(\mathbf{y})$$

for $\mathbf{t} = (t_1,\dots,t_n)$ such that $t_i > 0$, $i = 1,\dots,n$, and taking supremum. □

Remark 1.3.1. Note that for $n = 2$ the statements of Lemmas 1.3.1 – 1.3.3 follow from Theorem 1.1.1.

To prove the main results of this Section first we use the technique by E. Ushakova from her Ph.D. thesis [171] (see also [145]).

The first main theorem in this Section reads:

Theorem 1.3.1. *Let $1 < p \le q < \infty$ and the weight function v be of product type (1.3.2). Then inequality (1.3.1) holds for all measurable functions f on $\mathbb{R}^n_+$ with some finite constant C, which is independent on f, if and only if $A_{M_n} < \infty$, where*

$$A_{M_n} := \sup_{t_i>0, i=1,\dots,n} V(t_1,\dots,t_n)^{\frac{1}{q}} W_1(t_1)^{\frac{1}{p'}}\dots W_n(t_n)^{\frac{1}{p'}} \tag{1.3.10}$$

and

$$W_i(t_i) := \int\limits_0^{t_i} w_i(x_i)^{1-p'}dx_i, \qquad i = 1,\dots,n.$$

Moreover, $C \approx A_{M_n}$ with constants of equivalence depending only on the parameters p, q and the dimension n.

Proof. The necessary part of the proof follows from Lemma 1.3.1 while the sufficiency can be obtained from the n-dimensional extension of Theorem 1.1.3 and from the following Lemma □

Lemma 1.3.4. *Let*

$$A_{W_n} := A_{W_n}(s_1,\dots,s_n) := \sup_{t_i>0, i=1,\dots,n} W_1(t_1)^{\frac{s_1-1}{p}}\dots W_n(t_n)^{\frac{s_n-1}{p}}$$

$$\times\left(\int\limits_{t_1}^{\infty}\dots\int\limits_{t_n}^{\infty} v(\mathbf{x})W_1(x_1)^{\frac{q(p-s_1)}{p}}\dots W_n(x_n)^{\frac{q(p-s_n)}{p}}d\mathbf{x}\right)^{\frac{1}{q}},$$

where $s_i \in (1,p)$, $i = 1,\dots,n$. Then

$$A_{W_n} \le cA_{M_n}. \tag{1.3.11}$$

Proof. Let $n = 2$ and $s_1 = s_2 = \frac{1+p}{2}$. Then

$$A_{W_2} = \sup_{t_1,t_2>0} W_1(t_1)^{\frac{1}{2p'}} W_2(t_2)^{\frac{1}{2p'}} \left(\int_{t_1}^{\infty}\int_{t_2}^{\infty} v(x_1,x_2) W_1(x_1)^{\frac{q}{2p'}} W_2(x_2)^{\frac{q}{2p'}} dx_2 dx_1 \right)^{\frac{1}{q}}.$$

Since

$$W_i(x_i)^{\frac{q}{2p'}} = \frac{q}{2p'} \int_0^{x_i} w_i(y_i)^{1-p'} W_i(y_i)^{\frac{q}{2p'}-1} dy_i, \quad i = 1, 2,$$

we have that

$$\begin{aligned} W_1(x_1)^{\frac{q}{2p'}} W_2(x_2)^{\frac{q}{2p'}} &\approx \left(\left[\int_0^{t_1} + \int_{t_1}^{x_1} \right] w_1(y_1)^{1-p'} W_1(y_1)^{\frac{q}{2p'}-1} dy_1 \right) \\ &\times \left(\left[\int_0^{t_2} + \int_{t_2}^{x_2} \right] w_2(y_2)^{1-p'} W_2(y_2)^{\frac{q}{2p'}-1} dy_2 \right) \\ &= \int_0^{t_1} w_1(y_1)^{1-p'} W_1(y_1)^{\frac{q}{2p'}-1} dy_1 \int_0^{t_2} w_2(y_2)^{1-p'} W_2(y_2)^{\frac{q}{2p'}-1} dy_2 \\ &+ \int_{t_1}^{x_1} w_1(y_1)^{1-p'} W_1(y_1)^{\frac{q}{2p'}-1} dy_1 \int_{t_2}^{x_2} w_2(y_2)^{1-p'} W_2(y_2)^{\frac{q}{2p'}-1} dy_2 \\ &+ \int_0^{t_1} w_1(y_1)^{1-p'} W_1(y_1)^{\frac{q}{2p'}-1} dy_1 \int_{t_2}^{x_2} w_2(y_2)^{1-p'} W_2(y_2)^{\frac{q}{2p'}-1} dy_2 \\ &+ \int_{t_1}^{x_1} w_1(y_1)^{1-p'} W_1(y_1)^{\frac{q}{2p'}-1} dy_1 \int_0^{t_2} w_2(y_2)^{1-p'} W_2(y_2)^{\frac{q}{2p'}-1} dy_2 \\ &:= I_{11} + I_{22} + I_{12} + I_{21}. \end{aligned}$$

Thus,

$$\begin{aligned} &\int_{t_1}^{\infty}\int_{t_2}^{\infty} v(x_1,x_2) W_1(x_1)^{\frac{q}{2p'}} W_2(x_2)^{\frac{q}{2p'}} dx_2 dx_1 \\ &\approx \int_{t_1}^{\infty}\int_{t_2}^{\infty} v(x_1,x_2) \left[I_{11} + I_{22} + I_{12} + I_{21} \right] dx_2 dx_1 \end{aligned}$$

$$:= J_{11} + J_{22} + J_{12} + J_{21}.$$

Clearly it yields that

$$W_1(t_1)^{\frac{1}{2p'}} W_2(t_2)^{\frac{1}{2p'}} [J_{11}]^{\frac{1}{q}} \ll A_{M_2}.$$

Further,

$$J_{22} = \int_{t_1}^{\infty}\int_{t_2}^{\infty} v(x_1,x_2) \left(\int_{t_1}^{x_1} w_1(y_1)^{1-p'} W_1(y_1)^{\frac{q}{2p'}-1} dy_1 \right)$$

$$\times \left(\int_{t_2}^{x_2} w_2(y_2)^{1-p'} W_2(y_2)^{\frac{q}{2p'}-1} dy_2 \right) dx_2 dx_1$$

$$= \int_{t_1}^{\infty}\int_{t_2}^{\infty} w_1(y_1)^{1-p'} W_1(y_1)^{\frac{q}{2p'}-1} w_2(y_2)^{1-p'} W_2(y_2)^{\frac{q}{2p'}-1}$$

$$\times \left(\int_{y_1}^{\infty}\int_{y_2}^{\infty} v(x_1,x_2) dx_2 dx_1 \right) dy_2 dy_1$$

$$\le A_{M_2}^q \int_{t_1}^{\infty} W_1(y_1)^{-\frac{q}{2p'}-1} w_1(y_1)^{1-p'} dy_1 \int_{t_2}^{\infty} W_2(y_2)^{-\frac{q}{2p'}-1} w_2(y_2)^{1-p'} dy_2$$

$$\ll A_{M_2}^q W_1(t_1)^{-\frac{q}{2p'}} W_2(t_2)^{-\frac{q}{2p'}}.$$

Hence,

$$W_1(t_1)^{\frac{1}{2p'}} W_2(t_2)^{\frac{1}{2p'}} [J_{11}]^{\frac{1}{q}} \ll A_{M_2}.$$

The terms with J_{12} and J_{21} are estimated analogously. The method works for any $n > 2$ by induction and the proof is complete. □

Remark 1.3.2. The condition $A_{M_n} < \infty$ may be regarded as a natural end point of the conditions given in the n-dimensional version of Theorem 1.1.3 and also as a natural generalization of the usual Muckenhoupt–Bradley condition in one dimension.

The alternative criterion for the Hardy inequality (1.3.1) to hold with product type weight w satisfying (1.3.2) in the case $1 < p \le q < \infty$ is stated by the following

Theorem 1.3.2. *Let $1 < p \le q < \infty$ and the weight function v be of product type (1.3.2). Then inequality (1.3.1) holds for all measurable functions f on $\mathbb{R}^n_+$ with some finite constant C, which is independent on f, if and only if $A_{PS_n} < \infty$, where*

$$A_{PS_n} := \sup_{t_i>0, i=1,\dots,n} W_1(t_1)^{-\frac{1}{p}} \dots W_n(t_n)^{-\frac{1}{p}}$$

$$\times \left(\int_0^{t_1} \dots \int_0^{t_n} v(\mathbf{x}) W_1(x_1)^q \dots W_n(x_n)^q d\mathbf{x} \right)^{\frac{1}{q}}. \qquad (1.3.12)$$

Moreover, $C \approx A_{PS_n}$ with constants of equivalence depending only on the parameters p, q and n.

Proof. The necessary part follows from Lemma 1.3.2. The proof of the sufficiency can be obtained from Theorem 1.3.1 and the following Lemma. □

Lemma 1.3.5. *We have*

$$A_{M_n} \ll A_{PS_n}. \qquad (1.3.13)$$

Proof. Let $n = 2$. Suppose first that $W_1(\infty) = W_2(\infty) = \infty$. Then

$$\int_{t_1}^{\infty} \int_{t_2}^{\infty} v(x_1, x_2) dx_2 dx_1 = \int_{t_2}^{\infty} \int_{t_1}^{\infty} v(x_1, x_2) W_1(x_1)^q W_1(x_1)^{-q} dx_1 dx_2$$

$$= q \int_{t_2}^{\infty} \int_{t_1}^{\infty} v(x_1, x_2) W_1(x_1)^q \left(\int_{x_1}^{\infty} W_1(y_1)^{-q-1} dW_1(y_1) \right) dx_1 dx_2$$

$$= q \int_{t_2}^{\infty} \int_{t_1}^{\infty} W_1(y_1)^{-q-1} \left(\int_{t_1}^{y_1} v(x_1, x_2) W_1(x_1)^q dx_1 \right) dW_1(y_1) dx_2$$

$$\leq q \int_{t_1}^{\infty} W_1(y_1)^{-q-1} \int_{t_2}^{\infty} W_2(x_2)^q W_2(x_2)^{-q} \left(\int_0^{y_1} v(x_1, x_2) W_1(x_1)^q dx_1 \right) dx_2 dW_1(y_1)$$

$$\leq q^2 \int_{t_1}^{\infty} \int_{t_2}^{\infty} W_1(y_1)^{-q-1} W_2(y_2)^{-q-1}$$

$$\times \left(\int_0^{y_2} \int_0^{y_1} v(x_1, x_2) W_1(x_1)^q W_2(x_2)^q dx_1 dx_2 \right) dW_2(y_2) dW_1(y_1)$$

$$\leq q^2 A_{PS_2}^q \int_{t_1}^{\infty} \int_{t_2}^{\infty} W_1(y_1)^{-\frac{q}{p'}-1} W_2(y_2)^{-\frac{q}{p'}-1} dW_2(y_2) dW_1(y_1)$$

$$= (p')^2 A_{PS_2}^q W_1(t_1)^{-\frac{q}{p'}} W_2(t_2)^{-\frac{q}{p'}}.$$

Thus, we get that

$$W_1(t_1)^{\frac{1}{p'}}W_2(t_2)^{\frac{1}{p'}}\left(\int_{t_1}^{\infty}\int_{t_2}^{\infty} v(x_1,x_2)dx_2dx_1\right)^{\frac{1}{q}} \ll A_{PS_2}.$$

Further, suppose that $W_i(\infty)<\infty$ for all $i=1,2$. Note that

$$W_i(x_i)^{-q} = W_i(\infty)^{-q} + q\int_{x_i}^{\infty} W_i(y_i)^{-q-1}dW_i(y_i), \qquad i=1,2. \tag{1.3.14}$$

Therefore,

$$\int_{t_1}^{\infty}\int_{t_2}^{\infty} v(x_1,x_2)dx_2dx_1$$

$$= \int_{t_2}^{\infty}\int_{t_1}^{\infty} v(x_1,x_2)W_1(x_1)^qW_2(x_2)^qW_1(x_1)^{-q}W_2(x_2)^{-q}dx_1dx_2$$

$$= W_1(\infty)^{-q}W_2(\infty)^{-q}\int_{t_1}^{\infty}\int_{t_2}^{\infty} v(x_1,x_2)W_1(x_1)^qW_2(x_2)^qdx_2dx_1$$

$$+qW_1(\infty)^{-q}\int_{t_1}^{\infty}\int_{t_2}^{\infty} v(x_1,x_2)W_1(x_1)^qW_2(x_2)^q\left(\int_{x_2}^{\infty} W_2(y_2)^{-q-1}dW_2(y_2)\right)dx_2dx_1$$

$$+qW_2(\infty)^{-q}\int_{t_2}^{\infty}\int_{t_1}^{\infty} v(x_1,x_2)W_1(x_1)^qW_2(x_2)^q\left(\int_{x_1}^{\infty} W_1(y_1)^{-q-1}dW_1(y_1)\right)dx_1dx_2$$

$$+q^2\int_{t_2}^{\infty}\int_{t_1}^{\infty} v(x_1,x_2)W_1(x_1)^qW_2(x_2)^q$$

$$\times\left(\int_{x_1}^{\infty}\int_{x_2}^{\infty} W_1(y_1)^{-q-1}W_2(y_2)^{-q-1}dW_2(y_2)dW_1(y_1)\right)dx_1dx_2$$

$$:= J_{11}+J_{12}+J_{21}+J_{22}.$$

Obviously, it holds the next estimate

$$J_{11} \le W_1(\infty)^{-q}W_2(\infty)^{-q}\int_{0}^{\infty}\int_{0}^{\infty} v(x_1,x_2)W_1(x_1)^qW_2(x_2)^qdx_2dx_1$$

$$\le A^q_{PS_2} W_1(\infty)^{-\frac{q}{p'}} W_2(\infty)^{-\frac{q}{p'}}.$$

By changing the order of integration we have that

$$J_{12} \le q W_1(\infty)^{-q} \int_0^\infty \int_{t_2}^\infty v(x_1,x_2) W_1(x_1)^q W_2(x_2)^q$$

$$\times \left(\int_{x_2}^\infty W_2(y_2)^{-q-1} dW_2(y_2) \right) dx_2 dx_1 \le q W_1(\infty)^{-q} \int_0^\infty \int_{t_2}^\infty W_2(y_2)^{-q-1}$$

$$\times \left(\int_0^{y_2} v(x_1,x_2) W_1(x_1)^q W_2(x_2)^q dx_2 \right) dW_2(y_2) dx_1 = q W_1(\infty)^{-q} \int_{t_2}^\infty W_2(y_2)^{-q-1}$$

$$\times \left(\int_0^{y_2} \int_0^\infty v(x_1,x_2) W_1(x_1)^q W_2(x_2)^q dx_2 dx_1 \right) dW_2(y_2)$$

$$\le q A^q_{PS_2} W_1(\infty)^{-\frac{q}{p'}} \int_{t_2}^\infty W_2(y_2)^{-\frac{q}{p'}-1} dW_2(y_2)$$

$$= p' A^q_{PS_2} \left[W_1(\infty)^{-\frac{q}{p'}} W_2(t_2)^{-\frac{q}{p'}} - W_1(\infty)^{-\frac{q}{p'}} W_2(\infty)^{-\frac{q}{p'}} \right].$$

Analogously,

$$J_{21} \le p' A^q_{PS_2} \left[W_1(t_1)^{-\frac{q}{p'}} W_2(\infty)^{-\frac{q}{p'}} - W_1(\infty)^{-\frac{q}{p'}} W_2(\infty)^{-\frac{q}{p'}} \right].$$

By changing the order of integration we have for J_{22} that

$$J_{22} \le q^2 \int_{t_1}^\infty \int_{t_2}^\infty \left(\int_0^{y_1} \int_0^{y_2} v(x_1,x_2) W_1(x_1)^q W_2(x_2)^q dx_2 dx_1 \right)$$

$$\times W_1(y_1)^{-q-1} W_2(y_2)^{-q-1} dW_2(y_2) dW_1(y_1)$$

$$\le (p')^2 A^q_{PS_2} \left[W_1(t_1)^{-\frac{q}{p'}} W_2(t_2)^{-\frac{q}{p'}} - W_1(\infty)^{-\frac{q}{p'}} W_2(t_2)^{-\frac{q}{p'}} \right.$$

$$\left. - W_1(t_1)^{-\frac{q}{p'}} W_2(\infty)^{-\frac{q}{p'}} + W_1(\infty)^{-\frac{q}{p'}} W_2(\infty)^{-\frac{q}{p'}} \right].$$

Therefore, it follows that

$$W_1(t_1)^{\frac{1}{p'}} W_2(t_2)^{\frac{1}{p'}} \left[J_{11} + J_{12} + J_{21} + J_{22} \right]^{\frac{1}{q}} \le (p')^{\frac{2}{q}} A_{PS_2}.$$

Consider now one of the mixed cases when $W_1(\infty) = \infty$ and $W_2(\infty) < \infty$. Write

$$\int_{t_1}^{\infty}\int_{t_2}^{\infty} v(x_1,x_2)dx_2dx_1 = \int_{t_2}^{\infty}\int_{t_1}^{\infty} v(x_1,x_2)W_1(x_1)^q W_1(x_1)^{-q}dx_1dx_2$$

$$= q\int_{t_2}^{\infty}\int_{t_1}^{\infty} v(x_1,x_2)W_1(x_1)^q\left(\int_{x_1}^{\infty} W_1(y_1)^{-q-1}dW_1(y_1)\right)dx_1dx_2$$

$$\le q\int_{t_2}^{\infty}\int_{t_1}^{\infty} W_1(y_1)^{-q-1}\left(\int_{0}^{y_1} v(x_1,x_2)W_1(x_1)^q dx_1\right)dW_1(y_1)dx_2.$$

Further, by using (3.1.14) with $i = 2$ we get that

$$\int_{t_1}^{\infty}\int_{t_2}^{\infty} v(x_1,x_2)dx_2dx_1$$

$$\le q\int_{t_1}^{\infty} W_1(y_1)^{-q-1}\int_{t_2}^{\infty}\left(\int_{0}^{y_1} v(x_1,x_2)W_1(x_1)^q dx_1\right)W_2(x_2)^q W_2(x_2)^{-q}dx_2dW_1(y_1)$$

$$= qW_2(\infty)^{-q}\int_{t_1}^{\infty} W_1(y_1)^{-q-1}\int_{t_2}^{\infty}\left(\int_{0}^{y_1} v(x_1,x_2)W_1(x_1)^q dx_1\right)W_2(x_2)^q dx_2dW_1(y_1)$$

$$+q^2\int_{t_1}^{\infty} W_1(y_1)^{-q-1}\int_{t_2}^{\infty}\left(\int_{0}^{y_1} v(x_1,x_2)W_1(x_1)^q dx_1\right)W_2(x_2)^q$$

$$\times\left(\int_{x_2}^{\infty} W_2(y_2)^{-q-1}dW_2(y_2)\right)dx_2dW_1(y_1)$$

$$\le qW_2(\infty)^{-q}\int_{t_1}^{\infty} W_1(y_1)^{-q-1}\left(\int_{0}^{\infty}\int_{0}^{y_1} v(x_1,x_2)W_1(x_1)^q W_2(x_2)^q dx_1dx_2\right)dW_1(y_1)$$

$$+q^2\int_{t_1}^{\infty} W_1(y_1)^{-q-1}\int_{t_2}^{\infty} W_2(y_2)^{-q-1}$$

$$\times\left(\int_{0}^{y_1}\int_{0}^{y_2} v(x_1,x_2)W_1(x_1)^q W_2(x_2)^q dx_2dx_1\right)dW_2(y_2)dW_1(y_1)$$

$$\leq (p')^2 A_{PS_2}^q W_1(t_1)^{-\frac{q}{p'}} W_2(t_2)^{-\frac{q}{p'}}.$$

Hence, it yields that

$$W_1(t_1)^{\frac{1}{p'}} W_2(t_2)^{\frac{1}{p'}} \left(\int_{t_1}^{\infty} \int_{t_2}^{\infty} v(x_1, x_2) dx_2 dx_1 \right)^{\frac{1}{q}} \leq (p')^{\frac{2}{q}} A_{PS_2}.$$

The case $W_1(\infty) < \infty$, $W_2(\infty) = \infty$ can be proved analogously. The proof for $n = 2$ is complete. For any $n > 2$ the statement of the lemma follows by induction and the proof is complete. □

Further, we discuss inequality (1.3.1) with the left–hand side weight function of product type. In particular, in Theorems (1.3.4) and (1.3.5) we state a Muckenhoupt–type and Persson–Stepanov–type criteria for inequality (1.3.1) to hold in the case $1 < p \leq q < \infty$ with the left– hand side weight w to be of product type (1.3.3). The proofs of these results are analogous to the proofs of Theorems 1.3.1, 1.3.2 and based on some statements formulated below. The first of them is dual to n –dimensional extension of Theorem 1.1.3 and reads:

Theorem 1.3.3. *Let $1 < q' \leq p' < \infty$, $s_i \in (1, q')$, $i = 1, \dots, n$, and the weight function w be of product type* (1.3.3). *Then inequality* (1.3.8) *holds for all measurable functions g if and only if $A^*_{W_n} < \infty$, where*

$$A^*_{W_n} := A^*_{W_n}(s_1, \dots, s_n) : = \sup_{t_i > 0, i=1,\dots,n} V_1(t_1)^{\frac{s_1 - 1}{q'}} \dots V_n(t_n)^{\frac{s_n - 1}{q'}}$$

$$\times \left(\int_0^{t_1} \dots \int_0^{t_n} w(\mathbf{x})^{1-p'} V_1(x_1)^{\frac{p'(q'-s_1)}{q'}} \dots V_n(x_n)^{\frac{p'(q'-s_n)}{q'}} d\mathbf{x} \right)^{\frac{1}{p'}},$$

$$V(t_i) := \int_{t_i}^{\infty} v_i(x_i) dx_i, \qquad i = 1, \dots, n.$$

Moreover, $C \approx A^*_{W_n}$ with constants of equivalence depending only on the parameters p, q and n.

The following two auxiliary statements are similar to Lemmas 1.3.4 and 1.3.5, respectively.

Lemma 1.3.6. *Let*

$$A^*_{M_n} : = \sup_{\substack{t_i > 0, \\ i=1,\dots,n}} V_1(t_1)^{\frac{1}{q}} \dots V_n(t_n)^{\frac{1}{q}} W(t_1, \dots, t_n)^{\frac{1}{p'}}.$$

Then

$$A^*_{W_n} \ll A^*_{M_n}. \tag{1.3.15}$$

Lemma 1.3.7. *Let*

$$A^*_{PS_n} := \sup_{\substack{t_i>0,\\ i=1,\dots,n}} V_1(t_1)^{-\frac{1}{q'}} \dots V_n(t_n)^{-\frac{1}{q'}}$$

$$\times \left(\int_{t_1}^{\infty} \dots \int_{t_n}^{\infty} w(\mathbf{x})^{1-p'} V_1(x_1)^{p'} \dots V_n(x_n)^{p'} d\mathbf{x} \right)^{\frac{1}{p'}}.$$

Then

$$A^*_{M_n} \ll A^*_{PS_n}.$$

Now by passing to the dual inequality (1.3.8) of (1.3.1) we can get Muckenhoupt–type and Persson-Stepanov–type criteria for (1.3.1) with the left–hand side weight w of product type (1.3.3). The necessity in the proofs of these results follow from Lemmas 1.3.1 and 1.3.3, while the sufficient parts can be proved in the similar ways as in Theorems 1.3.1 and 1.3.2 but by using Theorem 1.3.3, Lemmas 1.3.6 and 1.3.7 instead of Theorem 1.3.3, Lemmas 1.3.4 and 1.3.5, respectively.

Theorem 1.3.4. *Let $1 < p \le q < \infty$ and the weight function v be of product type* (1.3.2)*. Then inequality* (1.3.1) *holds for all measurable functions f on $\mathbb{R}^n_+$ with some finite constant C, which is independent on f, if and only if $A^*_{M_n} < \infty$. Moreover, $C \approx A^*_{M_n}$ with constants of equivalence depending only on the parameters p, q and n.*

Theorem 1.3.5. *Let $1 < p \le q < \infty$ and the weight function v be of product type* 1.3.2*. Then inequality* (1.3.1) *holds for all measurable functions f on $\mathbb{R}^n_+$ with some finite constant C, which is independent on f, if and only if $A^*_{PS_n} < \infty$. Moreover, $C \approx A^*_{PS_n}$ with constants of equivalence depending only on the parameters p, q and n.*

We finish this Section by giving an alternative and direct proof of Theorem 1.3.1 which is based on the induction and the arguments given in [120].

Direct Proof of Sufficiency of Theorem 1.3.1. Suppose that the theorem is true for the $n-1$ dimensional case. We will show that it is true for n dimension. We argue as in the proof of Theorem 1.1.6. First suppose that $S := \int_0^\infty w_n^{1-p'}(x_n)dy = \infty$. Let $\{a_k\}_{k=-\infty}^{+\infty}$ be a sequence of positive numbers for which the equality

$$2^k = \int_0^{a_k} w_n^{1-p'}(x_n)dx_n \tag{1.3.16}$$

holds for $k \in \mathbb{Z}$. It is clear that $\mathbb{R}_+ = \cup_{k\in\mathbb{Z}}[a_k, a_{k+1})$. Moreover, it is easy to verify that

$$2^k = \int_{a_k}^{a_{k+1}} w_n^{1-p'}(x_n)dx_n. \tag{1.3.17}$$

Let $f \geq 0$. We have that

$$I_n := \int_0^\infty \cdots \int_0^\infty v(x_1 \cdots x_n)\big(H_n(x_1 \cdots x_n)\big)^q dx_1 \cdots dx_n$$

$$\leq \sum_{k\in\mathbb{Z}} \underbrace{\int_0^\infty \cdots \int_0^\infty}_{n-1} \int_{a_k}^{a_{k+1}} v(x_1 \cdots x_{n-1}, x_n)$$

$$\times\left(\int_0^{x_1} \cdots \int_0^{x_{n-1}} \int_0^{x_n} f(t_1, \ldots, t_{n-1}, t_n) dt_1 \cdots dt_{n-1} dt_n\right)^q dx_1 \cdots dx_{n-1} dx_n$$

$$\leq \sum_{k\in\mathbb{Z}} \underbrace{\int_0^\infty \cdots \int_0^\infty}_{n-1} \left(\int_{a_k}^{a_{k+1}} v(x_1, \ldots, x_{n-1}, x_n) dx_n\right)$$

$$\times\left(\int_0^{x_1} \cdots \int_0^{x_{n-1}} \left(\int_0^{a_{k+1}} f(t_1, \ldots, t_{n-1}, t_n) dt_n\right) dt_1 \cdots dt_{n-1}\right)^q dx_1 \cdots dx_{n-1}$$

$$= \sum_{k\in\mathbb{Z}} \underbrace{\int_0^\infty \cdots \int_0^\infty}_{n-1} V_k(x_1, \ldots, x_{n-1})$$

$$\times\left(\int_0^{x_1} \cdots \int_0^{x_{n-1}} F_k(t_1, \ldots, t_{n-1}) dt_1 \cdots dt_{n-1}\right)^q dx_1 \cdots dx_{n-1},$$

where

$$V_k(x_1, \ldots, x_{n-1}) := \int_{a_k}^{a_{k+1}} v(x_1, \ldots, x_{n-1}, x_n) dx_n;$$

$$F_k(t_1, \ldots, t_{n-1}) := \int_0^{a_{k+1}} f(t_1, \ldots, t_{n-1}, t_n) dt_n.$$

It is clear that

$$A_{M_n}^q \geq \sup_{\substack{t_1, \ldots, t_{n-1} > 0, \\ j\in\mathbb{Z}}} \left(\underbrace{\int_{t_1}^\infty \cdots \int_{t_{n-1}}^\infty}_{n-1} \int_{a_j}^{a_{j+1}} v(x_1, \ldots, x_{n-1}, x_n) dx_1 \cdots dx_{n-1} dx_n\right)$$

$$\times\Big(\int_0^{t_1}\cdots\int_0^{t_{n-1}}\int_0^{a_j} w^{1-p'}(x_1,\dots,x_{n-1},x_n)dx_1\cdots dx_{n-1}dx_n\Big)^{q/p'}.$$

Further,

$$I_n \le cA^q_{M_n}\sum_{j\in\mathbb{Z}}\Big[\underbrace{\int_0^\infty\cdots\int_0^\infty}_{n-1} w_1(x_1)\cdots w_{n-1}(x_{n-1})$$

$$\times\Big(\int_0^{a_j} w_n^{1-p'}(x_n)dx_n\Big)^{1-p}(F_k(x_1,\dots,x_{n-1}))^p dx_1\cdots dx_{n-1}\Big]^{q/p}$$

$$\le cA^q_{M_n}\Big[\underbrace{\int_0^\infty\cdots\int_0^\infty}_{n-1} w_1(x_1)\cdots w_{n-1}(x_{n-1})\sum_{j\in\mathbb{Z}}\Big(\int_0^{a_j} w_n^{1-p'}(x_n)dx_n\Big)^{1-p}$$

$$\times\Big(\sum_{k=-\infty}^{j}\int_{a_k}^{a_{k+1}} f(x_1,\dots,x_{n-1},x_n)dx_n\Big)^p dx_1\cdots dx_{n-1}\Big]^{q/p}.$$

On the other hand, (1.3.16) and (1.3.17) yield that

$$\sum_{k=m}^{+\infty}\Big(\int_0^{x_k} w_n^{1-p'}(x_n)dx_n\Big)^{1-p}\Big(\sum_{k=-\infty}^{m}\int_{a_k}^{a_{k+1}} w_n^{1-p'}(x_n)dx_n\Big)^{p-1}$$

$$=\sum_{k=m}^{+\infty}\Big(\int_0^{a_k} w_n^{1-p'}(x_n)dx_n\Big)^{1-p}\Big(\int_0^{a_{m+1}} w_n^{1-p'}(x_n)dy\Big)^{p-1}$$

$$=\Big(\sum_{k=m}^{+\infty}2^{k(1-p)}\Big)2^{(m+1)(p-1)}\le c$$

for all $m\in\mathbb{Z}$. Hence by Theorem 1.1.4 and Hölder's inequality we have

$$I_n \le cA^q_{M_n}\Big[\underbrace{\int_0^\infty\cdots\int_0^\infty}_{n-1} w_1(x_1),\dots,w_{n-1}(x_{n-1})$$

$$\times\sum_{j\in\mathbb{Z}}\Big(\int_{a_j}^{a_{j+1}} w_n^{1-p'}(x_n)dx_n\Big)^{1-p}\Big(\int_{a_j}^{a_{j+1}} f(x_1,\dots,x_{n-1},x_n)dx_n\Big)^p dx_1\cdots dx_{n-1}\Big]^{q/p}$$

$$\leq cA_{M_n}^q \Bigg[\underbrace{\int_0^\infty \cdots \int_0^\infty}_{n-1} w_1(x_1) \cdots w_{n-1}(x_{n-1})$$

$$\times \sum_{j\in\mathbb{Z}} \Bigg(\int_{a_j}^{a_{j+1}} w_n(x_n) f^p(x_1, \ldots, x_{n-1}, x_n) dx_n \Bigg) dx_1 \cdots dx_{n-1} \Bigg]^{q/p}$$

$$= cA_{M_n}^q \|f\|_{L^p(\mathbb{R}_+^n)}^q.$$

The proof of the case $S < \infty$ is similar to that of $S = \infty$ (see also the proof of Theorem 1.1.6). □

1.4 The Multidimensional Case: $1 < q < p < \infty$

In this Section we will prove the similar results as in the previous section but in the case $1 < q < p < \infty$. Let us introduce the following n–dimensional versions of the Mazya–Rozin and Persson–Stepanov conditions in this case:

$$B_{MR_n} := \left(\int_{\mathbb{R}_+^n} V(\mathbf{t})^{\frac{r}{q}} W_1(t_1)^{\frac{r}{q'}} \ldots W_n(t_n)^{\frac{r}{q'}} dW_1(t_1) \ldots dW_n(t_n) \right)^{\frac{1}{r}},$$

$$B_{PS_n} := \left(\int_{\mathbb{R}_+^n} \left(\int_0^{t_1} \ldots \int_0^{t_n} v(\mathbf{x}) W_1(x_1)^q \ldots W_n(x_n)^q d\mathbf{x} \right)^{\frac{r}{q}} \right.$$

$$\left. \times W_1(t_1)^{-\frac{r}{q}} \ldots W_n(t_n)^{-\frac{r}{q}} dW_1(t_1) \ldots dW_n(t_n) \right)^{\frac{1}{r}}.$$

The following comparison between these constants is useful later on but also of independent interest.

Lemma 1.4.1. *We have*

$$B_{PS_n} \ll B_{MR_n}. \tag{1.4.1}$$

Proof. It yields that

$$\int_0^{t_1} \ldots \int_0^{t_n} v(\mathbf{x}) W_1(x_1)^q \ldots W_n(x_n)^q d\mathbf{x} = q^n \int_0^{t_1} \ldots \int_0^{t_n} v(\mathbf{x})$$

$$\times \left(\int_0^{x_1} \ldots \int_0^{x_n} W_1(y_1)^{q-1} \ldots W_n(y_n)^{q-1} dW_n(y_n) \ldots dW_1(y_1) \right) d\mathbf{x}$$

$$\leq q^n \int_0^{t_1} \dots \int_0^{t_n} V(\mathbf{y}) W_1(y_1)^{q-1} \dots W_n(y_n)^{q-1} dW_n(y_n) \dots dW_1(y_1)$$

$\big[$ applying Hölder's inequality with the exponents r/q and p/q $\big]$

$$= q^n \int_0^{t_1} \dots \int_0^{t_n} \left\{ V(\mathbf{y}) W_1(y_1)^{(q-1)+\frac{q}{2p}} \dots W_n(y_n)^{(q-1)+\frac{q}{2p}} \right\}$$

$$\times W_1(y_1)^{-\frac{q}{2p}} \dots W_n(y_n)^{-\frac{q}{2p}} dW_n(y_n) \dots dW_1(y_1)$$

$$\leq q^n \left(\int_0^{t_1} \dots \int_0^{t_n} V(\mathbf{y})^{\frac{r}{q}} W_1(y_1)^{(q-1+\frac{q}{2p})\frac{r}{q}} \dots W_n(y_n)^{(q-1+\frac{q}{2p})\frac{r}{q}} dW_n(y_n) \dots \right.$$

$$\left. \times W_1(y_1)^{\frac{q}{r}} \left(\int_0^{t_1} \dots \int_0^{t_n} W_1(y_1)^{-\frac{1}{2}} \dots W_n(y_n)^{-\frac{1}{2}} dW_n(y_n) \dots W_1(y_1) \right) \right)^{\frac{q}{p}}$$

$$= q^n 2^{\frac{qn}{p}} \left(\int_0^{t_1} \dots \int_0^{t_n} V(\mathbf{y})^{\frac{r}{q}} W_1(y_1)^{\frac{r}{q'}+\frac{r}{2p}} \dots \right.$$

$$\left. \times W_n(y_n)^{\frac{r}{q'}+\frac{r}{2p}} dW_n(y_n) \dots dW_1(y_1) \right)^{\frac{q}{r}} W_1(t_1)^{\frac{q}{2p}} \dots W_n(t_n)^{\frac{q}{2p}}.$$

Hence, we obtain that

$$B^r_{PS_n} \leq q^{\frac{rn}{q}} 2^{\frac{rn}{p}} \int_{\mathbb{R}^n_+} \left(\int_0^{t_1} \dots \int_0^{t_n} V(\mathbf{y})^{\frac{r}{q}} W_1(y_1)^{\frac{r}{q'}+\frac{r}{2p}} \dots \right.$$

$$\left. \times W_n(y_n)^{\frac{r}{q'}+\frac{r}{2p}} dW_n(y_n) \dots dW_1(y_1) \right)$$

$$\times W_1(t_1)^{\frac{r}{2p}-\frac{r}{q}} \dots W_n(t_n)^{\frac{r}{2p}-\frac{r}{q}} dW_1(t_1) \dots dW_n(t_n).$$

Therefore, by changing the order of integration, we get that

$$B^r_{PS_n} \leq q^{\frac{rn}{q}} 2^{\frac{rn}{p}} \int_{\mathbb{R}^n_+} V(\mathbf{y})^{\frac{r}{q}} W_1(y_1)^{\frac{r}{q'}+\frac{r}{2p}} \dots W_n(y_n)^{\frac{r}{q'}+\frac{r}{2p}}$$

$$\times \left(\int_{y_1}^{\infty} \dots \int_{y_n}^{\infty} W_1(t_1)^{\frac{r}{2p}-\frac{r}{q}} \dots W_n(t_n)^{\frac{r}{2p}-\frac{r}{q}} dW_n(t_n) \dots dW_1(t_1) \right)$$

$$\times dW_1(y_1) \dots dW_n(y_n) \leq q^{\frac{rn}{q}} 2^{\frac{rn}{p}} \left(\frac{2p}{r} \right)^n B^r_{MR_n}$$

and the required estimate (1.4.1) is proved. □

Next we will state a similar comparison between the following dual versions of the constants B_{MR_n} and B_{PS_n} :

$$B^*_{MR_n} := \left(\int\limits_{\mathbb{R}^n_+} W(\mathbf{t})^{\frac{r}{p'}} V_1(t_1)^{\frac{r}{p}} \dots V_n(t_n)^{\frac{r}{p}} d\left[-V_1(t_1)\right] \dots d\left[-V_n(t_n)\right] \right)^{\frac{1}{r}},$$

$$B^*_{PS_n} := \left(\int\limits_{\mathbb{R}^n_+} \left(\int\limits_{t_1}^{\infty} \dots \int\limits_{t_n}^{\infty} w(\mathbf{x})^{1-p'} V_1(x_1)^{p'} \dots V_n(x_n)^{p'} d\mathbf{x} \right)^{\frac{r}{p'}} \right.$$
$$\left. \times V_1(t_1)^{-\frac{r}{p'}} \dots V_n(t_n)^{-\frac{r}{p'}} d\left[-V_1(t_1)\right] \dots d\left[-V_n(t_n)\right] \right)^{\frac{1}{r}}.$$

Lemma 1.4.2. *It yields that*

$$B^*_{PS_n} \le c B^*_{MR_n}. \tag{1.4.2}$$

Proof. The proof is similar to that of Lemma 1.4.1 so we omit the details. □

The following Theorems state necessary and sufficient conditions for the validity of (1.3.1) in the case $1 < q < p < \infty$ with weights satisfying some of the following additional conditions:

$$W_1(\infty) = \dots = W_n(\infty) = \infty \tag{1.4.3}$$

or

$$V_1(0) = \dots = V_n(0) = \infty \tag{1.4.4}$$

Theorem 1.4.1. *Let $1 < q < p < \infty$ and $1/r = 1/q - 1/p$. Suppose that the weight function v satisfies conditions* (1.3.2) *and* (1.4.3)*. Then inequality* (1.3.1) *holds for all measurable functions f on $\mathbb{R}^n_+$ with some finite constant C, which is independent on f, if and only if $B_{MR_n} < \infty$. Moreover, $C \approx B_{MR_n}$ with constants of equivalence depending only on the parameters p, q and the dimension n.*

Proof. *Necessity.* Suppose that inequality (1.3.1) holds with $C < \infty$ and put

$$f(\mathbf{y}) = V(\mathbf{y})^{\frac{r}{pq}} W_1(y_1)^{\frac{r}{pq'}} w_1(y_1)^{1-p'} \dots W_n(y_n)^{\frac{r}{pq'}} w_n(y_n)^{1-p'}.$$

It is easy to see that $\left(\int_{\mathbb{R}^n_+} f^p(\mathbf{x}) w(\mathbf{x}) d\mathbf{x}\right)^{\frac{1}{p}} = B^{\frac{r}{p}}_{MR_n}$. On the left–hand side we have

$$\left(\int\limits_{\mathbb{R}^n_+} \left(H_n f\right)^q (\mathbf{x}) v(\mathbf{x}) d\mathbf{x} \right)^{\frac{1}{q}}$$

$$= \left(\int_{\mathbb{R}^n_+} \left(\int_0^{x_1} \dots \int_0^{x_n} f(\mathbf{t}) d\mathbf{t} \right) \left(\int_0^{x_1} \dots \int_0^{x_n} f(\mathbf{y}) d\mathbf{y} \right)^{q-1} v(\mathbf{x}) d\mathbf{x} \right)^{\frac{1}{q}}$$

$$= \left(\int_{\mathbb{R}^n_+} f(\mathbf{t}) \left(\int_{t_1}^{\infty} \dots \int_{t_n}^{\infty} \left(\int_0^{x_1} \dots \int_0^{x_n} f(\mathbf{y}) d\mathbf{y} \right)^{q-1} v(\mathbf{x}) d\mathbf{x} \right) d\mathbf{t} \right)^{\frac{1}{q}}$$

$$= \left(\int_{\mathbb{R}^n_+} V(\mathbf{t})^{\frac{r}{pq}} W_1(t_1)^{\frac{r}{pq'}} \dots W_n(t_n)^{\frac{r}{pq'}} \left(\int_{t_1}^{\infty} \dots \int_{t_n}^{\infty} \left(\int_0^{x_1} \dots \int_0^{x_n} V(\mathbf{y})^{\frac{r}{pq}} \right.\right.\right.$$

$$\left.\left.\left. \times\ W_1(y_1)^{\frac{r}{pq'}} \dots W_n(y_n)^{\frac{r}{pq'}} dW_n(y_n) \dots dW_1(y_1) \right)^{q-1} v(\mathbf{x}) d\mathbf{x} \right) dW_1(t_1) \dots dW_n(t_n) \right)^{\frac{1}{q}}$$

$$\geq \left(\int_{\mathbb{R}^n_+} V(\mathbf{t})^{\frac{r}{pq}+1} W_1(t_1)^{\frac{r}{pq'}} \dots W_n(t_n)^{\frac{r}{pq'}} \left(\int_0^{t_1} \dots \int_0^{t_n} V(\mathbf{y})^{\frac{r}{pq}} \right.\right.$$

$$\left.\left. \times\ W_1(y_1)^{\frac{r}{pq'}} \dots W_n(y_n)^{\frac{r}{pq'}} dW_n(y_n) \dots dW_1(y_1) \right)^{q-1} dW_1(t_1) \dots dW_n(t_n) \right)^{\frac{1}{q}}$$

[since the function V is non-increasing and $r/pq' + 1 = r/p'q$]

$$\geq \left(\int_{\mathbb{R}^n_+} \left(\int_0^{t_1} \dots \int_0^{t_n} W_1(y_1)^{\frac{r}{pq'}} \dots W_n(y_n)^{\frac{r}{pq'}} dW_n(y_n) \dots dW_1(y_1) \right)^{q-1} \right.$$

$$\left. \times\ V(\mathbf{t})^{\frac{r}{q}} W_1(t_1)^{\frac{r}{pq'}} \dots W_n(t_n)^{\frac{r}{pq'}} dW_1(t_1) \dots dW_n(t_n) \right)^{\frac{1}{q}} = \left(\frac{p'q}{r} \right)^{\frac{n}{q'}} B_{MR_n}^{\frac{r}{q}}$$

and the estimate $B_{MR_n}^{\frac{r}{q}} \leq C B_{MR_n}^{\frac{r}{p}}$ follows. Therefore, $B_{MR_n} \leq C < \infty$.

Sufficiency. Suppose that $B_{MR_n} < \infty$. On the strength of (1.4.3) we find that

$$\int_{\mathbb{R}^n_+} (H_n f)^q (\mathbf{x}) v(\mathbf{x}) d\mathbf{x}$$

$$= \int_{\mathbb{R}^n_+} (H_n f)^q (\mathbf{x}) W_1(x_1)^q W_1(x_1)^{-q} \dots W_n(x_n)^q W_n(x_n)^{-q} v(\mathbf{x}) d\mathbf{x}$$

$$= q^n \int_{\mathbb{R}^n_+} (H_n f)^q (\mathbf{x}) W_1(x_1)^q \dots W_n(x_n)^q$$

$$\times \left(\int_{x_1}^{\infty} \cdots \int_{x_n}^{\infty} W_1(y_1)^{-q-1} \ldots W_n(y_n)^{-q-1} dW_n(y_n) \ldots dW_1(y_1) \right) v(\mathbf{x}) d\mathbf{x}$$

$$\leq q^n \int_{\mathbb{R}^n_+} (H_n f)^q (\mathbf{y}) W_1(y_1)^{-q-1} \ldots W_n(y_n)^{-q-1}$$

$$\times \left(\int_0^{y_1} \cdots \int_0^{y_n} v(\mathbf{x}) W_1(x_1)^q \ldots W_n(x_n)^q d\mathbf{x} \right) dW_n(y_n) \ldots dW_1(y_1)$$

$$= q^n \int_{\mathbb{R}^n_+} \left\{ (H_n f)^q (\mathbf{y}) W_1(y_1)^{-q} \ldots W_n(y_n)^{-q} \right\} \left\{ W_1(y_1)^{-1} \ldots W_n(y_n)^{-1} \right.$$

$$\left. \times \left(\int_0^{y_1} \cdots \int_0^{y_n} v(\mathbf{x}) W_1(x_1)^q \ldots W_n(x_n)^q d\mathbf{x} \right) \right\} dW_n(y_n) \ldots dW_1(y_1)$$

[by using Hölder's inequality with exponents p/q and r/q]

$$\ll B^q_{PS_n} \left(\int_{\mathbb{R}^n_+} (H_n f)^p (\mathbf{y}) W_1(y_1)^{-p} \ldots W_n(y_n)^{-p} dW_n(y_n) \ldots dW_1(y_1) \right)^{\frac{q}{p}}.$$

Moreover, according to Theorem 1.3.2,

$$\left(\int_{\mathbb{R}^n_+} (H_n f)^p (\mathbf{y}) W_1(y_1)^{-p} \ldots W_n(y_n)^{-p} dW_n(y_n) \ldots dW_1(y_1) \right)^{\frac{q}{p}}$$

$$\ll \left(\int_{\mathbb{R}^n_+} f^p(\mathbf{x}) w_1(x_1) \ldots w_n(x_n) d\mathbf{x} \right)^{\frac{q}{p}}.$$

By combining these inequalities we have that

$$\int_{\mathbb{R}^n_+} (H_n f)^q (\mathbf{x}) v(\mathbf{x}) d\mathbf{x} \ll B^q_{PS_n} \left(\int_{\mathbb{R}^n_+} f^p(\mathbf{x}) w_1(x_1) \ldots w_n(x_n) d\mathbf{x} \right)^{\frac{q}{p}}. \tag{1.4.5}$$

Therefore, in view of Lemma 1.4.1, inequality (1.3.1) holds and the proof is complete.
□

The corresponding result with the constant B_{PS_n} involved reads:

Theorem 1.4.2. *Let $1 < q < p < \infty$ and $1/r = 1/q - 1/p$. Suppose that the weight function v satisfies conditions (1.3.2) and (1.4.3). Then inequality (1.3.1) holds for all measurable functions f on $\mathbb{R}^n_+$ with some finite constant C, which is independent on f, if and only if $B_{PS_n} < \infty$. Moreover, $C \approx B_{PS_n}$ with constants of equivalence depending only on the parameters p, q and the dimension n.*

Proof. The necessity follows from Lemma 1.4.1 and Theorem 1.4.1. The sufficiency is proved by (1.4.5). □

Remark 1.4.1. Note that the sufficient parts of Theorems 1.4.1 and 1.4.2 in fact hold for all $0 < q < p < \infty$. Moreover, the necessary parts of these Theorems are correct even without assuming that condition (1.4.3) is satisfied.

By passing to the dual inequality (1.3.8) of (1.3.1) we can in a similar way as above (but now using Lemma 1.4.2 instead of Lemma 1.4.1) get the following results for the case $1 < q < p < \infty$ with the left–hand side weight w of product type (1.3.3).

Theorem 1.4.3. *Let $1 < q < p < \infty$ and $1/r = 1/q - 1/p$. Assume that the weight function w satisfies conditions (1.3.3) and (1.4.4). Then inequality (1.3.1) holds for all measurable functions f on $\mathbb{R}^n_+$ with some finite constant C, which is independent on f, if and only if $B^*_{MR_n} < \infty$. Moreover, $C \approx B^*_{MR_n}$ with constants of equivalence depending only on the parameters p, q and the dimension n.*

Theorem 1.4.4. *Let $1 < q < p < \infty$ and $1/r = 1/q - 1/p$. Suppose that the weight function w satisfies conditions (1.3.3) and (1.4.4). Then inequality (1.3.1) holds for all measurable functions f on $\mathbb{R}^n_+$ with some finite constant C, which is independent on f, if and only if $B^*_{PS_n} < \infty$. Moreover, $C \approx B^*_{PS_n}$ with constants of equivalence depending only on the parameters p, q and the dimension n.*

1.5 Multidimensional Pólya–Knopp Type Inequalities

In this Section we will apply the results of Theorems 1.3.2 and 1.4.2. Namely, we will characterize the inequality

$$\left(\int_{\mathbb{R}^n_+} (G_n f)^q (\mathbf{x}) v(\mathbf{x}) d\mathbf{x} \right)^{\frac{1}{q}} \leq C \left(\int_{\mathbb{R}^n_+} f^p(\mathbf{y}) w(\mathbf{y}) d\mathbf{y} \right)^{\frac{1}{p}} \tag{1.5.1}$$

in the case $0 < p \leq q < \infty$ and give a sufficient condition for (1.5.1) to hold in the case $0 < q < p < \infty$.

According to Jensen's inequality it holds for any $\mathbf{x} \in \mathbb{R}^n_+$ that

$$(G_n f)(\mathbf{x}) \leq \frac{1}{x_1 \ldots x_n} (H_n f)(\mathbf{x}), \tag{1.5.2}$$

where H_n is the usual Hardy operator defined in Section 1.3.

This fact allows us to find a upper estimate for the best constant of (1.5.1) via the inequality

$$\left(\int_{\mathbb{R}^n_+} (H_n f)^q(\mathbf{x})v(\mathbf{x})d\mathbf{x}\right)^{\frac{1}{q}} \le C\left(\int_{\mathbb{R}^n_+} f^p(\mathbf{y})w(\mathbf{y})d\mathbf{y}\right)^{\frac{1}{p}} \tag{1.5.3}$$

for the Hardy operator H_n, which was considered in the previous section and with a product type weight on one side. It is useful to rewrite (1.5.1) in the following way

$$\left(\int_{\mathbb{R}^n_+} (G_n g)^q(\mathbf{x})u(\mathbf{x})d\mathbf{x}\right)^{\frac{1}{q}} \le C\left(\int_{\mathbb{R}^n_+} g^p(\mathbf{x})d\mathbf{x}\right)^{\frac{1}{p}} \tag{1.5.4}$$

with $g(\mathbf{x}) = f(\mathbf{x})w(\mathbf{x})^{1/p}$ and

$$u(\mathbf{x}) := (G_n v)(\mathbf{x})^{-\frac{q}{p}}v(\mathbf{x}). \tag{1.5.5}$$

Further, for any $0 < s < q$ we put $\widetilde{p}\colon = p/s$, $\widetilde{q}\colon = q/s$ and after a new substitution $g(\mathbf{x}) = h(\mathbf{x})^{1/s}$ inequality (1.5.4) gets the form

$$\left(\int_{\mathbb{R}^n_+} (G_n h)^{\widetilde{q}}(\mathbf{x})u(\mathbf{x})d\mathbf{x}\right)^{1/\widetilde{q}} \le \widetilde{C}\left(\int_{\mathbb{R}^n_+} h^{\widetilde{p}}(\mathbf{x})d\mathbf{x}\right)^{1/\widetilde{p}}, \tag{1.5.6}$$

where $\widetilde{C} = C^s$. Therefore, in view of (1.5.2) we have that the inequality corresponding to (1.5.6) for the operator

$$(\widetilde{H}_n h)(\mathbf{x}) := \frac{1}{x_1 \dots x_n}(H_n h)(\mathbf{x}) \tag{1.5.7}$$

has the form

$$\left(\int_{\mathbb{R}^n_+} \left(\widetilde{H}_n h\right)^{\widetilde{q}}(\mathbf{x})u(\mathbf{x})d\mathbf{x}\right)^{1/\widetilde{q}} \le \bar{C}\left(\int_{\mathbb{R}^n_+} h^{\widetilde{p}}(\mathbf{x})d\mathbf{x}\right)^{1/\widetilde{p}}. \tag{1.5.8}$$

This is an inequality for the Hardy operator H_n with $1 < \widetilde{p}, \widetilde{q} < \infty$, $u(\mathbf{x}) = (x_1 \dots x_n)^{-\widetilde{q}}v(\mathbf{x})$ and with the product weight $w(\mathbf{x}) \equiv 1$. Now we are ready to state and prove our results for inequality (1.5.1). Our main result for the case $0 < p \le q < \infty$ reads:

Theorem 1.5.1. *Let $0 < p \le q < \infty$. Then inequality (1.5.1) holds for all positive measurable functions f on $\mathbb{R}_+^n$ if and only if $A_{G_n} < \infty$, where*

$$A_{G_n} := \sup_{\substack{t_i > 0 \\ i=1,\dots,n}} t_1^{-1/p} \dots t_n^{-1/p} \left(\int_0^{t_1} \dots \int_0^{t_n} u(\mathbf{x}) d\mathbf{x} \right)^{1/q} \tag{1.5.9}$$

with $u(x)$ defined by (1.5.5). Moreover, $C \approx A_{G_n}$ with constants of equivalence depending only on the parameters p, q and the dimension n.

Proof. *Sufficiency.* On the strength of (1.5.2) and Theorem 1.3.2 for $1 < \widetilde{p} \le \widetilde{q} < \infty$ inequality (1.5.8) holds if

$$\bar{A}_{G_n} : = \sup_{\substack{t_i 0 \\ i=1,\dots,n}} t_1^{-1/\widetilde{p}} \dots t_n^{-1/\widetilde{p}} \left(\int_0^{t_1} \dots \int_0^{t_n} u(\mathbf{x}) d\mathbf{x} \right)^{1/\widetilde{q}} < \infty.$$

Note that because of definitions of $\widetilde{p}$ and $\widetilde{q}$ it yields that

$$\left(\bar{A}_{G_n}\right)^{\frac{1}{s}} = A_{G_n}.$$

Therefore, according to the fact that $C = \widetilde{C}^{1/s}$ it follows that $A_{G_n} < \infty$ is a sufficient condition for the validity of inequality (1.5.1) in the case $0 < p \le q < \infty$.

Necessity. Suppose that (1.5.1) and, thus, (1.5.4) holds with $C < \infty$. Take a test function

$$g_{\mathbf{t}}(\mathbf{y}) = \chi_{[0,t_1]}(y_1) t_1^{-\frac{1}{p}} \dots \chi_{[0,t_n]}(y_n) t_n^{-\frac{1}{p}}$$

and put it into inequality (1.5.4). The function $g_{\mathbf{t}}(\mathbf{y})$ is such that the right–hand side of (1.5.4) is equal to 1. Therefore,

$$C \ge \left(\int_{\mathbb{R}_+^n} \left(G_n g_{\mathbf{t}}\right)^q (\mathbf{x}) u(\mathbf{x}) d\mathbf{x} \right)^{\frac{1}{q}} \ge t_1^{-\frac{1}{p}} \dots t_n^{-\frac{1}{p}} \left(\int_0^{t_1} \dots \int_0^{t_n} u(\mathbf{x}) d\mathbf{x} \right)^{\frac{1}{q}}.$$

Hence, by taking supremum over all t_i, $i = 1, \dots, n$, we have that $A_{G_n} < \infty$ and the proof is complete. □

Remark 1.5.1. Our proof above shows that Theorem 1.5.1 may be regarded as a limit case of the result in Theorem 1.3.2.

Remark 1.5.2. Note that for the case $n = 2$ we have here obtained another characterization of 1.5.1 than that in our Theorem 1.2.1. They are both endpoint characterization of the corresponding scales of Hardy inequalities with the Wedestig and Persson–Stepanov– type descriptions, respectively.

Moreover, inequality (1.5.2) and Theorem 1.4.2 allow us to obtain a sufficient condition for (1.5.1) to hold in the case $0 < q < p < \infty$. We state this result in the following form:

Theorem 1.5.2. *Let $0 < q < p < \infty$. Then inequality (1.5.1) holds if $B_{G_n} < \infty$, where*

$$B_{G_n} := \left(\int\limits_{\mathbb{R}^n_+} \left(\int\limits_0^{t_1} \dots \int\limits_0^{t_n} u(\mathbf{x}) d\mathbf{x} \right)^{\frac{r}{q}} t_1^{-\frac{r}{q}} \dots t_n^{-\frac{r}{q}} dt_1 \dots dt_n \right)^{\frac{1}{r}}.$$

Proof. The statement follows from Theorem 1.4.2 by using the same arguments as for the proof of a sufficiency part of Theorem 1.5.1. □

Remark 1.5.3. Note that the condition $B_{G_n} < \infty$ is also necessary for (1.5.1) to hold in the case $0 < q < p < \infty$ with the additional assumption that the weight function v is of product type. In this case we also have that $C \approx B_{G_n}$, where C is the best constant in (1.5.1).

1.6 Double Riemann–Liouville Transform Without Singularity

In this section we discuss the criteria guaranteeing the validity of the Hardy–type transforms for the Riemann–Liouville transforms with multiple kernels

$$(\mathcal{R}_{\alpha,\beta} f)(x,y) := \int\limits_0^x \int\limits_0^y \frac{f(t,\tau)}{(x-t)^{1-\alpha}(y-\tau)^{1-\beta}} dt d\tau$$

where $\alpha, \beta > 1$, provided that the right–hand side weight function satisfies the doubling condition.

The next statement is due to V. Stepanov [163] (see also [165])and F. J. Martin–Reyes and E. Sawyer [114], which gives two–weight criteria for the Riemann–Liouville transform

$$\mathcal{R}_\alpha f(x) := \int\limits_0^x \frac{f(y)}{(x-y)^{1-\alpha}} dy, \quad x > 0,$$

where $\alpha > 1$.

Theorem 1.6.1. *Let $1 < p \le q < \infty$, $\alpha > 1$. Then the operator $\mathcal{R}_\alpha$ acts boundedly from $L^p_w(\mathbb{R}_+)$ to $L^q_v(\mathbb{R}_+)$ if and only if the following two conditions*

$$A^{(1)}_{SMS} := \sup_{t>0} \left(\int\limits_t^\infty \frac{v(x)}{(x-t)^{(1-\alpha)q}} dx \right)^{1/q} \left(\int\limits_0^t w^{1-p'}(x) dx \right)^{1/p'} < \infty$$

and

$$A_{SMS}^{(2)} := \sup_{t>0} \Big(\int_t^{\infty} v(x)dx \Big)^{1/q} \Big(\int_0^t \frac{w^{1-p'}(x)}{(t-x)^{(1-\alpha)p'}} dx \Big)^{1/p'} < \infty$$

hold.

Moreover, there exist positive constants c_1 *and* c_2 *depending only on* α, p *and* q *such that*

$$c_1 \max\{A_{SMS}^{(1)}, A_{SMS}^{(2)}\} \le \|\mathcal{R}_\alpha\| \le c_2 \max\{A_{SMS}^{(1)}, A_{SMS}^{(2)}\}.$$

In order to formulate the main results of this section we need the following definition:

Definition 1.6.1. We say that a measurable positive function $\rho : \mathbb{R}_+^2 \to \mathbb{R}^1$ satisfies the doubling condition uniformly with respect to $x \in \mathbb{R}_+$ ($\rho \in DC(y)$) if there exists a constant $b > 1$ such that for all $t > 0$ and almost every $x > 0$,

$$\int_0^{2t} \rho(x,y)dy \le b \int_{J_t} \rho(x,y)dy$$

holds, where J_t denotes the following two types of intervals: $[0,t]$ and $[t,2t]$. Analogously, is defined the class $DC(x)$.

Note that if the weight ρ is locally integrable on $\mathbb{R}_+^2$, then $\rho \in DC(y)$ is equivalent to the following condition: there exists a constant $b > 1$ such that, for all intervals of finite length $I \subset \mathbb{R}_+$ and $t > 0$, the inequality

$$\int_I \int_0^{2t} \rho(x,y)dxdy \le b \int_I \int_{J_t} \rho(x,y)dxdy$$

holds.

Lemma 1.6.1. *Let* $\rho \in DC(y)$ *(resp.* $\rho \in DC(y)$ *). Then there exists a constant* $\eta > 1$ *such that for all* $t > 0$ *and a.e.* $x \in \mathbb{R}$ *the following inequality holds*

$$\int_0^{2t} \rho(x,y)dy \ge \eta \int_0^t \rho(x,y)dy \tag{1.6.1}$$

$$\Big(\text{resp.} \quad \int_0^{2\tau} \rho(x,y)dx \ge \eta \int_0^{\tau} \rho(x,y)dx \Big).$$

Proof. We have that

$$\int_0^{2t} \rho(x,y)dy = \int_0^{t} \rho(x,y)dy + \int_t^{2t} \rho(x,y)dy$$

$$\geq \int_0^{t} \rho(x,y)dy + (1/b)\int_0^{2t} \rho(x,y)dy.$$

Hence

$$\int_0^{2t} \rho(x,y)dy \geq \frac{b}{b-1}\int_0^{t} \rho(x,y)dy.$$

Now we take $\eta = b/(b-1)$ and we are done. □

Theorem 1.6.2. *Let $1 < p \leq q < \infty$ and let $\alpha, \beta \geq 1$. Suppose that $w^{1-p'} \in DC(y)$. Then the operator $\mathcal{R}_{\alpha,\beta}$ is bounded from $L^p_w(\mathbb{R}^2_+)$ to $L^q_v(\mathbb{R}^2_+)$ if and only if*

(i)

$$P_1 := \sup_{a,b>0}\left(\int_0^a\int_0^b \frac{w^{1-p'}(x,y)}{(a-x)^{(1-\alpha)q}}dxdy\right)^{1/p'}\left(\int_a^\infty\int_b^\infty \frac{v(x,y)}{y^{(1-\beta)q}}dxdy\right)^{1/q} < \infty;$$

(ii)

$$P_2 := \sup_{a,b>0}\left(\int_0^a\int_0^b w^{1-p'}(x,y)dxdy\right)^{1/p'}\left(\int_a^\infty\int_b^\infty \frac{v(x,y)}{(x-a)^{(1-\alpha)q}y^{(1-\beta)q}}dxdy\right)^{1/q} < \infty.$$

Moreover, $\|\mathcal{R}_{\alpha,\beta}\| \approx \max\{P_1, P_2\}$.

Proof. *Sufficiency.* Let us introduce the notation:

$$v_1(x,y) := \frac{v(x,y)}{y^{(1-\beta)q}};\ \ \widetilde{v}_{1,j}(x) := \int_{2^j}^{2^{j+1}} v_1(x,y)dy;$$

$$F_j(t) := \int_0^{2^{j+1}} f(t,\tau)d\tau;\ \ P := \max\{P_1, P_2\}.$$

Let $f \geq 0$. Then by using Theorem 1.6.1 we find that

$$I := \iint_{\mathbb{R}^2_+} v(x,y)(\mathcal{R}_{\alpha,\beta}f)^q(x,y)dxdy$$

$$\leq \iint\limits_{\mathbb{R}^2_+} v_1(x,y)\Big(\int\limits_0^x (x-t)^{\alpha-1}\Big(\int\limits_0^y f(t,\tau)d\tau\Big)dt\Big)^q dxdy$$

$$\leq \sum_{j\in\mathbb{Z}} \int\limits_0^\infty \widetilde{v}_{1,j}(x)\Big(\int\limits_0^x (x-t)^{\alpha-1}F_j(t)dt\Big)^q dx$$

$$\leq cP^q \sum_{j\in\mathbb{Z}} \Big[\int\limits_0^\infty \Big(\int\limits_0^{2^j} w^{1-p'}(x,y)dy\Big)^{1-p} F_j^p(x)dx\Big]^{q/p}$$

$$\leq cP^q \Big[\int\limits_0^\infty \sum_{j\in\mathbb{Z}} \Big(\int\limits_0^{2^j} w^{1-p'}(x,y)dy\Big)^{1-p} F_j^p(x)dx\Big]^{q/p}.$$

On the other hand, we have that

$$\sup_{\substack{x>0\\ k\in\mathbb{Z}}} J(x,k) := \sup_{\substack{x>0\\ k\in\mathbb{Z}}} \sum_{j=k}^\infty \Big(\int\limits_0^{2^j} w^{1-p'}(x,y)dy\Big)^{1-p}$$
$$\times\Big(\sum_{j=-\infty}^k \int\limits_{2^j}^{2^{j+1}} w^{1-p'}(x,y)dy\Big)^{p-1} < \infty.$$

Indeed, (1.6.1) and the condition $w^{1-p'} \in DC(y)$ lead us to the inequalities:

$$J(x,k) = \sum_{j=k}^\infty \Big(\int\limits_0^{2^j} w^{1-p'}(x,y)dy\Big)^{1-p}$$
$$\times\Big(\int\limits_0^{2^{k+1}} w^{1-p'}(x,y)dy\Big)^{p-1} \leq \Big(\sum_{j=k}^\infty \eta^{(j-k)(1-p)}\Big)$$
$$\times\Big(\int\limits_0^{2^k} w^{1-p'}(x,y)dy\Big)^{1-p}\Big(\int\limits_0^{2^{k+1}} w^{1-p'}(x,y)dy\Big)^{p-1} \leq c,$$

where the positive constant c does not depend on x and k. Consequently, taking into account Theorem 1.1.4 and Hölder's inequality we find that

$$\Big[\int\limits_0^\infty \sum_{j\in\mathbb{Z}} \Big(\int\limits_0^{2^j} w^{1-p'}(x,y)dy\Big)^{1-p} F_j^p(x)dx\Big]^{q/p}$$

$$\leq c\Big[\int_0^\infty \sum_{j\in\mathbb{Z}}\Big(\int_0^{2^j} w^{1-p'}(x,y)dy\Big)^{1-p}\Big(\sum_{k=-\infty}^{j}\int_{2^k}^{2^{k+1}} f(x,t)dt\Big)^p dx\Big]^{q/p}$$

$$\leq c\Big[\int_0^\infty \sum_{k\in\mathbb{Z}}\Big(\int_{2^k}^{2^{k+1}} w^{1-p'}(x,\tau)d\tau\Big)^{1-p}\Big(\int_{2^k}^{2^{k+1}} f(x,t)dt\Big)^p dx\Big]^{q/p}$$

$$\leq c\Big[\int_0^\infty \sum_{j\in\mathbb{Z}}\Big(\int_{2^j}^{2^{j+1}} w^{1-p'}(x,y)dy\Big)^{1-p}$$

$$\times\Big(\int_{2^j}^{2^{j+1}} w(x,t)f^p(x,t)dt\Big)\Big(\int_{2^j}^{2^{j+1}} w^{1-p'}(x,t)dt\Big)^{p-1} dx\Big]^{q/p} \leq c\|f\|^q_{L^p(\mathbb{R}^2_+)}.$$

Necessity. Let $f \geq 0$ and let $a, b > 0$. It is easy to see that

$$I \geq \int_0^\infty\int_0^\infty v(x,y)\Big(\int_0^x\int_0^{y/2} \frac{f(t,\tau)}{(x-t)^{1-\alpha}(y-\tau)^{1-\beta}}dtd\tau\Big)^q dxdy$$

$$\geq c\Big(\int_a^\infty\int_b^\infty v_1(x,y)dxdy\Big)\Big(\int_0^a\int_0^{b/2} \frac{f(t,\tau)}{(a-t)^{1-\alpha}}dtd\tau\Big)^q.$$

By using the latter inequality and the boundedness of $\mathcal{R}_{\alpha,\beta}$ on the class of functions $f_{a,b}(x,y) = \chi_{(0,a)}(x)\chi_{(0,b/2)}(y)w^{1-p'}(x,y)(a-t)^{(\alpha-1)(q-1)}$, $a, b > 0$, we find that

$$\Big(\int_0^a\int_0^{b/2} \frac{w^{1-p'}(x,y)}{(a-x)^{(1-\alpha)q}}dxdy\Big)^q \int_a^\infty\int_b^\infty v_1(x,y)dxdy$$

$$\leq I \leq c\Big(\int_0^a\int_0^{b/2} \frac{w^{1-p'}(x,y)}{(a-x)^{(1-\alpha)q}}dxdy\Big)^{q/p} < \infty.$$

Hence, this inequality and the condition $w^{1-p'} \in CD(y)$ give us the condition $P_1 < \infty$.

Taking into account the arguments used above and the fact that the operator $\mathcal{R}_{\alpha,\beta}$ is bounded from $L^p_w(\mathbb{R}^2_+)$ to $L^q_v(\mathbb{R}^2_+)$ if and only if the operator

$$(\mathcal{W}_{\alpha,\beta}f)(x,y) = \int_x^\infty\int_y^\infty (t-x)^{\alpha-1}(\tau-y)^{\beta-1}f(t,\tau)dtd\tau$$

is bounded from $L^{q'}_{v^{1-q'}}(\mathbb{R}^2_+)$ to $L^{p'}_{w^{1-p'}}(\mathbb{R}^2_+)$, we obtain that $P_2 < \infty$. The proof is complete. □

Analogously, it can be proved the next statement:

Theorem 1.6.3. *Let $1 < p \le q < \infty$ and let $\alpha, \beta \ge 1$. Suppose that $w^{1-p'} \in DC(x)$. Then the operator $\mathcal{R}_{\alpha,\beta}$ is bounded from $L^p_w(\mathbb{R}^2_+)$ to $L^q_v(\mathbb{R}_+)$ if and only if*

(i)

$$\widetilde{P}_1 := \sup_{a,b>0} \bigg(\int_0^a \int_0^b \frac{w^{1-p'}(x,y)}{(b-y)^{(1-\beta)q}} dxdy \bigg)^{1/p'} \bigg(\int_a^\infty \int_b^\infty \frac{v(x,y)}{x^{(1-\alpha)q}} dxdy \bigg)^{1/q} < \infty;$$

(ii)

$$\widetilde{P}_2 := \sup_{a,b>0} \bigg(\int_0^a \int_0^b w^{1-p'}(x,y) dxdy \bigg)^{1/p'} \bigg(\int_a^\infty \int_b^\infty \frac{v(x,y)}{(y-b)^{(1-\beta)q} x^{(1-\alpha)q}} dxdy \bigg)^{1/q} < \infty.$$

Moreover, $\|\mathcal{R}_{\alpha,\beta}\| \approx \max\{\widetilde{P}_1, \widetilde{P}_2\}$.

Corollary 1.6.1. *Let $1 < p \le q < \infty$. Suppose that $w^{1-p'} \in DC(x)$ or $w^{1-p'} \in DC(y)$. Then the double Hardy operator H_2 is bounded from $L^p_w(\mathbb{R}^2_+)$ to $L^q_v(\mathbb{R}^2_+)$ if and only if* (1.1.2) *holds.*

1.7 Further Results

We begin to discuss some refined multidimensional Hardy–type inequalities via superquadracity for $p \ge 2$ and their duals. We shall see that these inequalities hold in the reversed direction when $1 < p \le 2$. In particular, this implies the surprising fact that we have equality for $p = 2$, which is completely different from the usual situation with Hardy–type inequalities when we can have equality only for $p = 1$ (see, e.g., our Remark 1.7.12). The results obtained are based mainly on some new results for superquadratic and subquadratic functions. See [2] and [3] (c. f. also [1]). In particular, our results further extend the recent results in [134] to a multidimensional setting (see [136]).

In the sequel, we use the following notation:

$$\mathbf{x} := (x_1, \dots, x_n), \;\; \mathbf{y} := (y_1, \dots, y_n), \;\; \mathbf{0} := (0, \dots, 0) \in \mathbb{R}^n, \;\; \mathbf{1} := (1, \dots, 1) \in \mathbb{R}^n.$$

Furthermore, the relations $<$, $\le$, $>$, and $\ge$ are, as usual, defined componentwise. For example, for $\mathbf{x}, \mathbf{y} \in \mathbb{R}^n$, we write $\mathbf{x} < \mathbf{y}$ if $x_i < y_i$, $i = 1, \dots, n$. Moreover, $\mathbf{0} < \boldsymbol{b} \le \boldsymbol{\infty}$ means that $0 < b_i \le \infty$, $i = 1, \dots, n$. In addition, we introduce a notation for $n-$cells, that is, axis parallel rectangular blocks in $\mathbb{R}^n$. For $\boldsymbol{a}, \boldsymbol{b} \in \mathbb{R}^n$, $\boldsymbol{a} < \boldsymbol{b}$, let

$$(\boldsymbol{a}, \boldsymbol{b}) := (a_1, b_1) \times \cdots \times (a_n, b_n) = \{\boldsymbol{x} \in \mathbb{R}^n : \boldsymbol{a} < \boldsymbol{x} < \boldsymbol{b}\},$$

$$(\boldsymbol{a},\boldsymbol{b}] := (a_1,b_1]\times\cdots\times(a_n,b_n] = \{\boldsymbol{x}\in\mathbb{R}^n : \boldsymbol{a}<\boldsymbol{x}\le\boldsymbol{b}\},$$

and analogously also for $[\boldsymbol{a},\boldsymbol{b})$ and $[\boldsymbol{a},\boldsymbol{b}]$. In particular, we have $\mathbb{R}^n_+ = (\mathbf{0},\boldsymbol{\infty})$, $(\mathbf{0},\boldsymbol{\infty}] = \{\mathbf{0}<\boldsymbol{x}\le\boldsymbol{\infty}\}$, and $[\mathbf{0},\boldsymbol{\infty}) = \{\boldsymbol{x}\in\mathbb{R}^n : \boldsymbol{x}\ge\mathbf{0}\}$. Furthermore, all functions are assumed to be measurable and expressions of the form $0\cdot\infty$, $\frac{\infty}{\infty}$, and $\frac{0}{0}$ are taken to be equal to zero. Moreover, $u(\boldsymbol{x})$ denotes a weight function, i.e., a nonnegative and measurable function on $\mathbb{R}^n$, and we define a corresponding weight function $v(\boldsymbol{t})$ by

$$w(\boldsymbol{t}) = \begin{cases} t_1\cdots t_n \int_{t_1}^{b_1}\cdots\int_{t_n}^{b_n} \frac{v(\boldsymbol{x})}{x_1^2\cdots x_n^2}d\boldsymbol{x}, \boldsymbol{t}\in(\mathbf{0},\mathbf{b}) \\ \frac{1}{t_1\cdots t_n}\int_{b_1}^{t_1}\cdots\int_{b_n}^{t_n} v(\boldsymbol{x})d\boldsymbol{x} \ <\infty, \boldsymbol{t}\in(\mathbf{b},\boldsymbol{\infty}) \end{cases} \tag{1.7.1}$$

First, we state a definition and some results in [2], which are germane to the proofs of our propositions below.

Definition 1.7.1 ([2, Definition 2.1]). A function $\varphi:[0,\infty)\to\mathbb{R}$ is superquadratic provided that for all $x\ge 0$ there exists a constant $C_x\in\mathbb{R}$ such that

$$\varphi(y)-\varphi(x)-\varphi(|y-x|)\ge C_x(y-x) \quad \text{for all } y\ge 0.$$

We say that f is subquadratic if $-f$ is superquadratic.

Lemma 1.7.1 ([2, Theorem 2.3]). *Let (Ω,μ) be a probability measure space. The inequality*

$$\varphi\left(\int_\Omega f(s)d\mu(s)\right)\le\int_\Omega\varphi(f(s))d\mu(s)-\int_\Omega\varphi\left(\left|f(s)-\int_\Omega f(s)d\mu(s)\right|\right)d\mu(s) \tag{1.7.2}$$

holds for all probability measures μ and all nonnegative $\mu-$ integrable functions f if and only if φ is superquadratic. Moreover, (1.7.2) *holds in the reversed direction if and only if φ is subquadratic.*

Lemma 1.7.2 ([2, Lemma 3.1]). *Suppose $\varphi:[0,\infty)\to\mathbb{R}$ is continuously differentiable and $\varphi(0)\le 0$. If φ' is superadditive or $\frac{\varphi'(x)}{x}$ is nondecreasing, then φ is superquadratic.*

Remark 1.7.1. According to Lemmas 1.7.1 and 1.7.2 we have that if $p\ge 2$ in Lemma 1.7.1, then

$$\left(\int_\Omega f(s)d\mu(s)\right)^p\le\int_\Omega(f(s))^p d\mu(s)-\int_\Omega\left|f(s)-\int_\Omega f(s)d\mu(s)\right|^p d\mu(s)$$

holds and the reversed inequality holds when $1<p\le 2$ (see also [1, Example 1, p. 1448]).

Proposition 1.7.1. *Let* $\boldsymbol{b} \in (\boldsymbol{0}, \boldsymbol{\infty})$, $v : (\boldsymbol{0}, \boldsymbol{b}) \to \mathbb{R}$ *be a weight function which is locally integrable in* $(\boldsymbol{0}, \boldsymbol{b})$ *and* $w(\boldsymbol{x})$ *be defined by* (1.7.1). *Suppose* $I = (a, c)$, $0 \le a < c \le \infty$, $\varphi : I \to \mathbb{R}$, *and* $f : (\boldsymbol{0}, \boldsymbol{b}) \to \mathbb{R}$ *is an integrable function, such that* $f(\boldsymbol{x}) \in I$, *for all* $\boldsymbol{x} \in (\boldsymbol{0}, \boldsymbol{b})$.

(i) *If* φ *is superquadratic, then the following inequality holds:*

$$\int_0^{b_1} \cdots \int_0^{b_n} v(\boldsymbol{x}) \varphi \left(\frac{1}{x_1 \cdots x_n} \int_0^{x_1} \cdots \int_0^{x_n} f(\boldsymbol{t}) d\boldsymbol{t} \right) \frac{d\boldsymbol{x}}{x_1 \cdots x_n}$$

$$+ \int_0^{b_1} \cdots \int_0^{b_n} \int_{t_1}^{b_1} \cdots \int_{t_n}^{b_n} \varphi \left(\left| f(\boldsymbol{t}) - \frac{1}{x_1 \cdots x_n} \int_0^{x_1} \cdots \int_0^{x_n} f(\boldsymbol{t}) d\boldsymbol{t} \right| \right) \frac{v(\boldsymbol{x})}{x_1^2 \cdots x_n^2} d\boldsymbol{x} d\boldsymbol{t}$$

$$\le \int_0^{b_1} \cdots \int_0^{b_n} w(\boldsymbol{x}) \varphi(f(\boldsymbol{x})) \frac{d\boldsymbol{x}}{x_1 \cdots x_n}. \tag{1.7.3}$$

(ii) *If* φ *is subquadratic, then* (1.7.3) *holds in the reversed direction.*

Remark 1.7.2. If we consider Proposition 1.7.1 for $v(\boldsymbol{x}) \equiv 1$, then we have

$$w(\boldsymbol{x}) = x_1 \cdots x_n \int_{x_1}^{b_1} \cdots \int_{x_n}^{b_n} \frac{d\boldsymbol{t}}{t_1^2 \cdots t_n^2} = \prod_{i=1}^{n} \left(1 - \frac{x_i}{b_i} \right), \ \boldsymbol{x} \in (\mathbf{0}, \mathbf{b}),$$

so (1.7.3) reads as follows: If φ is a superquadratic, then

$$\int_0^{b_1} \cdots \int_0^{b_n} \varphi \left(\frac{1}{x_1 \cdots x_n} \int_0^{x_1} \cdots \int_0^{x_n} f(\boldsymbol{t}) d\boldsymbol{t} \right) \frac{d\boldsymbol{x}}{x_1 \cdots x_n}$$

$$+ \int_0^{b_1} \cdots \int_0^{b_n} \int_{t_1}^{b_1} \cdots \int_{t_n}^{b_n} \varphi \left(\left| f(\boldsymbol{t}) - \frac{1}{x_1 \cdots x_n} \int_0^{x_1} \cdots \int_0^{x_n} f(\boldsymbol{t}) d\boldsymbol{t} \right| \right) \frac{d\boldsymbol{x}}{x_1^2 \cdots x_n^2} d\boldsymbol{t}$$

$$\le \int_0^{b_1} \cdots \int_0^{b_n} \varphi(f(\boldsymbol{x})) \prod_{i=1}^{n} \left(1 - \frac{x_i}{b_i} \right) \frac{d\boldsymbol{x}}{x_1 \cdots x_n}, \tag{1.7.4}$$

and (1.7.4) holds in the reversed direction when φ is subquadratic.

Proof. (i) Let φ be superquadratic. Then, by applying the refined Jensen's inequality (1.7.2) to the first term on the left–hand side of (1.7.4) and then the Fubini theorem repeatedly, we have that

$$\int_0^{b_1} \cdots \int_0^{b_n} v(\boldsymbol{x}) \varphi \left(\frac{1}{x_1 \cdots x_n} \int_0^{x_1} \cdots \int_0^{x_n} f(\boldsymbol{t}) d\boldsymbol{t} \right) \frac{d\boldsymbol{x}}{x_1 \cdots x_n}$$

$$\leq \int_0^{b_1} \cdots \int_0^{b_n} \frac{v(\boldsymbol{x})}{x_1^2 \cdots x_n^2} \left(\int_0^{x_1} \cdots \int_0^{x_n} \varphi(f(\boldsymbol{t}))d\boldsymbol{t} \right) d\boldsymbol{x}$$

$$- \int_0^{b_1} \cdots \int_0^{b_n} \frac{v(\boldsymbol{x})}{x_1^2 \cdots x_n^2} \int_0^{x_1} \cdots \int_0^{x_n} \varphi \left(\left| f(\boldsymbol{t}) - \frac{1}{x_1 \cdots x_n} \int_0^{x_1} \cdots \int_0^{x_n} f(\boldsymbol{t})d\boldsymbol{t} \right| \right) d\boldsymbol{t}d\boldsymbol{x}$$

$$= \int_0^{b_1} \cdots \int_0^{b_n} \varphi(f(\boldsymbol{t})) \int_{t_1}^{b_1} \cdots \int_{t_n}^{b_n} \frac{v(\boldsymbol{x})}{x_1^2 \cdots x_n^2} d\boldsymbol{x}d\boldsymbol{t}$$

$$- \int_0^{b_1} \cdots \int_0^{b_n} \int_{t_1}^{b_1} \int_{t_n}^{b_n} \varphi \left(\left| f(\boldsymbol{t}) - \frac{1}{x_1 \cdots x_n} \int_0^{x_1} \cdots \int_0^{x_n} f(\boldsymbol{t})d\boldsymbol{t} \right| \right) \frac{v(\boldsymbol{x})}{x_1^2 \cdots x_n^2} d\boldsymbol{x}d\boldsymbol{t}$$

$$= \int_0^{b_1} \cdots \int_0^{b_n} w(\boldsymbol{t})\varphi(f(\boldsymbol{t})) \frac{d\boldsymbol{t}}{t_1 \cdots t_n}$$

$$- \int_0^{b_1} \cdots \int_0^{b_n} \int_{t_1}^{b_1} \cdots \int_{t_n}^{b_n} \varphi \left(\left| f(\boldsymbol{t}) - \frac{1}{x_1 \cdots x_n} \int_0^{x_1} \cdots \int_0^{x_n} f(\boldsymbol{t})d\boldsymbol{t} \right| \right) \frac{v(\boldsymbol{x})}{x_1^2 \cdots x_n^2} d\boldsymbol{x}d\boldsymbol{t},$$

from which (1.7.3) follows.

(ii) Similar to the proof of (i) and by making the same calculations with φ subquadratic we see that only the inequality sign will be reversed. □

Proposition 1.7.2. *Let $\boldsymbol{b} \in [\boldsymbol{0}, \boldsymbol{\infty})$, $v : (\boldsymbol{b}, \boldsymbol{\infty}) \to \mathbb{R}$ be a weight function which is locally integrable in $(\boldsymbol{0}, \boldsymbol{b})$ and $w(\boldsymbol{x})$ be defined by (1.7.1). Suppose $I = (a, c)$, $0 \leq a < c \leq \infty$, $\varphi : I \to \mathbb{R}$, and $f : (\boldsymbol{b}, \boldsymbol{\infty}) \to \mathbb{R}$ is an integrable function, such that $f(\boldsymbol{x}) \in I$, for all $\boldsymbol{x} \in (\boldsymbol{b}, \boldsymbol{\infty})$.*

(i) If φ is superquadratic, then the following inequality holds:

$$\int_{b_1}^{\infty} \cdots \int_{b_n}^{\infty} v(\boldsymbol{x})\varphi \left(x_1 \cdots x_n \int_{x_1}^{\infty} \cdots \int_{x_n}^{\infty} f(\boldsymbol{t}) \frac{d\boldsymbol{t}}{t_1^2 \cdots t_n^2} \right) \frac{d\boldsymbol{x}}{x_1 \cdots x_n}$$

$$+ \int_{b_1}^{\infty} \cdots \int_{b_n}^{\infty} \int_{b_1}^{t_1} \cdots \int_{b_n}^{t_n} \varphi \left(\left| f(\boldsymbol{t}) - x_1 \cdots x_n \int_{x_1}^{\infty} \cdots \int_{x_n}^{\infty} f(\boldsymbol{t}) \frac{d\boldsymbol{t}}{t_1^2 \cdots t_n^2} \right| \right) v(\boldsymbol{x})d\boldsymbol{x} \frac{d\boldsymbol{t}}{t_1^2 \cdots t_n^2}$$

$$\leq \int_{b_1}^{\infty} \cdots \int_{b_n}^{\infty} w(\boldsymbol{x})\varphi(f(\boldsymbol{x})) \frac{d\boldsymbol{x}}{x_1 \cdots x_n}. \tag{1.7.5}$$

(ii) If φ is subquadratic, then the inequality sign in (1.7.5) is reversed.

$$I_2=\int_0^{a_1}\cdots\int_0^{a_n}\int_{t_1}^{a_1}\cdots\int_{t_n}^{a_n}\Bigg|f(t_1^{\frac{p}{k_1-1}},\dots,t_n^{\frac{p}{k_n-1}})\prod_{i=1}^n t_i^{\frac{p}{k_i-1}-1}$$

$$-\frac{1}{x_1\cdots x_n}\int_0^{x_1}\cdots\int_0^{x_n}f(t_1^{\frac{p}{k_1-1}},\dots,t_n^{\frac{p}{k_n-1}})\prod_{i=1}^n t_i^{\frac{p}{k_i-1}-1}d\boldsymbol{t}\Bigg|^p\frac{d\boldsymbol{x}}{x_1^2\cdots x_n^2}d\boldsymbol{t}$$

$$=\left(\prod_{i=1}^n\frac{k_i-1}{p}\right)^{p+1}\int_0^{b_1}\cdots\int_0^{b_n}\int_{s_1^{\frac{k_1-1}{p}}}^{a_1}\cdots\int_{s_n^{\frac{k_n-1}{p}}}^{a_n}\Bigg|\prod_{i=1}^n\frac{p}{k_i-1}s_i^{1-\frac{k_i-1}{p}}f(\boldsymbol{s})$$

$$-\frac{1}{x_1\cdots x_n}\int_0^{x^{\frac{p}{k_1-1}}}\cdots\int_0^{x^{\frac{p}{k_n-1}}}f(\boldsymbol{s})d\boldsymbol{s}\Bigg|^p\frac{d\boldsymbol{x}}{x_1^2\cdots x_n^2}\prod_{i=1}^n s_i^{\frac{k_i-1}{p}-1}d\boldsymbol{s}$$

$$=\left(\prod_{i=1}^n\frac{k_i-1}{p}\right)^{p+2}\int_0^{b_1}\cdots\int_0^{b_n}\int_{s_1}^{b_1}\cdots\int_{s_n}^{b_n}\Bigg|\prod_{i=1}^n\frac{p}{k_i-1}s_i^{1-\frac{k_i-1}{p}}f(\boldsymbol{s})$$

$$-\frac{1}{y_1^{\frac{k_1-1}{p}}\cdots y_n^{\frac{k_n-1}{p}}}\int_0^{y_1}\cdots\int_0^{y_n}f(\boldsymbol{s})d\boldsymbol{s}\Bigg|^p\prod_{i=1}^n y_i^{\frac{1-k_i}{p}-1}d\boldsymbol{y}\prod_{i=1}^n s_i^{\frac{k_i-1}{p}-1}d\boldsymbol{s}$$

$$=\left(\prod_{i=1}^n\frac{k_i-1}{p}\right)^{p+2}\int_0^{b_1}\cdots\int_0^{b_n}\int_{s_1}^{b_1}\cdots\int_{s_n}^{b_n}\Bigg|\prod_{i=1}^n\frac{p}{k_i-1}\left(\frac{s_i}{y_i}\right)^{1-\frac{k_i-1}{p}}f(\boldsymbol{s})$$

$$-\frac{1}{y_1\cdots y_n}\int_0^{y_1}\cdots\int_0^{y_n}f(\boldsymbol{s})d\boldsymbol{s}\Bigg|^p\prod_{i=1}^n y_i^{p-k_i-\frac{k_i-1}{p}}d\boldsymbol{y}\prod_{i=1}^n s_i^{\frac{k_i-1}{p}-1}d\boldsymbol{s}. \qquad (1.7.10)$$

and

$$I_3=\int_0^{a_1}\cdots\int_0^{a_n}f^p(x_1^{\frac{p}{k_i-1}},\dots,x_n^{\frac{p}{k_n-1}})\prod_{i=1}^n x_i^{p\left(\frac{p}{k_i-1}-1\right)}\left(1-\frac{x_i}{a_i}\right)\frac{d\boldsymbol{x}}{x_1\cdots x_n}$$

$$=\left(\prod_{i=1}^n\frac{k_i-1}{p}\right)\int_0^{b_1}\cdots\int_0^{b_n}\prod_{i=1}^n\left(1-\left[\frac{y_i}{b_i}\right]^{\frac{k_i-1}{p}}\right)y_i^{p-k_i}f^p(\boldsymbol{y})d\boldsymbol{y}. \qquad (1.7.11)$$

The proof of (1.7.7) follows by combining (1.7.8)–(1.7.11).

(ii) The proof for the case $1<p\le 2$ is similar to that of (i). The only difference is that in this case all the inequalities signs are reversed. $\square$

In the next result we state the dual of Theorem 1.7.1.

Theorem 1.7.2. *Let* $1 < p < \infty$, $\boldsymbol{k} = (k_1, \cdots k_n) \in \mathbb{R}^n$ *be such that* $k_i < 1$, $i = 1, 2, \ldots, n$, $0 \le \boldsymbol{b} < \infty$, *and let the function* f *be locally integrable on* $(\boldsymbol{b}, \infty)$ *and such that* $0 < \int_{b_1}^{\infty} \cdots \int_{b_n}^{\infty} \prod_{i=1}^{n} x_i^{p-k_i} f^p(\boldsymbol{x}) d\boldsymbol{x} < \infty$.

(iii) *If* $p \ge 2$, *then*

$$\int_{b_1}^{\infty} \cdots \int_{b_n}^{\infty} \prod_{i=1}^{n} x_i^{-k_i} \left(\int_{x_1}^{\infty} \cdots \int_{x_n}^{\infty} f(\boldsymbol{t}) d\boldsymbol{t} \right)^p d\boldsymbol{x}$$

$$+ \left(\prod_{i=1}^{n} \frac{1-k_i}{p} \right) \int_{b_1}^{\infty} \cdots \int_{b_n}^{\infty} \int_{b_1}^{t_1} \cdots \int_{b_n}^{t_n} \left| \prod_{i=1}^{n} \frac{p}{1-k_i} \left(\frac{t_i}{x_i} \right)^{\frac{1-k_i}{p}+1} f(\boldsymbol{t}) \right.$$

$$\left. - \frac{1}{x_1 \cdots x_n} \int_{x_1}^{\infty} \cdots \int_{x_n}^{\infty} f(\boldsymbol{t}) d\boldsymbol{t} \right|^p \prod_{i=1}^{n} x_i^{\frac{1-k_i}{p}+p-k_i} d\boldsymbol{x} \prod_{i=1}^{n} t_i^{\frac{k_i-1}{p}-1} d\boldsymbol{t}$$

$$\le \left(\prod_{i=1}^{n} \frac{p}{1-k_i} \right)^p \int_{b_1}^{\infty} \cdots \int_{b_n}^{\infty} \prod_{i=1}^{n} \left(1 - \left[\frac{b_i}{x_i} \right]^{\frac{1-k_i}{p}} \right) x_i^{p-k_i} f^p(\boldsymbol{x}) d\boldsymbol{x}. \tag{1.7.12}$$

(iv) *If* $1 < p \le 2$, *then inequality* (1.7.12) *holds in the reversed direction.*

Remark 1.7.4. Note that for the case $n = 1$ Theorem 1.7.2 reduces to Theorem 3.2 in [134].

Proof. (iii) By applying Proposition 1.7.2 with the superquadratic function $\varphi(x) = x^p$, $p \ge 2$ and $v(\boldsymbol{x}) \equiv 1$ (cf. Remark 1.7.2), we find that

$$\int_{b_1}^{\infty} \cdots \int_{b_n}^{\infty} \left(x_1 \cdots x_n \int_{x_1}^{\infty} \cdots \int_{x_n}^{\infty} f(\boldsymbol{t}) \frac{d\boldsymbol{t}}{t_1^2 \cdots t_n^2} \right)^p \frac{d\boldsymbol{x}}{x_1 \cdots x_n}$$

$$+ \int_{b_1}^{\infty} \cdots \int_{b_n}^{\infty} \int_{b_1}^{t_1} \cdots \int_{b_n}^{t_n} \left| f(\boldsymbol{t}) - x_1 \cdots x_n \int_{x_1}^{\infty} \cdots \int_{x_n}^{\infty} f(\boldsymbol{t}) \frac{d\boldsymbol{t}}{t_1^2 \cdots t_n^2} \right|^p d\boldsymbol{x} \frac{d\boldsymbol{t}}{t_1^2 \cdots t_n^2}$$

$$\le \int_{b_1}^{\infty} \cdots \int_{b_n}^{\infty} \prod_{i=1}^{n} \left(1 - \frac{b_i}{x_i} \right) f^p(\boldsymbol{x}) \frac{d\boldsymbol{x}}{x_1 \cdots x_n}. \tag{1.7.13}$$

Again, denote the first and second terms on the left–hand side of (1.7.13) by I_1 and I_2 and the term on the right–hand side by I_3, respectively.

Then, in (1.7.13) replace the parameter b_i by $a_i = b_i^{\frac{1-k_i}{p}}$, $i = 1, 2, \ldots, n$, and the function f by $\boldsymbol{x} \mapsto f(x_1^{\frac{p}{1-k_1}}, \ldots, x_n^{\frac{p}{1-k_n}}) \prod_{i=1}^{n} x_i^{\frac{p}{1-k_i}+1}$.

The remaining part of the proof follows by making similar calculations as in the proof of (i) in the Theorem 1.7.1, so we omit the details.

(iv) The proof for this case is similar to that above, now all inequality signs are just reversed. The proof is complete. □

Now we give some examples.

Example 1.7.1. In Theorem 1.7.1 if we let $1 < p < \infty$, $k_1 = \dots = k_n = k$, where $k > 1 \in \mathbb{R}$, $0 < \boldsymbol{b} \leq \infty$, and the function f be locally integrable on $(\boldsymbol{0}, \boldsymbol{b})$ such that $0 < \int_0^{b_1} \cdots \int_0^{b_n} \prod_{i=1}^n x_i^{p-k} f^p(\boldsymbol{x}) d\boldsymbol{x} < \infty$, then for $p \geq 2$, inequality (1.7.7) reads

$$\int_0^{b_1} \cdots \int_0^{b_n} \prod_{i=1}^n x_i^{-k} \left(\int_0^{x_1} \cdots \int_0^{x_n} f(\boldsymbol{t}) d\boldsymbol{t} \right)^p d\boldsymbol{x}$$

$$+ \left(\frac{k-1}{p} \right)^n \int_0^{b_1} \cdots \int_0^{b_n} \int_{t_1}^{b_1} \cdots \int_{t_n}^{b_n} \left| \left(\frac{p}{k-1} \right)^n \prod_{i=1}^n \left(\frac{t_i}{x_i} \right)^{1-\frac{k-1}{p}} f(\boldsymbol{t}) \right.$$

$$\left. - \frac{1}{x_1 \cdots x_n} \int_0^{x_1} \cdots \int_0^{x_n} f(\boldsymbol{t}) d\boldsymbol{t} \right|^p \prod_{i=1}^n x_i^{p-k-\frac{k-1}{p}} d\boldsymbol{x} \prod_{i=1}^n t_i^{\frac{k-1}{p}-1} d\boldsymbol{t}$$

$$\leq \left(\frac{p}{k-1} \right)^{np} \int_0^{b_1} \cdots \int_0^{b_n} \prod_{i=1}^n \left(1 - \left[\frac{x_i}{b_i} \right]^{\frac{k-1}{p}} \right) x_i^{p-k} f^p(\boldsymbol{x}) d\boldsymbol{x}. \tag{1.7.14}$$

The sign of the inequality in (1.7.14) is reversed if $1 < p \leq 2$.

Remark 1.7.5. By applying Example 1.7.1 with $p = 2$ we obtain the following remarkable identity: If $k > 1$, then

$$\int_0^{b_1} \cdots \int_0^{b_n} \prod_{i=1}^n x_i^{-k} \left(\int_0^{x_1} \cdots \int_0^{x_n} f(\boldsymbol{t}) d\boldsymbol{t} \right)^2 d\boldsymbol{x}$$

$$+ \left(\frac{k-1}{2} \right)^n \int_0^{b_1} \cdots \int_0^{b_n} \int_{t_1}^{b_1} \cdots \int_{t_n}^{b_n} \left| \left(\frac{2}{k-1} \right)^n \prod_{i=1}^n \left(\frac{t_i}{x_i} \right)^{1-\frac{k-1}{2}} f(\boldsymbol{t}) \right.$$

$$\left. - \frac{1}{x_1 \cdots x_n} \int_0^{x_1} \cdots \int_0^{x_n} f(\boldsymbol{t}) d\boldsymbol{t} \right|^2 \prod_{i=1}^n x_i^{2-k-\frac{k-1}{2}} d\boldsymbol{x} \prod_{i=1}^n t_i^{\frac{k-1}{2}-1} d\boldsymbol{t}$$

$$\leq \left(\frac{2}{k-1} \right)^{2n} \int_0^{b_1} \cdots \int_0^{b_n} \prod_{i=1}^n \left(1 - \left[\frac{x_i}{b_i} \right]^{\frac{k-1}{2}} \right) x_i^{2-k} f^2(\boldsymbol{x}) d\boldsymbol{x}.$$

Note that for the case $n = 1$ this identity coincides with that pointed out in Remark 3.1 in [134].

Remark 1.7.6. For the special case $k = p$ in Example 1.7.1, inequality (1.7.14) takes the form: If $p \geq 2$, then

$$\int_0^{b_1} \cdots \int_0^{b_n} \left(\frac{1}{x_1 x_2 \cdots x_n} \int_0^{x_1} \cdots \int_0^{x_n} f(\boldsymbol{t}) d\boldsymbol{t} \right)^p d\boldsymbol{x}$$

$$+ \left(\frac{p-1}{p} \right)^n \int_0^{b_1} \cdots \int_0^{b_n} \int_{t_1}^{b_1} \cdots \int_{t_n}^{b_n} \left| \left(\frac{p}{p-1} \right)^n \prod_{i=1}^n \left(\frac{t_i}{x_i} \right)^{\frac{1}{p}} f(\boldsymbol{t}) \right.$$

$$\left. - \frac{1}{x_1 \cdots x_n} \int_0^{x_1} \cdots \int_0^{x_n} f(\boldsymbol{t}) d\boldsymbol{t} \right|^p \prod_{i=1}^n x_i^{-\frac{p-1}{p}} d\boldsymbol{x} \prod_{i=1}^n t_i^{-\frac{1}{p}} d\boldsymbol{t}$$

$$\leq \left(\frac{p}{p-1} \right)^{np} \int_0^{b_1} \cdots \int_0^{b_n} \prod_{i=1}^n \left(1 - \left[\frac{x_i}{b_i} \right]^{\frac{p-1}{p}} \right) f^p(\boldsymbol{x}) d\boldsymbol{x}. \tag{1.7.15}$$

The sign of the inequality in (1.7.15) is reversed if $1 < p \leq 2$.

Note that in the one-dimensional case ($n = 1$) and $b_1 = b_2 = \cdots b_n = \infty$, these inequalities coincide with those given in Example 4.3 in [134].

Example 1.7.2. In Theorem 1.7.2 if we let $1 < p < \infty$, $k_1 = \cdots = k_n = k$, where $k < 1 \in \mathbb{R}$, $0 \leq \boldsymbol{b} < \infty$, and the function f be locally integrable in $(\boldsymbol{b}, \boldsymbol{\infty})$ and such that $0 < \int_{b_1}^{\infty} \cdots \int_{b_n}^{\infty} \prod_{i=1}^n x_i^{p-k} f^p(\boldsymbol{x}) d\boldsymbol{x} < \infty$. Then for $p \geq 2$,we obtain that

$$\int_{b_1}^{\infty} \cdots \int_{b_n}^{\infty} \prod_{i=1}^n x_i^{-k} \left(\int_{x_1}^{\infty} \cdots \int_{x_n}^{\infty} f(\boldsymbol{t}) d\boldsymbol{t} \right)^p d\boldsymbol{x}$$

$$+ \left(\frac{1-k}{p} \right)^n \int_{b_1}^{\infty} \cdots \int_{b_n}^{\infty} \int_{b_1}^{t_1} \cdots \int_{b_n}^{t_n} \left| \left(\frac{p}{1-k} \right)^n \prod_{i=1}^n \left(\frac{t_i}{x_i} \right)^{\frac{1-k}{p}+1} f(\boldsymbol{t}) \right.$$

$$\left. - \frac{1}{x_1 \cdots x_n} \int_{x_1}^{\infty} \cdots \int_{x_n}^{\infty} f(\boldsymbol{t}) d\boldsymbol{t} \right|^p \prod_{i=1}^n x_i^{\frac{1-k}{p}+p-k} d\boldsymbol{x} \prod_{i=1}^n t_i^{\frac{k-1}{p}-1} d\boldsymbol{t}$$

$$\leq \left(\frac{p}{1-k} \right)^{np} \int_{b_1}^{\infty} \cdots \int_{b_n}^{\infty} \prod_{i=1}^n \left(1 - \left[\frac{b_i}{x_i} \right]^{\frac{1-k}{p}} \right) x_i^{p-k} f^p(\boldsymbol{x}) d\boldsymbol{x}. \tag{1.7.16}$$

Inequality (1.7.16) holds in the reversed direction if $1 < p \leq 2$.

Remark 1.7.7. In Example 1.7.2 by setting $p = 2$, inequality (1.7.16) can be replaced by the following remarkable identity: If $k < 1$, then

$$\int_{b_1}^{\infty} \cdots \int_{b_n}^{\infty} \prod_{i=1}^{n} x_i^{-k} \left(\int_{x_1}^{\infty} \cdots \int_{x_n}^{\infty} f(\boldsymbol{t}) d\boldsymbol{t} \right)^2 d\boldsymbol{x}$$

$$+ \left(\frac{1-k}{2} \right)^n \int_{b_1}^{\infty} \cdots \int_{b_n}^{\infty} \int_{b_1}^{t_1} \cdots \int_{b_n}^{t_n} \left(\left(\frac{2}{1-k} \right)^n \prod_{i=1}^{n} \left(\frac{t_i}{x_i} \right)^{\frac{1-k}{2}+1} f(\boldsymbol{t}) \right.$$

$$\left. - \frac{1}{x_1 \cdots x_n} \int_{x_1}^{\infty} \cdots \int_{x_n}^{\infty} f(\boldsymbol{t}) d\boldsymbol{t} \right)^2 \prod_{i=1}^{n} x_i^{\frac{1-k}{2}+2-k} d\boldsymbol{x}$$

$$\prod_{i=1}^{n} t_i^{\frac{k-1}{2}-1} d\boldsymbol{t}$$

$$= \left(\frac{2}{1-k} \right)^{2n} \int_{b_1}^{\infty} \cdots \int_{b_n}^{\infty} \prod_{i=1}^{n} \left(1 - \left[\frac{b_i}{x_i} \right]^{\frac{1-k}{2}} \right) x_i^{2-k} f^2(\boldsymbol{x}) d\boldsymbol{x}. \tag{1.7.17}$$

In particular, for the one dimensional case ($n = 1$) and $b_1 = b_2 = \cdots = b_n = 0$ inequality (1.7.17) reduces to the identity in Remark 3.2 in [134].

We finish this Section by using a similar technique to prove some usual (not refined) multidimensional weighted Hardy and Pólya–Knopp type inequalities and their duals via Jensen's inequality. Our results are mainly taken from [132] and, in particular, they unify and further extend several results of this type in the literature including the fairly new results in [31], [79] and [133] (see also [28], [29], [30], [33], [80] and [110]). Our approach is to mainly use a convexity argument, which is completely different from the classical one used by Hardy and others (see, e.g., [67] and [102]). This direct way to prove Hardy inequality was first discovered in 1965 by Godunova [57] and further extended to a multidimensional setting in her paper [58] in 1968 (see also [100]). This idea was rediscovered independently in 1977 by Imoru [73] (see also [80]).

First we state some propositions, which are crucial for the proofs of our main results but also of independent interest.

Proposition 1.7.3. *Let $\boldsymbol{b} \in (\boldsymbol{0}, \boldsymbol{\infty}]$, $-\infty \le a < c \le \infty$ and let Φ be a positive function on $[a, c]$. Suppose that $v(\boldsymbol{x})$ is a weight function and $w(\boldsymbol{x})$ is defined by (1.7.1).*

(i) *If* Φ *is convex, then*

$$\int_0^{b_1}\cdots\int_0^{b_n} v(\boldsymbol{x})\Phi\left(\frac{1}{x_1\dots x_n}\int_0^{x_1}\cdots\int_0^{x_n} f(\mathbf{t})\,dt_1\dots dt_n\right)\frac{dx_1\dots dx_n}{x_1\dots x_n}$$

$$\leq \int_0^{b_1}\cdots\int_0^{b_n} w(\boldsymbol{x})\,\Phi(f(\boldsymbol{x}))\frac{dx_1\dots dx_n}{x_1\dots x_n} \tag{1.7.18}$$

holds for every function f *on* $(\mathbf{0},\mathbf{b})$ *such that* $a < f(\boldsymbol{x}) < c$.

(ii) *If* Φ *is concave, then*

$$\int_0^{b_1}\cdots\int_0^{b_n} v(\boldsymbol{x})\Phi\left(\frac{1}{x_1\dots x_n}\int_0^{x_1}\cdots\int_0^{x_n} f(\mathbf{t})\,dt_1\dots dt_n\right)\frac{dx_1\dots dx_n}{x_1\dots x_n}$$

$$\geq \int_0^{b_1}\cdots\int_0^{b_n} w(\boldsymbol{x})\,\Phi(f(\boldsymbol{x}))\frac{dx_1\dots dx_n}{x_1\dots x_n}, \tag{1.7.19}$$

holds for every function f *on* $(\mathbf{0},\mathbf{b})$ *such that* $a < f(\boldsymbol{x}) < c$.

Remark 1.7.8. For the special case $u(\boldsymbol{x}) \equiv 1$ in Proposition 1.7.3, inequalities (1.7.18) and (1.7.19) imply the corresponding inequalities (2.1) and (2.2) in Theorem 2.1 in [79].

Remark 1.7.9. Note that in the case $n = 1$ inequality (1.7.18) in Proposition 1.7.3 coincides with inequality (4) in Theorem 1 in [31].

Proof. (i) By applying Jensen's inequality and Fubini's theorem to the left–hand side of (1.7.18) we obtain that

$$\int_0^{b_1}\cdots\int_0^{b_n} v(\boldsymbol{x})\Phi\left(\frac{1}{x_1\dots x_n}\int_0^{x_1}\cdots\int_0^{x_n} f(\mathbf{t})\,dt_1\dots t_n\right)\frac{dx_1\dots dx_n}{x_1\dots x_n}$$

$$\leq \int_0^{b_1}\cdots\int_0^{b_n} v(\boldsymbol{x})\left(\frac{1}{x_1\dots x_n}\int_0^{x_1}\cdots\int_0^{x_n}\Phi(f(\mathbf{t}))dt_1\dots t_n\right)\frac{dx_1\dots dx_n}{x_1\dots x_n}$$

$$= \int_0^{b_1}\cdots\int_0^{b_n}\Phi(f(\mathbf{t}))\left(\int_{t_1}^{b_1}\cdots\int_{t_n}^{b_n} v(\boldsymbol{x})\,x_1^{-2}\dots x_n^{-2}dx_1\dots dx_n\right)dt_1\dots dt_n$$

$$= \int_0^{b_1}\cdots\int_0^{b_n} w(\mathbf{t})\Phi(f(\mathbf{t}))\frac{dt_1\dots dt_n}{t_1\dots t_n}.$$

(ii) The proof is completely similar to the proof of (i). In this case, since the function Φ is concave and, hence, the inequality sign is reversed. The proof is complete. $\square$

Proposition 1.7.4. *Let* $\mathbf{b} \in [\mathbf{0},\boldsymbol{\infty})$, $-\infty \leq a < c \leq \infty$ *and let* Φ *be a positive function on* $[a, c]$. *Assume that* $v(\boldsymbol{x})$ *is a weight function and* $w(\boldsymbol{x})$ *is defined by* (1.7.1).

(i) *If* Φ *is convex, then*

$$\int_{b_1}^{\infty} \cdots \int_{b_n}^{\infty} u(\boldsymbol{x})\Phi\left(x_1 \ldots x_n \int_{x_1}^{\infty} \cdots \int_{x_n}^{\infty} f(\boldsymbol{t}) \frac{dt_1 \ldots dt_n}{t_1^2 \ldots t_n^2}\right) \frac{dx_1 \ldots dx_n}{x_1 \ldots x_n}$$

$$\leq \int_{b_1}^{\infty} \cdots \int_{b_n}^{\infty} v(\boldsymbol{x}) \Phi(f(\boldsymbol{x})) \frac{dx_1 \ldots dx_n}{x_1 \ldots x_n} \tag{1.7.20}$$

holds for every function f *on* $[\mathbf{b},\boldsymbol{\infty})$ *such that* $a < f(\boldsymbol{x}) < c$.

(ii) *If* Φ *is concave, then*

$$\int_{b_1}^{\infty} \cdots \int_{b_n}^{\infty} u(\boldsymbol{x})\Phi\left(x_1 \ldots x_n \int_{x_1}^{\infty} \cdots \int_{x_n}^{\infty} f(\boldsymbol{t}) \frac{dt_1 \ldots dt_n}{t_1^2 \ldots t_n^2}\right) \frac{dx_1 \ldots dx_n}{x_1 \ldots x_n}$$

$$\geq \int_{b_1}^{\infty} \cdots \int_{b_n}^{\infty} v(\boldsymbol{x}) \Phi(f(\boldsymbol{x})) \frac{dx_1 \ldots dx_n}{x_1 \ldots x_n} \tag{1.7.21}$$

holds for every function f *on* $[\mathbf{b},\boldsymbol{\infty})$ *such that* $a < f(\boldsymbol{x}) < c$.

Remark 1.7.10. For the case $n = 1$ inequality (1.7.20) in Proposition 1.7.4 coincides with inequality (6) in Theorem 2 in [31].

Proof. (i) By applying Jensen's inequality and Fubini's theorem to the left–hand side of (1.7.20) we find that

$$\int_{b_1}^{\infty} \cdots \int_{b_n}^{\infty} v(\boldsymbol{x})\Phi\left(x_1 \ldots x_n \int_{x_1}^{\infty} \cdots \int_{x_n}^{\infty} f(\boldsymbol{t}) \frac{dt_1 \ldots dt_n}{t_1^2 \ldots t_n^2}\right) \frac{dx_1 \ldots dx_n}{x_1 \ldots x_n}$$

$$\leq \int_{b_1}^{\infty} \cdots \int_{b_n}^{\infty} v(\boldsymbol{x})\left(x_1 \ldots x_n \int_{x_1}^{\infty} \cdots \int_{x_n}^{\infty} \Phi(f(\boldsymbol{t})) \frac{dt_1 \ldots dt_n}{t_1^2 \ldots t_n^2}\right) \frac{dx_1 \ldots dx_n}{x_1 \ldots x_n}$$

$$= \int_{b_1}^{\infty} \cdots \int_{b_n}^{\infty} \Phi(f(\boldsymbol{t}))\left(\int_{b_1}^{t_1} \cdots \int_{b_n}^{t_n} v(\boldsymbol{x})\, dx_1 \ldots dx_n\right) \frac{dt_1 \ldots dt_n}{t_1^2 \ldots t_n^2}$$

$$= \int_{b_1}^{\infty} \cdots \int_{b_n}^{\infty} w(\boldsymbol{t}) \Phi \left(f \left(\boldsymbol{t} \right) \right) \frac{dt_1 \dots dt_n}{t_1 \dots t_n}.$$

(ii) The proof is completely similar to the proof of (i). In this case, by assumption the function Φ is concave and, hence, the inequality sign is reversed. □

We now state the following multidimensional version of the Hardy–type inequalities in [31]:

Theorem 1.7.3. (i) *Let* $m > 1$, $\boldsymbol{b} \in (\boldsymbol{0}, \boldsymbol{\infty}]$, *and*

$$0 < \int_0^{b_1} \cdots \int_0^{b_n} x_1^{p-m} \cdots x_n^{p-m} f^p(\boldsymbol{x}) dx_1 \cdots dx_n < \infty.$$

If $p \geq 1$, *then*

$$\int_0^{b_1} \cdots \int_0^{b_n} x_1^{-m} \cdots x_n^{-m} \left(\int_0^{x_1} \cdots \int_0^{x_n} f \left(\boldsymbol{t} \right) dt_1 \cdots dt_n \right)^p dx_1 \cdots dx_n \leq \left(\frac{p}{m-1} \right)^{pn}$$

$$\times \int_0^{b_1} \cdots \int_0^{b_n} \prod_{j=1}^n \left[1 - \left(\frac{x_j}{b_j} \right)^{\frac{m-1}{p}} \right] x_1^{p-m} \cdots x_n^{p-m} f^p(\boldsymbol{x}) dx_1 \cdots dx_n, \tag{1.7.22}$$

where f *be a nonnegative function on* $(\boldsymbol{0}, \mathbf{b})$. *If* $0 < p \leq 1$, *then* (1.7.22) *holds in the reversed direction.*

(ii) *Let* $m < 1$, $\boldsymbol{b} \in [\boldsymbol{0}, \boldsymbol{\infty})$, *and* f *be a nonnegative function on* $[\boldsymbol{b}, \boldsymbol{\infty})$ *such that*

$$0 < \int_{b_1}^{\infty} \cdots \int_{b_n}^{\infty} x_1^{p-m} \dots x_n^{p-m} f^p(\boldsymbol{x}) dx_1 \dots dx_n < \infty.$$

If $p \geq 1$, *then*

$$\int_{b_1}^{\infty} \cdots \int_{b_n}^{\infty} x_1^{-m} \dots x_n^{-m} \left(\int_{x_1}^{\infty} \cdots \int_{x_n}^{\infty} f \left(\boldsymbol{t} \right) dt_1 \cdots dt_n \right)^p dx_1 \cdots dx_n \leq \left(\frac{p}{1-m} \right)^{pn}$$

$$\times \int_{b_1}^{\infty} \cdots \int_{b_n}^{\infty} \prod_{j=1}^n \left[1 - \left(\frac{b_j}{x_j} \right)^{\frac{1-m}{p}} \right] x_1^{p-m} \cdots x_n^{p-m} f^p(\boldsymbol{x}) dx_1 \cdots dx_n. \tag{1.7.23}$$

If $0 < p \leq 1$, *then* (1.7.23) *holds in the reversed direction.*

(iii) *The constant* $\left(\frac{p}{|m-1|} \right)^{pn}$ *is the best possible in both inequalities* (1.7.22) *and* (1.7.23).

Remark 1.7.11. For the case $n = 1$ inequalities (1.7.22) and (1.7.23) in Theorem 1.7.3 coincide with inequalities (8) and (9) in Corollary 1 in [31].

Remark 1.7.12. A similar result for the case $p > 1$ can be found in [30] but our proof is completely different and simpler. Note that when $p = 1$ we have equality both in (1.7.22) and (1.7.23), which can also be seen directly by using Fubini's theorem.

Remark 1.7.13. For the special case $m = p$ inequality (1.7.23) coincides with inequality (2.3) in Corollary 2.2 in [79]. In particular, if $\boldsymbol{b} = \infty$ we obtain that

$$\int_0^\infty \cdots \int_0^\infty \left(\frac{1}{x_1 \dots x_n} \int_0^{x_1} \cdots \int_0^{x_n} f(\mathbf{t})\, dt_1 \dots dt_n \right)^p dx_1 \dots dx_n$$

$$\leq \left(\frac{p}{p-1} \right)^{pn} \int_0^\infty \cdots \int_0^\infty f^p(\boldsymbol{x}) dx_1 \dots dx_n, \tag{1.7.24}$$

for $p > 1$, $n \in \mathbb{N}$, which is a multidimensional generalization of the classical Hardy inequality (1) with the sharp constant $\left(\frac{p}{p-1}\right)^{pn}$.

Proof. (i) Let $p > 1$. By using Proposition 1.7.3(i) with the convex function $\Phi(\boldsymbol{x}) = \boldsymbol{x}^p$ and the weight function $u(\boldsymbol{x}) \equiv 1$ we obtain that

$$\int_0^{b_1} \cdots \int_0^{b_n} \left(\frac{1}{x_1 \dots x_n} \int_0^{x_1} \cdots \int_0^{x_n} f(\mathbf{t})\, dt_1 \dots dt_n \right)^p \frac{dx_1 \dots dx_n}{x_1 \dots x_n}$$

$$\leq \int_0^{b_1} \cdots \int_0^{b_n} \left(1 - \frac{x_1}{b_1}\right) \cdots \left(1 - \frac{x_n}{b_n}\right) f^p(x_1, \dots, x_n) \frac{dx_1 \dots dx_n}{x_1 \dots x_n}. \tag{1.7.25}$$

Now, replace b_j by $a_j = b_j^{\frac{m-1}{p}}$ and the function f by $f\left(\boldsymbol{x}^{\frac{p}{m-1}}\right) x_1^{\frac{p}{m-1}-1} \dots x_n^{\frac{p}{m-1}-1}$. Thereafter, use the substitutions $s_j = t_j^{\frac{p}{m-1}}$ and $y_j = x_j^{\frac{p}{m-1}}$, respectively. Then, the left–hand side of (1.7.25) is equal to

$$\int_0^{b_1} \cdots \int_0^{b_n} \left(\frac{1}{x_1 \dots x_n} \int_0^{x_1} \cdots \int_0^{x_n} f(\mathbf{t})\, dt_1 \dots dt_n \right)^p \frac{dx_1 \dots dx_n}{x_1 \dots x_n}$$

$$= \int_0^{a_1} \cdots \int_0^{a_n} \left(\frac{1}{x_1 \dots x_n} \int_0^{x_1} \cdots \right.$$

$$\cdots\int_0^{x_n} f\left(\boldsymbol{t}^{\frac{p}{m-1}}\right) t_1^{\frac{p}{m-1}-1}\cdots t_n^{\frac{p}{m-1}-1}dt_1\dots dt_n\Big)^p \frac{dx_1\dots dx_n}{x_1\dots x_n}$$

$$=\left(\frac{m-1}{p}\right)^{pn}\int_0^{a_1}\cdots\int_0^{a_n}\Big(\frac{1}{x_1\dots x_n}\int_0^{x_1^{\frac{p}{m-1}}}\cdots$$

$$\cdots\int_0^{x_n^{\frac{p}{m-1}}} f(\boldsymbol{s})\,ds_1\dots ds_n\Big)^p \frac{dx_1\dots dx_n}{x_1\dots x_n}$$

$$=\left(\frac{m-1}{p}\right)^{pn+n}\int_0^{b_1}\cdots\int_0^{b_n}\Big(\int_0^{y_1}\cdots$$

$$\cdots\int_0^{y_n} f(\boldsymbol{s})\,ds_1\dots ds_n\Big)^p y_1^{-\left(\frac{m-1}{p}\right)p}\cdots y_n^{-\left(\frac{m-1}{p}\right)p}\frac{dy_1\dots dy_n}{y_1\dots y_n}$$

$$=\left(\frac{m-1}{p}\right)^{pn+n}\int_0^{b_1}\cdots\int_0^{b_n}\Big(\int_0^{y_1}\cdots$$

$$\cdots\int_0^{y_n} f(\boldsymbol{s})\,ds_1\dots ds_n\Big)^p y_1^{-m}\cdots y_n^{-m}dy_1\dots dy_n. \tag{1.7.26}$$

Similarly, by using the above substitutions, the right–hand side of (1.7.25) becomes

$$\int_0^{a_1}\cdots\int_0^{a_n}\prod_{j=1}^n\left(1-\frac{x_j}{a_j}\right) f^p(\boldsymbol{x}^{\frac{p}{m-1}})x_1^{p(\frac{p}{m-1}-1)}\cdots x_n^{p(\frac{p}{m-1}-1)}\frac{dx_1\dots dx_n}{x_1\dots x_n}$$

$$=\left(\frac{m-1}{p}\right)^n\int_0^{b_1}\cdots\int_0^{b_n}\prod_{j=1}^n\left[1-\left(\frac{y_j}{b_j}\right)^{\frac{m-1}{p}}\right] f^p(\boldsymbol{y})y_1^{p-m+1}\cdots$$

$$\cdots y_n^{p-m+1}\frac{dy_1\dots dy_n}{y_1\dots y_n}$$

$$=\left(\frac{m-1}{p}\right)^n\int_0^{b_1}\cdots\int_0^{b_n}\prod_{j=1}^n\left[1-\left(\frac{y_j}{b_j}\right)^{\frac{m-1}{p}}\right] y_1^{p-m}\cdots$$

$$\cdots y_n^{p-m}f^p(\boldsymbol{y})dy_1\dots dy_n. \tag{1.7.27}$$

Inequality (1.7.22) follows by combining (1.7.26) and (1.7.27). Now, for the case $p = 1$, $m > 1$, we have by Fubini's theorem that

$$\int_0^{b_1}\cdots\int_0^{b_n} x_1^{-m}\dots x_n^{-m}\left(\int_0^{x_1}\cdots\int_0^{x_n} f(\boldsymbol{t})\,dt_1\dots dt_n\right)dx_1\dots dx_n$$

$$=\int_0^{b_1}\cdots\int_0^{b_n} f(\boldsymbol{t})\left(\int_{t_1}^{b_1}\cdots\int_{t_n}^{b_n} x_1^{-m}\dots x_n^{-m}dx_1\dots dx_n\right)dt_1\dots dt_n$$

$$=\left(\frac{1}{m-1}\right)^n\int_0^{b_1}\cdots\int_0^{b_n}\prod_{j=1}^n\left[1-\left(\frac{x_j}{b_j}\right)^{m-1}\right]t_1^{-m}\dots t_n^{-m}dt_1\dots dt_n.$$

For the case $0 < p < 1$ we just use Proposition 1.7.3 (ii) with the concave function $\Phi(\boldsymbol{x}) = \boldsymbol{x}^p$ and (1.7.25) holds in the reversed direction. By combining (1.7.26) and (1.7.27) we have completed the proof of (i).

(ii) Let $p > 1$ we use Proposition 1.7.4 (i) with the convex function $\Phi(\mathbf{x}) = \mathbf{x}^p$ and the weight function $v(\boldsymbol{x}) = 1$. Hence,

$$\int_{b_1}^{\infty}\cdots\int_{b_n}^{\infty}\left(x_1\dots x_n\int_{x_1}^{\infty}\cdots\int_{x_n}^{\infty} f(\boldsymbol{t})\frac{dt_1\dots dt_n}{t_1^2\dots t_n^2}\right)^p\frac{dx_1\dots dx_n}{x_1\dots x_n}$$

$$\le\int_{b_1}^{\infty}\cdots\int_{b_n}^{\infty}\prod_{j=1}^n\left(1-\frac{b_j}{x_j}\right)f^p(\boldsymbol{x})\frac{dx_1\dots dx_n}{x_1\dots x_n}. \tag{1.7.28}$$

Now, replace b_j by $a_j = a_j^{\frac{1-m}{p}}$ and f by $f\left(\boldsymbol{x}^{\frac{p}{1-m}}\right)x_1^{\frac{p}{1-m}+1}\dots x_n^{\frac{p}{1-m}+1}$. Thereafter, use the substitutions $s_j = t_j^{\frac{p}{1-m}}$ and $y_j = x_j^{\frac{p}{1-m}}$, respectively. The remaining part of the proof is similar to the corresponding part in the proof of (i).

For the case $p = 1$ we use Fubini's theorem as in the proof of (i) and we find that (1.7.23) holds with equality.

The proof of the case $0 < p < 1$ is completely similar to the proof of the case $p > 1$. Here we just use Proposition 1.7.4 (ii) to see that now (1.7.28) holds in the reversed direction.

(iii) Now we prove the sharpness of the constant $\left(\frac{p}{m-1}\right)^{pn}$ for inequality (1.7.22). Define the function f by $f(\boldsymbol{x}) = \prod_{j=1}^n f(x_j)$, $\boldsymbol{x} \in (\mathbf{0}, \boldsymbol{b})$. Then, by making a simple calculation the left–hand side of (1.7.22) yields

$$\int_0^{b_1}\cdots\int_0^{b_n} x_1^{-m}\dots x_n^{-m}\left(\int_0^{x_1}\cdots\int_0^{x_n} f(t_1)f(t_2)\dots f(t_n)dt_1\dots dt_n\right)^p dx_1\dots dx_n$$

$$= \int_0^{b_1} x_1^{-m}\left(\int_0^{x_1} f(t_1)\,dt_1\right)^p dx_1 \int_0^{b_2} x_2^{-m}\left(\int_0^{x_1} f(t_2)\,dt_2\right)^p dx_2 \cdots$$

$$\cdots \int_0^{b_n} x_1^{-m}\left(\int_0^{x_n} f(t_n)\,dt_n\right)^p dx_n$$

$$\leq \left(\frac{p}{m-1}\right)^{pn} \int_0^{b_1}\left[1-\left(\frac{x_1}{b_1}\right)^{\frac{m-1}{p}}\right] x_1^{p-m} f^p(x_1)dx_1 \cdots$$

$$\cdots \int_0^{b_n}\left[1-\left(\frac{x_n}{b_n}\right)^{\frac{m-1}{p}}\right] x_n^{p-m} f^p(x_n)dx_n$$

$$= \left(\frac{p}{m-1}\right)^{pn} \int_0^{b_1}\cdots\int_0^{b_n} \prod_{j=1}^{n}\left[1-\left(\frac{x_j}{b_j}\right)^{\frac{m-1}{p}}\right] x_1^{p-m}\ldots x_n^{p-m} f^p(\boldsymbol{x})dx_1\ldots dx_n.$$

It is well known that Hardy inequality in the one-dimensional case has the sharp constant $\left(\frac{p}{m-1}\right)^p$. Hence, the constant $\left(\frac{p}{m-1}\right)^{pn}$ in (1.7.22) is sharp.

The proof of the sharpness of the constant $\left(\frac{p}{1-m}\right)^{pn}$ in (1.7.23) is similar if we define the function f by $f(\boldsymbol{x}) = \prod_{j=1}^{n} f(x_j)$, $\boldsymbol{x} \in (\boldsymbol{b}, \infty)$. □

Remark 1.7.14. By applying Theorem 1.7.3 (ii) and (iii) with $b = 0$, $m = p$, $0 < p < 1$, we obtain the following reversed Hardy-type inequality (c.f. inequality (1.7.24)):

$$\int_0^{\infty}\cdots\int_0^{\infty}\left(\frac{1}{x_1\ldots x_n}\int_{x_1}^{\infty}\cdots\int_{x_n}^{\infty} f(\boldsymbol{t})dt_1\ldots dt_n\right)^p dx_1\ldots dx_n$$

$$\geq \left(\frac{p}{1-p}\right)^{pn}\int_0^{\infty}\cdots\int_0^{\infty} f^p(\boldsymbol{x})dx_1\ldots dx_n,$$

for $0 < p < 1$, $n \in \mathbb{N}$.

For $n = 1$ this is just Theorem 337 in [68].

Next we point out that our simple convexity argument directly gives also the following results of Čižmešija and Pečarić [30]:

Theorem 1.7.4. *Let $\gamma \in \mathbb{R}$, and let f be a positive function.*

(i) *If* $\alpha > 0$, $\boldsymbol{b} \in (\boldsymbol{0}, \boldsymbol{\infty}]$, *and*

$$0 < \int_0^{b_1} \cdots \int_0^{b_n} x_1^{\gamma-1} \dots x_n^{\gamma-1} f(\boldsymbol{x}) dx_1 \dots dx_n < \infty,$$

then

$$\int_0^{b_1} \cdots \int_0^{b_n} x_1^{\gamma-1} \cdots x_n^{\gamma-1} \exp\left(\prod_{j=1}^n \frac{\alpha^n}{x_j^\alpha} \int_0^{x_1} \cdots \right.$$

$$\left. \cdots \int_0^{x_n} t_1^{\alpha-1} \dots t_n^{\alpha-1} \log f(\boldsymbol{t}) dt_1 \dots dt_n \right) dx_1 \dots dx_n$$

$$\le e^{\frac{\gamma}{\alpha} n} \int_0^{b_1} \cdots \int_0^{b_n} \prod_{j=1}^n \left[1 - \left(\frac{x_j}{b_j} \right)^\alpha \right] x_1^{\gamma-1} \dots x_n^{\gamma-1} f(\boldsymbol{x}) dx_1 \dots dx_n. \tag{1.7.29}$$

(ii) *If* $\alpha < 0$, $\boldsymbol{b} \in [\boldsymbol{0}, \boldsymbol{\infty})$, *and*

$$0 < \int_{b_1}^{\infty} \cdots \int_{b_n}^{\infty} x_1^{\gamma-1} \dots x_n^{\gamma-1} f(\boldsymbol{x}) dx_1 \dots dx_n < \infty,$$

then

$$\int_{b_1}^{\infty} \cdots \int_{b_n}^{\infty} x_1^{\gamma-1} \dots x_n^{\gamma-1} \exp\left(\prod_{j=1}^n \frac{(-\alpha)^n}{x_j^\alpha} \int_{x_1}^{\infty} \cdots \right.$$

$$\left. \cdots \int_{x_n}^{\infty} t_1^{\alpha-1} \dots t_n^{\alpha-1} \log f(\boldsymbol{t}) dt_1 \dots dt_n \right) dx_1 \dots dx_n$$

$$\le e^{\frac{\gamma}{\alpha} n} \int_{b_1}^{\infty} \cdots \int_{b_n}^{\infty} \prod_{j=1}^n \left[1 - \left(\frac{b_j}{x_j} \right)^{-\alpha} \right] x_1^{\gamma-1} \dots x_n^{\gamma-1} f(\boldsymbol{x}) dx_1 \dots dx_n. \tag{1.7.30}$$

The constant $e^{\frac{\gamma}{\alpha} n}$ *is the best possible in both inequalities* (1.7.29) *and* (1.7.30).

Proof. (i) By using Proposition 1.7.3(i) with the convex function $\Phi(\boldsymbol{x}) = \exp \boldsymbol{x}$ and $u(\boldsymbol{x}) \equiv 1$ we find that

$$\int_0^{b_1} \cdots \int_0^{b_n} \exp\left(\frac{1}{x_1 \cdots x_n} \int_0^{x_1} \cdots \int_0^{x_n} f(\boldsymbol{t}) dt_1 \dots dt_n \right) \frac{dx_1 \dots dx_n}{x_1 \dots x_n}$$

$$\leq \int_0^{b_1} \cdots \int_0^{b_n} \prod_{j=1}^n \left(1 - \frac{x_j}{b_j}\right) \exp f(\boldsymbol{x}) \frac{dx_1 \ldots dx_n}{x_1 \ldots x_n}. \tag{1.7.31}$$

Now, replace b_j by $a_j = b_j^\alpha$, f by $\log\left(x_1^{\frac{\gamma}{\alpha}} \ldots x_n^{\frac{\gamma}{\alpha}} f\left(\boldsymbol{x}^{\frac{1}{\alpha}}\right)\right)$ and, thereafter, use the substitutions $s_j = t_j^{\frac{1}{\alpha}}$ and $y_j = x_j^{\frac{1}{\alpha}}$ in (1.7.31). Then a simple calculation yields the desired result.

(ii) The proof follows by using Proposition 1.7.4 (i) in a similar way with $\Phi(\boldsymbol{x}) = \exp \boldsymbol{x}$ and $u(\boldsymbol{x}) \equiv 1$. Hence, inequality (1.7.20) reads

$$\int_{b_1}^{\infty} \cdots \int_{b_n}^{\infty} \exp\left(x_1 \ldots x_n \int_{x_1}^{\infty} \cdots \int_{x_n}^{\infty} f(\boldsymbol{t}) \frac{dt_1 \ldots dt_n}{t_1^2 \ldots t_n^2}\right) \frac{dx_1 \ldots dx_n}{x_1 \ldots x_n}$$

$$\leq \int_{b_1}^{\infty} \cdots \int_{b_n}^{\infty} \prod_{j=1}^n \left(1 - \frac{b_j}{x_j}\right) \exp f(\boldsymbol{x}) \frac{dx_1 \ldots dx_n}{x_1 \ldots x_n}. \tag{1.7.32}$$

Replace b_j by $a_j = b_j^{-\alpha}$, f by $\log\left(x_1^{-\frac{\gamma}{\alpha}} \ldots x_n^{-\frac{\gamma}{\alpha}} f\left(\boldsymbol{x}^{-\frac{1}{\alpha}}\right)\right)$ and, afterwards, use the substitutions $s_j = t_j^{-\frac{1}{\alpha}}$ and $y_j = x_j^{-\frac{1}{\alpha}}$ in (1.7.32). Then a simple calculation yields the desired result.

The proof of the sharpness of the constant can also be done by again using product functions as in the proof of Theorem 1.7.3 and the well-known fact that the constant $e^{\frac{\gamma}{\alpha}n}$ is sharp in the one-dimensional case (cf. also [29]). □

Next, we point out the following multidimensional versions of the so called Cochran-Lee and Love inequalities:

Remark 1.7.15. For the case $\boldsymbol{b} = \infty$, inequality (1.7.29) in Theorem 1.7.4 reads

$$\int_0^{\infty} \cdots \int_0^{\infty} x_1^{\gamma-1} \ldots x_n^{\gamma-1} \exp\left(\frac{\alpha^n}{x_1^\alpha \ldots x_n^\alpha} \int_0^{x_1} \cdots \right.$$

$$\left. \cdots \int_0^{x_n} t_1^{\alpha-1} \ldots t_n^{\alpha-1} \log f(\boldsymbol{t}) dt_1 \ldots dt_n \right) dx_1 \ldots dx_n$$

$$\leq e^{\frac{\gamma}{\alpha}n} \int_0^{\infty} \cdots \int_0^{\infty} x_1^{\gamma-1} \ldots x_n^{\gamma-1} f(\boldsymbol{x}) dx_1 \ldots dx_n, \tag{1.7.33}$$

which coincides in the special case $n = 1$ with an inequality by Cochran and Lee (see [110] and also inequality (16) in [31]).

On the other hand, inequality (1.7.30) in Theorem 1.7.4 rewritten with $\boldsymbol{b} = \boldsymbol{0}$ reads

$$\int_0^\infty \cdots \int_0^\infty x_1^{\gamma-1} \dots x_n^{\gamma-1} \exp\left(\frac{(-\alpha)^n}{x_1^\alpha \dots x_n^\alpha} \int_{x_1}^\infty \cdots \right.$$

$$\left. \cdots \int_{x_n}^\infty t_1^{\alpha-1} \dots t_n^{\alpha-1} \log f(\boldsymbol{t}) dt_1 \dots dt_n \right) dx_1 \dots dx_n$$

$$\leq e^{\frac{\gamma}{\alpha} n} \int_0^\infty \cdots \int_0^\infty x_1^{\gamma-1} \dots x_n^{\gamma-1} f(\boldsymbol{x}) dx_1 \dots dx_n \tag{1.7.34}$$

which in the special case $n = 1$ coincides also to an inequality by Love (see [110] and also inequality (17) in [31]).

Thus, inequality (1.7.33) and its dual (1.7.34) are multidimensional generalizations of the Cochran-Lee and Love inequalities, respectively.

Remark 1.7.16. For the case $n = 1$ inequalities (1.7.29) and (1.7.30) in Theorem 1.7.4 take the forms

$$\int_0^b x^{\gamma-1} \exp\left(\frac{\alpha}{x^\alpha} \int_0^x t^{\alpha-1} \log f(t) dt \right) dx$$

$$\leq e^{\frac{\gamma}{\alpha}} \int_0^b \left[1 - \left(\frac{x}{b} \right)^\alpha \right] x^{\gamma-1} f(x) dx$$

and

$$\int_b^\infty x^{\gamma-1} \exp\left(- \frac{\alpha}{x^\alpha} \int_x^\infty t^{\alpha-1} \log f(t) dt \right) dx$$

$$\leq e^{\frac{\gamma}{\alpha}} \int_b^\infty \left[1 - \left(\frac{b_1}{x_1} \right)^{-\alpha} \right] x^{\gamma-1} f(x) dx,$$

respectively, which coincide with inequalities (12) and (13) in Corollary 2 in [31].

Remark 1.7.17. In this paper we have, in particular, proved that Jensen's inequality implies multidimensional forms of Hardy and Pólya-Knopp's inequalities. It is well known that Jensen's inequality implies also generalization of several other classical inequalities including, e.g., those by Hölder, Young, Minkowski, Beckenbach–Dresher, Carlson, Hilbert, Levin, Hardy–Littlewood–Pólya and the AG-mean inequality or, more generally, that the scale of power means P_α or even general Gini means $G_{\alpha,\beta}$ increases in both α and β ($-\infty < \alpha, \beta < \infty$). See, e.g., the book of Niculescu–Persson [131] and the references given there.

1.8 Notes and Comments on Chapter 1

Theorem 1.1.3 and also the limiting results in Section 1.2 are based on the PhD thesis [174] by A. Wedestig (see also [175]). Theorem 1.1.6 and the results in Section 1.6 are due to A. Meskhi [120], [121] (see also the surveys [94] and [95]). Moreover, Sections 1.3 and 1.4 are based on the PhD thesis by E. Ushakova [171] (see also [145]). Another (direct) proof of Theorem 1.3.1 is based on the arguments used in [120]. The results and techniques in Section 1.7 are mainly taken from [132] and [136]. Moreover, the PhD theses by S. Barza [9] and M. Johansson [77] have also influenced our results and ideas in this chapter. In particular, [9] together with the paper [144] were important to connect the results in this chapter with those in second part of our next Chapter. For another type of multidimensional Hardy–type inequalities we refer to the review article [32] and for some recent results concerning Hardy–type inequalities involving both general measures and scales of conditions we refer to [137] and [138] and the references given in these papers. Finally, for weighted criteria for discrete Hardy inequality we refer the reader to [20], [21], [48], [100] and [132].

Chapter 2

Weighted Boundedness Criteria for Integral Transforms with Product Kernels

In this chapter necessary and sufficient conditions for a weight function v governing the boundedness of the Volterra, truncated and ball fractional integral operators with product kernels from L^p to L^q_v are established, where $1 < p \leq q < \infty$. The result for the operator with product positive kernels

$$Kf(x,y) = \int_0^x \int_0^y k_1(x,t)k_2(y,\tau)f(t,\tau)dtd\tau, \ x,y \in (0,\infty),$$

enables us to derive a trace inequality, for example, for the multiple Riemann-Liouville transform

$$\mathcal{R}_{\alpha,\beta}f(x,y) = \int_0^x \int_0^y (x-t)^{\alpha-1}(y-\tau)^{\beta-1}f(t,\tau)dtd\tau$$

with $\alpha, \beta > 1/p$.

It is shown that to derive two–weighted boundedness criteria for multiple integral transforms it is not enough using the iteration process and, moreover, corresponding inequalities for integral transforms with single kernel does not lead us to desired results. An example of weight function presented here shows that the condition which arises applying twice one-dimensional boundedness result to the operator

$$\mathcal{R}_{\alpha}(x) = \int_0^x (x-t)^{\alpha-1}f(t)dt$$

is different from the criterion guaranteeing the inequality

$$\|\mathcal{R}_{\alpha,\beta}f\|_{L^q_v(\mathbb{R}^2_+)} \le c\|f\|_{L^p(\mathbb{R}^2_+)}.$$

We provide also a special treatment of weighted boundedness criteria for integral transforms with product of mixed–type kernels.

Moreover, we prove some new multidimensional Hardy–type inequalities with general kernels via convexity. Finally, we prove and discuss some weighted integral inequalities for monotonic weights of several variables both in the case $1 < p \le q < \infty$ and the case $0 < q < p < \infty$. In particular, we present and discuss embeddings between some multidimensional Lorentz spaces introduced in [10]. The results are formulated in a fairly general frame with measures involved but can, in particular, be applied with multidimensional integral operators with general kernels involved. Some basic definitions and results concerning these multidimensional Lorentz spaces are included in an Appendix.

2.1 Integrals with General Product Kernels

In this section we introduce some classes of kernels.

Definition 2.1.1. We say that a kernel $k : \{(x,y) : 0 < y < x < \infty\} \longrightarrow \mathbb{R}_+$ belongs to V $(k \in V)$ if there exists a positive constant d_1 such that for all x, t, z with $0 < t < z < x < \infty$ the inequality

$$k(x,t) \le d_1 k(x,z)$$

holds.

Definition 2.1.2. A kernel k belongs to V_λ $(k \in V_\lambda)$ $(1 < \lambda < \infty)$ if there exists a positive constant d_2 such that for all $x, x > 0$, the inequality

$$\int\limits_{x/2}^{x} k^{\lambda'}(x,y)dy \le d_2 x k^{\lambda'}(x, x/2)$$

is fulfilled, where $\lambda' = \lambda/(\lambda-1)$.

It is easy to verify that if k_1 is a positive measurable function on $\mathbb{R}_+$ satisfying the condition

$$\int\limits_{0}^{x/2} k_1^{\lambda'}(t)dt \le d_3 x k_1^{\lambda'}(x/2), \quad \lambda' = \lambda/(\lambda-1),$$

with some positive constant d_3, then $k(x,y) = k_1(x-y)$ belongs to $V \cap V_\lambda$.

Example 2.1.1. Let $k_1(y) \equiv y^{\alpha-1}$, where $\alpha > 0$. If $1/\lambda < \alpha \le 1$, then the kernel $k(x,y) \equiv k_1(x-y)$ belongs to $V \cap V_\lambda$.

For other examples of the kernels in $V \cap V_\lambda$ see [119], [39, Section 2.7].

Let E be a measurable subset of $\mathbb{R}^n$. A measurable, almost everywhere positive function ρ (sometimes we will assume that ρ is also locally integrable) on E will be called a weight (or weight function). We say that $f \in L^p_\rho(E)$, where $1 \le p < \infty$, if

$$\|f\|_{L^p_\rho(E)} := \left(\int_E |f(x)|^p \rho(x) dx \right)^{1/p} < \infty.$$

If $\rho \equiv 1$, then we denote $L^p_\rho(E) := L^p(E)$.

Now we are ready to prove the main statements of this section.

Theorem 2.1.1. *Let $1 < p \le q < \infty$. Suppose that the kernels k_1 and k_2 belong to $V \cap V_p$. Then the following statements are equivalent:*
(i) *the operator K is bounded from $L^p(\mathbb{R}_+)$ to $L^q_v(\mathbb{R})_+$;*
(ii)

$$B := \sup_{a,b>0} \left(\int_a^\infty \int_b^\infty v(x,y) k_1^q(x, x/2) k_2^q(y, y/2) dx dy \right)^{1/q} (ab)^{1/p'} < \infty; \tag{2.1.1}$$

(iii)

$$\widetilde{B} := \sup_{k,j \in \mathbb{Z}} \left(\int_{2^k}^{2^{k+1}} \int_{2^j}^{2^{j+1}} v(x,y)(xy)^{q/p'} k_1^q(x, x/2) k_2^q(y, y/2) dx dy \right)^{1/q} < \infty. \tag{2.1.2}$$

Moreover, $\|K\| \approx B \approx \widetilde{B}$.

Proof. First we show that (ii) $\Rightarrow$ (i). Let $f \ge 0$. Represent $Kf(x,y)$ as a sum of four double integrals:

$$Kf(x,y) = \int_0^{x/2} \int_0^{y/2} (\cdots) dt d\tau + \int_0^{x/2} \int_{y/2}^{y} (\cdots) dt d\tau$$

$$+ \int_{x/2}^{x} \int_0^{y/2} (\cdots) dt d\tau + \int_{x/2}^{x} \int_{y/2}^{y} (\cdots) dt d\tau$$

$$:= K^{(1)} f(x,y) + K^{(2)} f(x,y) + K^{(3)} f(x,y) + K^{(4)} f(x,y).$$

The condition $k_1, k_2 \in V$ and Theorem 1.1.6 with $w \equiv 1$ lead to the conclusion

$$\|K^{(1)} f\|^q_{L^q_v(\mathbb{R}^2_+)} \le cB^q \|f\|^q_{L^p(\mathbb{R}^2_+)}.$$

By using Hölder's inequality, the conditions $k_1, k_2 \in V_p$ and (2.1.1) we find that

$$\|K^{(4)} f\|^q_{L^q_v(\mathbb{R}^2_+)} \le \int_0^\infty \int_0^\infty v(x,y) \Big(\int_{x/2}^x \int_{y/2}^y (f(t,\tau))^p dt d\tau \Big)^{q/p}$$

$$\times \Big(\int_{x/2}^x k_1^{p'}(x,t) dt \Big)^{q/p'} \Big(\int_{y/2}^y k_2^{p'}(y,\tau) d\tau \Big)^{q/p'} dx dy$$

$$\le c \sum_{k \in \mathbb{Z}} \sum_{j \in \mathbb{Z}} \int_{2^k}^{2^{k+1}} \int_{2^j}^{2^{j+1}} v(x,y) k_1^q(x, x/2) k_2^q(y, y/2) (xy)^{q/p'}$$

$$\times \Big(\int_{x/2}^x \int_{y/2}^y (f(t,\tau))^p dt d\tau \Big)^{q/p} dx dy$$

$$\le cB^q \sum_k \sum_j \Big(\int_{2^{k-1}}^{2^{k+1}} \int_{2^{j-1}}^{2^{j+1}} (f(t,\tau))^p dt d\tau \Big)^{q/p} \le cB^q \|f\|^q_{L^p(\mathbb{R}^2_+)}.$$

Due to Hölder's inequality with respect to the one-dimensional integral $\int_{y/2}^y (\cdots)$, the conditions $k_1 \in V$, $k_2 \in V_p$, one–dimensional Hardy inequality with power weights (see Theorem 1.1.2 for $v(x) = x^{-q/p'-1}$, $w(x) \equiv 1$) and (2.1.1) we have that

$$\|K^{(2)} f\|^q_{L^q(\mathbb{R}^2_+)} \le c \int_0^\infty \int_0^\infty v(x,y) k_1^q(x, x/2) \Big(\int_0^{x/2} \int_{y/2}^y f(t,\tau) k_2(y,\tau) dt d\tau \Big)^q dx dy$$

$$\le c \int_0^\infty \int_0^\infty v(x,y) k_1^q(x, x/2) \Big(\int_0^{x/2} \Big(\int_{y/2}^y (f(t,\tau))^p d\tau \Big)^{1/p}$$

$$\times \Big(\int_{y/2}^y k_2^{p'}(y,\tau) d\tau \Big)^{1/p'} dt \Big)^q dx dy$$

$$\le c \int_0^\infty \int_0^\infty v(x,y) k_1^q(x, x/2) k_2^q(y, y/2) y^{q/p'}$$

$$\times \Big(\int_0^{x/2} \Big(\int_{y/2}^y (f(t,\tau))^p d\tau \Big)^{1/p} dt \Big)^q dx dy$$

$$\leq c\sum_{k\in\mathbb{Z}}\sum_{j\in\mathbb{Z}}\Bigg(\int\limits_{2^k}^{2^{k+1}}\int\limits_{2^j}^{2^{j+1}} v(x,y)k_1^q(x,x/2)k_2^q(y,y/2)y^{q/p'}dxdy\Bigg)$$

$$\times\Bigg(\int\limits_0^{2^k}\Bigg(\int\limits_{2^{j-1}}^{2^{j+1}}(f(t,\tau))^p d\tau\Bigg)^{1/p}dt\Bigg)^q$$

$$\leq cB^q\sum_k\sum_j(2^{-q/p'})^k\Bigg(\int\limits_0^{2^k}\Bigg(\int\limits_{2^{j-1}}^{2^{j+1}}(f(t,\tau))^p d\tau\Bigg)^{1/p}dt\Bigg)^q$$

$$= cB^q\sum_j\int\limits_0^{\infty}x^{-q/p'-1}\Bigg(\int\limits_0^{x}\Bigg(\int\limits_{2^{j-1}}^{2^{j+1}}(f(t,\tau))^p d\tau\Bigg)^{1/p}dt\Bigg)^q dx$$

$$\leq cB^q\sum_j\Bigg(\int\limits_0^{\infty}\int\limits_{2^{j-1}}^{2^{j+1}}(f(x,\tau))^p d\tau dx\Bigg)^{q/p}\leq cB^q\|f\|^q_{L^p(\mathbb{R}^2_+)}.$$

Analogously, one can obtain that

$$\|K^{(3)}f\|^q_{L^q(\mathbb{R}^2_+)}\leq cB^q\|f\|^q_{L^p(\mathbb{R}^2_+)}.$$

Taking into account the estimates for $K^{(j)}f$, $j=1,2,3,4$, we finally conclude that

$$\|Kf\|^q_{L^q_v(\mathbb{R}^2_+)}\leq cB^q\|f\|^q_{L^p(\mathbb{R}^2_+)}.$$

To prove that (i)$\Rightarrow$ (iii) we take the test functions

$$f(x,y)=f_{kj}(x,y)=\chi_{(0,2^{k+1})}(x)\chi_{(0,2^{j+1})}(y).$$

Then we have that

$$\|Kf_{kj}\|_{L^q_v(\mathbb{R}^2_+)}\geq\Bigg(\int\limits_{2^k}^{2^{k+1}}\int\limits_{2^j}^{2^{j+1}}v(x,y)\Bigg(\int\limits_{x/2}^{x}\int\limits_{y/2}^{y}k_1(x,t)k_2(y,\tau)dtd\tau\Bigg)^q dxdy\Bigg)^{1/q}$$

$$\geq c\Bigg(\int\limits_{2^k}^{2^{k+1}}\int\limits_{2^j}^{2^{j+1}}v(x,y)k_1^q(x,x/2)k_2^q(y,y/2)(xy)^q dxdy\Bigg)^{1/q}.$$

On the other hand, $\|f_{kj}\|_{L^p(\mathbb{R}^2_+)}=c(2^k2^j)^{1/p}$. Consequently, by using the two–weight inequality for K, we have that $\widetilde{B}<\infty$.

Let us now show that (iii) $\Rightarrow$ (ii), i.e., that $B\leq c\widetilde{B}$. Denote

$$\Bigg(\int\limits_a^{\infty}\int\limits_b^{\infty}v(x,y)k_1^q(x,x/2)k_2^q(y,y/2)dxdy\Bigg)^{1/q}(ab)^{1/p'}:=cB(a,b).$$

We have that

$$B^q(a,b) \le (2^{n+1}2^{m+1})^{q/p'} \int_{2^n}^{\infty}\int_{2^m}^{\infty} v(x,y)k_1^q(x,x/2)k_2^q(y,y/2)dxdy$$

$$\le c(2^n2^m)^{q/p'} \sum_{k=n}^{+\infty}\sum_{j=m}^{\infty}\Bigg(\int_{2^k}^{2^{k+1}}\int_{2^j}^{2^{j+1}} v(x,y)k_1^q(x,x/2)k_2^q(y,y/2)dxdy\Bigg)$$

$$\le c\widetilde{B}^q(2^n2^m)^{q/p'} \sum_{k=n}^{+\infty}\sum_{j=m}^{\infty}(2^{-k}2^{-j})^{q/p'} \le c\widetilde{B}^q.$$

The proof is complete. □

By using Theorem 2.1.1 we observe that the next statement, for example, for the two-dimensional Riemann–Liouville operator

$$\mathcal{R}_{\alpha,\beta}f(x,y) := \int_0^x\int_0^y f(t,\tau)(x-t)^{\alpha-1}(y-\tau)^{\beta-1}dtd\tau$$

holds:

Theorem 2.1.2. *Let $1 < p \le q < \infty$, $\alpha, \beta > 1/p$. Then the following statements are equivalent:*

(i) *$\mathcal{R}_{\alpha,\beta}$ is bounded from $L^p(\mathbb{R}^2_+)$ to $L^q_v(\mathbb{R}^2_+)$;*

(ii)

$$B_1 := \sup_{a,b>0}\Bigg(\int_a^{\infty}\int_b^{\infty} v(x,y)x^{(\alpha-1)q}y^{(\beta-1)q}dxdy\Bigg)^{1/q}(ab)^{1/p'} < \infty;$$

(iii)

$$\overline{B}_1 := \sup_{k,j\in\mathbb{Z}}\Bigg(\int_{2^k}^{2^{k+1}}\int_{2^j}^{2^{j+1}} v(x,y)dxdy\Bigg)^{1/q} 2^{k(\alpha-1/p)}2^{j(\beta-1/p)} < \infty.$$

Moreover, $\|\mathcal{R}_{\alpha,\beta}\| \approx B_1 \approx \overline{B}_1$.

Proof. If $1/p < \alpha,\ \beta \le 1$, then the statement in the theorem follows directly from Theorem 2.1.1. Next, the inequality

$$\mathcal{R}_{\alpha,\beta}f(x,y) \le x^{\alpha-1}y^{\beta-1}H_2f(x,y),\quad f \ge 0,$$

is obvious for $\alpha, \beta > 1$, where H_2 is the double Hardy transform. Hence, by Theorem 1.1.6 with $w \equiv 1$ we have the desired result.

Next, let $\alpha > 1$ and $1/p < \beta < 1$. Then the following estimate holds

$$\mathcal{R}_{\alpha,\beta}f(x,y) \le x^{\alpha-1}\int_0^x\int_0^y f(t,\tau)(y-\tau)^{\beta-1}dtd\tau,\ f \ge 0.$$

Consequently, applying Theorem 2.1.1 with respect to the kernels $k_1(x,t) \equiv 1$, $k_2(y,\tau) = (y-\tau)^{\beta-1}$ and the weight function $x^{(\alpha-1)q}v(x,y)$ we conclude that the statement holds. The proof of the case $1/p < \alpha < 1$ and $\beta > 1$ is similar to that of the latter one. This completes the proof. □

Remark 2.1.1. Suppose that $v(x,y)$ is equal to $m2^{-m\alpha p}$ for $(x,y) \in [2^m, 2^{m+1}] \times (1, 1+1/m]$, $m = 1,2,\ldots$, $v(x,y) = 2^{-m\alpha p}$ for $(x,y) \in [2^m, 2^{m+1}] \times [1+1/m, 2]$, $m = 1,2,\cdots$, and $v(x,y) = x^{-\alpha p}y^{-\beta p}$ otherwise. Then for the weight function v, we have that

$$\widetilde{B} := \sup_j \left(\int_{2^j}^{2^{j+1}} \left[\sup_k \left(\int_{2^k}^{2^{k+1}} v(x,y)dx\right)2^{k(\alpha p-1)}\right]dy\right)2^{j(\beta p-1)} = \infty,$$

but the operator $\mathcal{R}_{\alpha,\beta}$ is bounded from $L^p(\mathbb{R}^2_+)$ to $L^p_v(\mathbb{R}^2_+)$. Indeed, we observe that $F_n(y)$ defined by

$$F_n(y) := \left(\int_{2^n}^{2^{n+1}} v(x,y)dx\right)2^{n(\alpha p-1)},\quad n \ge 1,$$

is equal to n when $y \in [1, 1+1/n]$, $F_n(y) = 1$ for $y \in (1+1/n, 2]$, and $F_n(y) \approx y^{-\beta p}$ otherwise. Consequently, $\sup\limits_{n\ge1} F_n(y) = m$ for $y \in [1+1/(m+1), 1+1/m]$, $m = 1,2,\ldots$. Hence,

$$\int_1^2 \sup_n F_n(y)dy = \sum_{m=1}^{\infty} m \int_{1+1/(m+1)}^{1+1/m} dy = \infty.$$

On the other hand, $\int_1^2 F_n(y)dy = 2 + 1/n$. Therefore its supremum with respect to all n, $n \ge 1$, is finite. Further, it is easy to see that

$$\sup_{k,n}\left(\int_{2^k}^{2^{k+1}}\int_{2^n}^{2^{n+1}} v(x,y)dxdy\right)2^{k(\alpha p-1)}2^{n(\beta p-1)}$$

$$\le \sup_{k\neq 0,n}\left(\int_{2^k}^{2^{k+1}}\int_{2^n}^{2^{n+1}} v(x,y)dxdy\right)2^{k(\alpha p-1)}2^{n(\beta p-1)}$$

$$+\sup_{n\geq 1}\bigg(\int_1^2\int_{2^n}^{2^{n+1}} v(x,y)dxdy\bigg)2^{\alpha p-1}2^{n(\beta p-1)}$$

$$+\sup_{n\leq 0}\bigg(\int_1^2\int_{2^n}^{2^{n+1}} v(x,y)dxdy\bigg)2^{\alpha p-1}2^{n(\beta p-1)}<\infty.$$

This example shows that the condition $\widetilde{B}<\infty$ which arises using twice the one-dimensional result for the operators $\mathcal{R}_\alpha$ and $\mathcal{R}_\beta$ respectively, is not necessary for the boundedness of $\mathcal{R}_{\alpha,\beta}$.

2.2 Truncated Potentials and Ball Fractional Integrals

In this section we obtain boundedness criteria of the operators:

$$J_{\alpha,\beta}f(x,y):=\int\limits_{|t|\leq 2|x|}\int\limits_{|\tau|\leq 2|y|} f(t,\tau)|x-t|^{\alpha-n}|y-\tau|^{\beta-n}dtd\tau,\quad \alpha,\beta>0,$$

$$B_{\alpha,\beta}f(x,y):=\int\limits_{|t|<|x|}\int\limits_{|\tau|<|y|} f(t,\tau)\frac{(|x|^2-|t|^2)^\alpha}{|x-t|^n}\frac{(|y|^2-|\tau|^2)^\beta}{|y-\tau|^n}dtd\tau,\quad \alpha,\beta>0,$$

from $L^p(\mathbb{R}^{2n})$ to $L^q_v(\mathbb{R}^{2n})$.

We denote $L^p:=L^p(\mathbb{R}^{2n})$, $L^q_v:=L^q_v(\mathbb{R}^{2n})$.

To prove the main results of this section we need the following statement which is a direct consequence of the two–weight inequality of the n–dimensional Hardy–type transform

$$Hf(x)=\int\limits_{\{t\in\mathbb{R}^n:|t|<|x|\}} f(t)dt,\quad x\in\mathbb{R}^n,$$

(see, e.g., [100], [39, Chapter 1]).

Proposition 2.2.1. *Let $1<p\leq q<\infty$. Then the inequality*

$$\|Hf\|_{L^q_v(\mathbb{R}^n)}\leq c\|f\|_{L^p(\mathbb{R}^n)}$$

holds if and only if

$$A:=\sup_{t>0}\bigg(\int\limits_{|y|>t} v(x)dx\bigg)^{1/q}t^{n/p'}.$$

Moreover, $\|H\|\approx A$.

Corollary 2.2.1. *Let $1\leq p\leq q\leq\infty$. Then the inequality*

$$\bigg(\int\limits_{\mathbb{R}^n}|x|^{-qn/p'-n}(Hf(x))^q dx\bigg)^{1/q}\leq c\bigg(\int\limits_{\mathbb{R}^n}(f(x))^p dx\bigg)^{1/p},$$

holds for all non–negative $f \in L^p(\mathbb{R}^n)$.

Let

$$\widetilde{H}f(x,y) := \int\limits_{\{t\in\mathbb{R}^n:|t|<|x|\}} \int\limits_{\{\tau\in\mathbb{R}^n:|\tau|<|y|\}} f(t,\tau)dtd\tau, \quad x,y \in \mathbb{R}^n.$$

By using Theorem 1.1.3, Fubini's theorem and the polar coordinates and passing to the polar coordinates we have the next proposition (see also [92] for the proof):

Proposition 2.2.2. *Let $1 < p \leq q < \infty$. Then the operator $\widetilde{H}$ is bounded from L^p to L^q_v if and only if*

$$\overline{A} := \sup_{a,b>0} \left(\int\limits_{|x|>a} \int\limits_{|y|>b} v(x,y)dxdy \right)^{1/q} (ab)^{n/p'} < \infty.$$

Moreover, $\|\widetilde{H}\| \approx \overline{A}$.

Theorem 2.2.1. *Let $1 < p \leq q < \infty$. Suppose that $\alpha, \beta > n/p$. Then the following statements are equivalent:*

(i) *The operator $J_{\alpha,\beta}$ is bounded from L^p to L^q_v;*

(ii)

$$D := \sup_{a,b>0} \left(\int\limits_{|x|>a} \int\limits_{|y|>b} v(x,y)|x|^{(\alpha-n)q}|y|^{(\beta-n)q}dxdy \right)^{1/q} (ab)^{n/p'} < \infty; \tag{2.2.1}$$

(iii)

$$D_1 := \sup_{k,j\in\mathbb{Z}} \left(\int\limits_{2^k<|x|<2^{k+1}} \int\limits_{2^j<|y|<2^{j+1}} v(x,y)dxdy \right)^{1/q} 2^{k(\alpha-n/p)}2^{j(\beta-n/p)}. \tag{2.2.2}$$

Moreover, $\|J_{\alpha,\beta}\| \approx D \approx D_1$.

Proof. Suppose that $f \geq 0$. Let us assume $n/p < \alpha, \beta \leq n$. To prove that (ii)⇒ (i) we represent $J_{\alpha,\beta}$ as follows:

$$\begin{aligned} J_{\alpha,\beta}f(x,y) = & \int\limits_{|t|\leq|x|/2} \int\limits_{|\tau|\leq|y|/2} (\cdots)dtd\tau + \int\limits_{|t|\leq|x|/2} \int\limits_{|y|/2<|\tau|\leq 2|y|} (\cdots)dtd\tau \\ & + \int\limits_{|x|/2<|t|\leq 2|x|} \int\limits_{|\tau|\leq|y|/2} (\cdots)dtd\tau + \int\limits_{|x|/2<|t|\leq 2|x|} \int\limits_{|y|/2<|\tau|\leq 2|y|} (\cdots)dtd\tau \\ := & J^{(1)}_{\alpha,\beta}f(x,y) + J^{(2)}_{\alpha,\beta}f(x,y) + J^{(3)}_{\alpha,\beta}f(x,y) + J^{(4)}_{\alpha,\beta}f(x,y). \end{aligned}$$

If $|t| < |x|/2$ and $|\tau| < |y|/2$, then $|x|/2 \le |x-t|$ and $|y|/2 \le |y-\tau|$. Consequently, by using Proposition 2.2.2, we find that

$$\|J^{(1)}_{\alpha,\beta} f\|^q_{L^q_v} \le cD^q \|f\|^q_{L^p}. \tag{2.2.3}$$

Hölder's inequality with respect to the integral $\int_{|x|/2<|t|<2|x|}\int_{|y|/2<|\tau|<2|y|}(\cdots)$ and the estimates

$$\int\limits_{|x|/2<|t|<2|x|} |x-t|^{(\alpha-n)p'} dt \le c|x|^{(\alpha-n)p'+n}, \quad \int\limits_{|y|/2<|\tau|<2|y|} |x-\tau|^{(\beta-n)p'} d\tau \le c|y|^{(\beta-n)p'+n}$$

lead us to the inequality

$$\|J^{(4)}_{\alpha,\beta} f\|^q_{L^q_v} \le cD^q \|f\|^q_{L^p}. \tag{2.2.4}$$

Due to Hölder's inequality with respect to the integral $\int_{|y|/2<|\tau|<2|y|}(\cdots)$, the inequality $|x|/2 \le |x-t|$ (which holds for $|t| < |x|/2$), Corollary 2.2.1 and Condition (2.2.1) we find that

$$\|J^{(2)}_{\alpha,\beta} f\|^q_{L^q} \le c \int\limits_{\mathbb{R}^n}\int\limits_{\mathbb{R}^n} v(x,y)|x|^{(\alpha-n)q}$$

$$\times\Bigg(\int\limits_{|t|\le|x|/2}\ \int\limits_{|y|/2<\tau\le 2|y|} f(t,\tau)|y-\tau|^{\beta-n} dt d\tau\Bigg)^q dxdy$$

$$\le c\int\limits_{\mathbb{R}^n}\int\limits_{\mathbb{R}^n} v(x,y)|x|^{(\alpha-n)q}\Bigg(\int\limits_{|t|\le|x|/2}\Bigg(\int\limits_{|y|/2<|\tau|\le 2|y|} (f(t,\tau))^p d\tau\Bigg)^{1/p}$$

$$\times\Bigg(\int\limits_{|y|/2<|\tau|\le|y|} |y-\tau|^{(\beta-n)p'} d\tau\Bigg)^{1/p'} dt\Bigg)^q dxdy$$

$$\le c\int\limits_{\mathbb{R}^n}\int\limits_{\mathbb{R}^n} v(x,y)|x|^{(\alpha-n)q}|y|^{(\beta-n)q}|y|^{nq/p'}$$

$$\times\Bigg(\int\limits_{|t|\le|x|/2}\Bigg(\int\limits_{|y|/2<|\tau|\le 2|y|} (f(t,\tau))^p d\tau\Bigg)^{1/p} dt\Bigg)^q dxdy$$

$$\le c\sum_k\sum_j\Bigg(\int\limits_{2^k<|t|\le 2^{k+1}}\ \int\limits_{2^j<|\tau|\le 2^{j+1}} v(x,y)|x|^{(\alpha-n)q}|y|^{(\beta-n)q}|y|^{nq/p'} dxdy\Bigg)$$

$$\times\Bigg(\int\limits_{|t|\le 2^k}\Bigg(\int\limits_{2^{j-1}<|\tau|\le 2^{j+1}} (f(t,\tau))^p d\tau\Bigg)^{1/p} dt\Bigg)^q$$

$$\le cD^q\sum_k\sum_j \int\limits_{2^k<|t|\le 2^{k+1}} |x|^{-nq/p'-n}$$

$$\times\Bigg(\int\limits_{|t|<|x|}\Bigg(\int\limits_{2^{j-1}<|\tau|\le 2^{j+1}}(f(t,\tau))^p d\tau\Bigg)^{1/p}dt\Bigg)^q dx$$

$$= cD^q\sum_j\int\limits_{\mathbb{R}^n}|x|^{-nq/p'-n}\Bigg(\int\limits_{|t|<|x|}\Bigg(\int\limits_{2^{j-1}<|t|\le 2^{j+1}}(f(t,\tau))^p d\tau\Bigg)^{1/p}dt\Bigg)^q dx$$

$$\le cD^q\sum_j\Bigg(\int\limits_{\mathbb{R}^n}\int\limits_{2^{j-1}<|\tau|\le 2^{j+1}}(f(x,\tau))^p d\tau dx\Bigg)^{q/p}\le cD^q\|f\|^q_{L^p}.$$

Thus

$$\|J^{(2)}_{\alpha,\beta}f\|^q_{L^q}\le cD^q\|f\|^q_{L^p}. \tag{2.2.5}$$

By using analogous arguments we find that

$$\|J^{(3)}_{\alpha,\beta}f\|^q_{L^q_v}\le cD^q\|f\|^q_{L^p}. \tag{2.2.6}$$

Summarizing the inequalities (2.2.3)–(2.2.6) we conclude that $J_{\alpha\beta}$ is bounded from L^p to L^q_v.

To prove that (i)$\Rightarrow$ (iii), we put the functions $f_{kj}(x,y)=\chi_{E_k}(x)\chi_{E_j}(y)$, where $E_n:=\{z:|z|\le 2^{n-1}\}$, into the inequality

$$\|J_{\alpha,\beta}f\|_{L^q_v}\le c\|f\|_{L^p}$$

and after simple calculation observe that $D_1<\infty$.

The proof of the fact (iii) $\Rightarrow$ (ii) is trivial (see also the part (iii)$\Rightarrow$ (ii) in the proof of Theorem 2.1.1). The proof is complete. □

Theorem 2.2.2. *Let $1<p\le q<\infty$. Suppose that $\alpha,\beta>n/p$. Then the following statements are equivalent:*

(i) *The operator $B_{\alpha,\beta}$ is bounded from L^p to L^q_v;*

(ii)

$$\overline{D}:=\sup_{a,b>0}\Bigg(\int\limits_{|x|>a}\int\limits_{|y|>b}v(x,y)|x|^{(2\alpha-n)q}|y|^{(2\beta-n)q}dxdy\Bigg)^{1/q}(ab)^{n/p'}<\infty;$$

(iii)

$$\overline{D}_1:=\sup_{k,j\in\mathbb{Z}}\Bigg(\int\limits_{2^k<|x|<2^{k+1}}\int\limits_{2^j<|y|<2^{j+1}}v(x,y)dxdy\Bigg)^{1/q}2^{k(2\alpha-n/p)}2^{j(2\beta-n/p)}<\infty.$$

Moreover, $\|B_{\alpha,\beta}\|\approx\overline{D}\approx\overline{D}_1$.

Proof. It is obvious that if $|t|<|x|$, then

$$(|x|^2-|t|^2)^\alpha=(|x|-|t|)^\alpha(|x|+|t|)^\alpha\le c|x|^\alpha|x-t|^\alpha,$$

where the positive constant c is independent of x. Analogously, we have that

$$(|y|^2-|\tau|^2)^\alpha\le c|y|^\alpha|y-\tau|^\alpha$$

for $|\tau|<|y|$.

Hence,

$$B_{\alpha,\beta}f(x,y)\le c|x|^\alpha|y|^\beta J_{\alpha,\beta}f(x,y),$$

where $f\ge 0$ and the constant c is independent of x and y. Consequently, by Theorem 2.2.1 we find that (ii)$\Rightarrow$ (i). To prove that (i)$\Rightarrow$ (iii), we take $f_{kj}(x,y)=\chi_{E_k}(x)\chi_{E_j}(y)$, where $E_n:=\{z:|z|<2^{n-1}\}$. Then

$$B_{\alpha,\beta}f(x,y)\ge\int\limits_{|t|<2^{k-1}}\int\limits_{|\tau|<2^{j-1}}\frac{(|x|^2-2^{(k-1)2})^\alpha}{|x-t|^n}\frac{(|y|^2-2^{(j-1)2})^\beta}{|y-\tau|^n}dtd\tau$$
$$\ge c|x|^{2\alpha}|y|^{2\beta}\int\limits_{|t|<2^{k-1}}\int\limits_{|\tau|<2^{j-1}}|x-t|^{-n}|y-\tau|^{-n}dtd\tau,$$

where $2^k\le|x|<2^{k+1}$, $2^j\le|y|<2^{j+1}$. Note that if $|t|<2^{k+1}$ and $|\tau|<2^{j+1}$, then $|x|/2\le|x-t|$ and $|y|/2\le|y-\tau|$. Hence,

$$B_{\alpha,\beta}f(x,y)\ge c|x|^{2\alpha-n}|y|^{2\beta-n}2^{kn}2^{jn}\ge c2^{2k\alpha}2^{2j\beta}.$$

From the boundedness of $B_{\alpha,\beta}$ we have that $\overline{D}_1<\infty$. The implication (iii)$\Rightarrow$ (ii) follows in the same manner as the inequality $B\le cB_1$ was proved in the proof of Theorem 2.1.1. The proof is complete. □

2.3 The Case of m–Multiple Kernels

An observation of the proofs of the main statements formulated in the previous sections enables us to obtain the trace inequality for the following integral transforms:

$$K^{(m)}f(x_1,\dots,x_m)=\int\limits_0^{x_1}\cdots\int\limits_0^{x_m}f(y_1,\dots,y_m)$$
$$\times\left(\prod_{i=1}^m k_i(x_i,y_i)\right)dy_1\cdots dy_m,\quad \alpha_j>0,\ \ x_j>0;$$

$$J^{(m)}f(x_1,\cdots x_m) = \int\limits_{\{y_1\in\mathbb{R}^n:|y_1|\le 2|x_1|\}} \cdots \int\limits_{\{y_m\in\mathbb{R}^n:|y_m|\le 2|x_m|\}} f(y_1,\ldots,y_m)$$
$$\times\Big(\prod_{i=1}^{m}|x_i-y_i|^{\alpha_i-n}\Big)dy_1\cdots dy_m,\ \ \alpha_j>0,\ x_j\in\mathbb{R}^n;$$

$$B^{(m)}f(x_1,\cdots x_m) = \int\limits_{\{y_1\in\mathbb{R}^n:|y_1|<|x_1|\}} \cdots \int\limits_{\{y^m\in\mathbb{R}^n:|y_m|<|x_m|\}} f(y_1,\ldots,y_m)$$
$$\times\Big(\prod_{i=1}^{m}\frac{(|x_i|^2-|y_i|^2)^{\alpha_i}}{|x_i-y_i|^n}\Big)dy_1\cdots dy_m,\ \ \alpha_j>0,\ x_j\in\mathbb{R}^n.$$

To avoid repeating the arguments used above, we present the statements without proofs.

Theorem 2.3.1. *Let $1<p\le q<\infty$, $k_j\in V\cap V_p$, $j=1,\ldots,m$. Then the following statements are equivalent:*

(i) *the operator $K^{(m)}$ is bounded from $L^p(\mathbb{R}^n_+)$ to $L^q_v(\mathbb{R}^n_+)$, where $\mathbb{R}^n_+ := \underbrace{\mathbb{R}_+\times\cdots\times\mathbb{R}_+}_{n}$;*

(ii)

$$B_m := \sup_{a_1,\ldots,a_m>0}\Big(\int_{a_1}^{\infty}\cdots\int_{a_m}^{\infty} v(x_1,\ldots,x_m)$$
$$\times\Big(\prod_{i=1}^{m}k_i(x_i,x_i/2)\Big)^q dx_1\cdots dx_m\Big)^{1/q}\big(\prod_{i=1}^{m}a_i\big)^{1/p'}<\infty;$$

(iii)

$$\widetilde{B}_m := \sup_{n_1,\ldots,n_m\in\mathbb{Z}}\Big(\int_{2^{n_1}}^{2^{n_1+1}}\cdots\int_{2^{n_m}}^{2^{n_m+1}} v(x_1,\ldots,x_m)\big(\prod_{i=1}^{m}x_i\big)^{q/p'}$$
$$\times\Big(\prod_{i=1}^{m}k_i(x_i,x_i/2)\Big)^q dx_1\cdots dx_m\Big)^{1/q}<\infty.$$

Moreover, $\|K^{(m)}\|\approx B_m\approx\widetilde{B}_m$.

For the operators $J^{(m)}$ and $B^{(m)}$ we have the next theorems:

Theorem 2.3.2. *Let $1<p\le q<\infty$. Suppose that $\alpha_1,\ldots,\alpha_m>n/p$. Then we have the following equivalent statements:*

(i) *the operator* $J^{(m)}$ *is bounded from* $L^p(\mathbb{R}^{mn})$ *to* $L^q_v(\mathbb{R}^{mn})$;

(ii)

$$D_m := \sup_{a_1,\dots,a_m>0} \Bigg(\int\limits_{|x_1|>a_1} \cdots \int\limits_{|x_m|>a_m} v(x_1,\dots,x_m) \times \Bigg(\prod_{i=1}^m |x_i|^{(\alpha_i-n)q}\Bigg) dx_1\cdots dx_m\Bigg)^{1/q} \Big(\prod_{i=i}^m a_i\Big)^{n/p'} < \infty;$$

(iii)

$$\widetilde{D}_m := \sup_{n_1,\dots,n_1\in\mathbb{Z}} \Bigg(\int\limits_{2^{n_1}<|x_1|<2^{n_m+1}} \cdots \int\limits_{2^{n_m}<|x_m|<2^{n_m+1}} v(x_1,\dots,x_m) \, dx_1\cdots dx_m\Bigg)^{1/q} \Bigg(\prod_{i=1}^m 2^{n_i(\alpha_i-n/p)}\Bigg) < \infty.$$

Moreover, $\|J^{(m)}\| \approx D_m \approx \widetilde{D}_m$.

Theorem 2.3.3. *Let* $1<p\le q<\infty$. *Suppose that* $\alpha_1,\dots,\alpha_m > n/p$. *Then we have the following equivalent statements:*

(i) *the operator* $B^{(m)}$ *is bounded from* $L^p(\mathbb{R}^{mn})$ *to* $L^q_v(\mathbb{R}^{mn})$;

(ii)

$$E_m := \sup_{a_1,\dots,a_m>0} \Bigg(\int\limits_{|x_1|>a_1} \cdots \int\limits_{|x_m|>a_m} v(x_1,\dots,x_m) \times \Bigg(\prod_{i=1}^m |x_i|^{(2\alpha_i-n)q}\Bigg) dx_1\cdots dx_m\Bigg)^{1/q} \Big(\prod_{i=1}^m a_i\Big)^{n/p'} < \infty;$$

(iii)

$$\widetilde{E}_m := \sup_{n_1,\dots,n_m\in\mathbb{Z}} \Bigg(\int\limits_{2^{n_1}<|x_1|<2^{n_1+1}} \cdots \int\limits_{2^{n_m}<|x_m|<2^{n_m+1}} v(x_1,\dots,x_m) \, dx_1\cdots dx_m\Bigg)^{1/q} \Bigg(\prod_{i=1}^m 2^{n_i(2\alpha_i-n/p)}\Bigg) < \infty.$$

Moreover, $\|B^{(m)}\| \approx E_m \approx \widetilde{E}_m$.

2.4 Multiple One–sided Potentials. Trace Inequality

The present section is devoted to the trace problem for the operator $\mathcal{R}_{\alpha,\beta}$ for $0 < \alpha < 1/p$ and $\beta > 1/p$. Criteria governing the trace inequality for the operators

$$(\mathcal{WR})_{\alpha,\beta}f(x,y) := \int_x^\infty \int_0^y (t-x)^{\alpha-1}(y-\tau)^{\beta-1}f(t,\tau)dtd\tau, \quad x,y>0,$$

and

$$(\mathcal{IR})_{\alpha,\beta}f(x,y) := \int_0^\infty \int_0^y |x-t|^{\alpha-1}(y-\tau)^{\beta-1}f(t,\tau)dtd\tau, \quad x,y>0,$$

are also derived.

Denote by $\mathcal{W}_\alpha$ the Weyl transform

$$\mathcal{W}_\alpha f(x) = \int_x^\infty (t-x)^{\alpha-1}f(t)dt, \quad x>0.$$

Recall that $\mathcal{R}_\alpha$ denotes the Riemann–Liouville transform defined on $\mathbb{R}_+$.

Before stating the main results of this section (Theorems 2.4.1– 2.4.6) we need to formulate some recent auxiliary statements of independent interest (Propositions 2.4.1–2.4.3).

The following statement was proved in [39, p. 131].

Proposition 2.4.1. *Let $1 < p < q < \infty$. Suppose that $0 < \alpha < 1/p$. Then the following statements are equivalent:*

(i) $\mathcal{R}_\alpha$ *is bounded from $L^p(\mathbb{R}_+)$ to $L^q_u(\mathbb{R}_+)$;*

(ii) $\mathcal{W}_\alpha$ *is bounded from $L^p(\mathbb{R}_+)$ to $L^q_u(\mathbb{R}_+)$;*

(iii)

$$S := \sup_{0\le h\le a} \left(\int_a^{a+h} u(x)dx \right)^{1/q} h^{\alpha-1/p} < \infty.$$

Moreover, there exist positive constants c_k, $k=1,\dots,4$, depending only on α, p and q such that $c_1S \le \|\mathcal{R}_\alpha\| \le c_2S$ and $c_3S \le \|\mathcal{W}_\alpha\| \le c_4S$.

The next statements were proved in [89]:

Proposition 2.4.2. *Let $1 < p < \infty$, $0 < \alpha < 1/p$. Then the operator $\mathcal{R}_\alpha$ is bounded from $L^p(\mathbb{R}_+)$ to $L^p_u(\mathbb{R}_+)$ if and only if $\mathcal{W}_\alpha u \in L^{p'}_{loc}(\mathbb{R}_+)$ and*

$$S_1 := \sup_{x\in\mathbb{R}_+} \left(\frac{\mathcal{W}_\alpha[\mathcal{W}_\alpha u]^{p'}(x)}{\mathcal{W}_\alpha u(x)} \right)^{1/p'} < \infty.$$

Moreover, there exist positive constants c_1 *and* c_2 *depending only on* α *and* p *such that* $c_1 S_1 \le \|\mathcal{R}_\alpha\| \le c_2 S_1$.

Proposition 2.4.3. *Let* $1 < p < \infty$, $0 < \alpha < 1/p$. *Then the operator* $\mathcal{W}_\alpha$ *is bounded from* $L^p(\mathbb{R}_+)$ *to* $L^p_u(\mathbb{R}_+)$ *if and only if* $\mathcal{R}_\alpha u \in L^{p'}_{loc}(\mathbb{R}_+)$ *and*

$$S_2 := \sup_{x\in\mathbb{R}_+} \left(\frac{\mathcal{R}_\alpha[\mathcal{R}_\alpha u]^{p'}(x)}{\mathcal{R}_\alpha u(x)} \right)^{1/p'} < \infty, \;\; p' = p/(p-1).$$

Moreover, there exist positive constants c_1 *and* c_2 *depending only on* α *and* p *such that* $c_1 S_2 \le \|\mathcal{W}_\alpha\| \le c_2 S_2$.

Now we formulate the main results of this section.

First we consider the case $p < q$.

Theorem 2.4.1. *Let* $1 < p < q < \infty$. *Suppose that* $0 < \alpha < 1/p$ *and* $\beta > 1/p$. *Then the operator* $\mathcal{R}_{\alpha,\beta}$ *is bounded from* $L^p(\mathbb{R}^2_+)$ *to* $L^q_v(\mathbb{R}^2_+)$ *if and only if*

$$A := \sup_{\substack{h,a,j \\ 0\le h\le a \\ j\in\mathbb{Z}}} \left(\int_a^{a+h} \int_{2^j}^{2^{j+1}} v(x,y)dxdy \right)^{1/q} h^{\alpha-1/p} 2^{j(\beta-1/p)} < \infty. \tag{2.4.1}$$

Moreover, $\|\mathcal{R}_{\alpha,\beta}\| \approx A$.

Proof. *Sufficiency.* Let $f \ge 0$. Represent $\mathcal{R}_{\alpha,\beta} f(x,y)$ as a sum of two two-dimensional integrals:

$$\mathcal{R}_{\alpha,\beta} f(x,y) = \int_0^x \int_0^{y/2} (\cdots)dtd\tau + \int_0^x \int_{y/2}^{y} (\cdots)dtd\tau := \mathcal{R}^{(1)}_{\alpha,\beta} f(x,y) + \mathcal{R}^{(2)}_{\alpha,\beta} f(x,y).$$

It easy to check that if $t < y/2$, then $(y-\tau)^{\beta-1} \le c_\beta y^{\beta-1}$, where $c_\beta = 2^{1-\beta}$ for $\beta < 1$, and $c_\beta = 1$ for $\beta \ge 1$. By using this estimate we find that

$$J_1 := \|\mathcal{R}^{(1)}_{\alpha,\beta} f\|^q_{L^q_v(\mathbb{R}^2_+)} \le c \iint_{\mathbb{R}^2_+} v(x,y) y^{(\beta-1)q} \left(\int_0^x \int_0^{y/2} \frac{f(t,\tau)}{(x-t)^{1-\alpha}} dtd\tau \right)^q dxdy$$

$$= c \sum_{j\in\mathbb{Z}} \int_0^\infty \int_{2^j}^{2^{j+1}} v(x,y) y^{(\beta-1)q} \left(\int_0^x \int_0^{y/2} \frac{f(t,\tau)}{(x-t)^{1-\alpha}} dtd\tau \right)^q dxdy$$

$$\le c \sum_{j\in\mathbb{Z}} \int_0^\infty \left(\int_{2^j}^{2^{j+1}} v(x,y) y^{(\beta-1)q} dy \right)$$

$$\times\Big(\int_0^x (x-t)^{\alpha-1}\Big(\int_0^{2^j} f(t,\tau)d\tau\Big)dt\Big)^q dx$$

$$=c\sum_{j\in\mathbb{Z}} 2^{-jq/p'}\int_0^\infty v_j(x)\Big(\int_0^x (x-t)^{\alpha-1}\Big(\int_0^{2^j} f(t,\tau)d\tau\Big)dt\Big)^q dx := J_1',$$

where

$$v_j(x) := \int_{2^j}^{2^{j+1}} v(x,y)y^{(\beta-1/p)q}dy, \quad j\in\mathbb{Z}.$$

Proposition 2.4.1 with respect to the one-dimensional weights v_j and the classical Hardy inequality (see Theorem 1.1.2 for $p=q$, $v(x)=x^{-p}$ and $w\equiv 1$) yield that

$$J_1' \le cA^q \sum_{j\in\mathbb{Z}} 2^{-jq/p'}\Big(\int_0^\infty\Big(\int_0^{2^j} f(t,\tau)d\tau\Big)^p dt\Big)^{q/p}$$

$$\le cA^q\Big(\int_0^\infty \sum_j 2^{j(1-p)}\Big(\int_0^{2^j} f(t,\tau)d\tau\Big)^p dt\Big)^{q/p}$$

$$\le cA^q\Big(\int_0^\infty\Big(\sum_j \int_{2^j}^{2^{j+1}} x^{-p}\Big(\int_0^x f(t,\tau)d\tau\Big)^p dx\Big)dt\Big)^{q/p}$$

$$= cA^q\Big(\int_0^\infty\Big(\int_0^\infty x^{-p}\Big(\int_0^x f(t,\tau)d\tau\Big)^p dx\Big)dt\Big)^{q/p}$$

$$\le cA^q\Big(\int_0^\infty\int_0^\infty f^p(t,\tau)dtd\tau\Big)^{q/p} = cA^q\|f\|^q_{L^p(\mathbb{R}^2_+)}.$$

Moreover, by Proposition 2.4.1 with respect to the weights $v_j(x)$, $j\in\mathbb{Z}$, we have that

$$J_2 := \|\mathcal{R}^{(2)}_{\alpha,\beta} f\|^q_{L^q_v(\mathbb{R}^2_+)} \le \int_0^\infty\int_0^\infty v(x,y)\Big(\int_0^x (x-t)^{\alpha-1}\Big(\int_{y/2}^y f^p(t,\tau)d\tau\Big)^{1/p}$$

$$\times\Big(\int_{y/2}^y (y-\tau)^{(\beta-1)p'}d\tau\Big)^{1/p'} dt\Big)^q dxdy$$

$$= c \sum_{j\in\mathbb{Z}} \int_0^{\infty} \int_{2^j}^{2^{j+1}} v(x,y) y^{q(\beta-1/p)}$$

$$\times \bigg(\int_0^x (x-t)^{\alpha-1} \bigg(\int_{y/2}^{y} f^p(t,\tau) d\tau \bigg)^{1/p} dt \bigg)^q dx dy$$

$$\leq c \sum_j \int_0^{\infty} v_j(x) \bigg(\int_0^x (x-t)^{\alpha-1} \bigg(\int_{2^{j-1}}^{2^{j+1}} f^p(t,\tau) d\tau \bigg)^{1/p} dt \bigg)^q dx$$

$$\leq cA^q \sum_j \bigg(\int_0^{\infty} \int_{2^{j-1}}^{2^{j+1}} f^p(t,\tau) dt d\tau \bigg)^{q/p} \leq cA^q \|f\|^q_{L^p(\mathbb{R}^2_+)}.$$

Combining the estimates for J_1 and J_2 we conclude that the operator $\mathcal{R}_{\alpha,\beta}$ is bounded from $L^p(\mathbb{R}^2_+)$ to $L^q_v(\mathbb{R}^2_+)$.

To prove *Necessity*, we take the test functions

$$f_{h,j}(t,\tau) = \chi_{(a-h,a)\times(2^{j-2},2^{j-1})}(t,\tau),$$

where $0 \leq h \leq a$ and $j \in \mathbb{Z}$. Then it is easy to see that

$$\|\mathcal{R}_{\alpha,\beta} f_{h,j}\|_{L^q_v(\mathbb{R}^2_+)} \geq \bigg(\int_a^{a+h} \int_{2^j}^{2^{j+1}} v(x,y)$$

$$\times \bigg(\int_{a-h}^{a} \int_{2^{j-2}}^{2^{j-1}} (x-t)^{\alpha-1} (y-\tau)^{\beta-1} dt d\tau \bigg)^q dx dy \bigg)^{1/q}$$

$$\geq c \bigg(\int_a^{a+h} \int_{2^j}^{2^{j+1}} v(x,y) dx dy \bigg)^{1/q} h^{\alpha} 2^{j\beta}.$$

On the other hand,

$$\|f_{h,j}\|_{L^p(\mathbb{R}^2_+)} = ch^{1/p} 2^{j/p}.$$

From the boundedness of $\mathcal{R}_{\alpha,\beta}$ we finally derive the desired result. The proof is complete. □

Theorem 2.4.2. *Let $1 < p < q < \infty$. Suppose that $0 < \alpha < 1/p$ and $\beta > 1/p$. Then the operator $(\mathcal{WR})_{\alpha,\beta}$ is bounded from $L^p(\mathbb{R}^2_+)$ to $L^q_v(\mathbb{R}^2_+)$ if and only if (2.4.1) holds. Moreover, $\|(\mathcal{WR})_{\alpha,\beta}\| \approx A$.*

Proof. To prove the sufficiency, we only need to repeat the arguments of the proof of Theorem 2.1.1 and use the boundedness of $\mathcal{W}_\alpha$ instead of the boundedness of $\mathcal{R}_\alpha$. Therefore the details are omitted.

For *necessity*, it is enough to show that the boundedness of $(\mathcal{WR})_{\alpha,\beta}$ implies the condition

$$A_1 := \sup \left(\int\limits_I \int\limits_{2^j}^{2^{j+1}} v(x,y)dxdy \right)^{1/q} |I|^{\alpha-1/p} 2^{j(\beta-1/p)} < \infty \tag{2.4.2}$$

where the supremum is taken with respect to all intervals $I \subset \mathbb{R}_+$ of finite length and for all $j \in \mathbb{Z}$ ($|I|$ is a length of I). Indeed, suppose that $I := (a,b)$ and $I_1 := (b,c)$, where $c-b = b-a$. Let us take the functions $f_{I,j}(x,y) = \chi_{I\times(2^{j-2},2^{j-1})}(x,y)$. Then a simple computation shows that $\|f\|_{L^p(\mathbb{R}^2_+)} = c|I|^{1/p}2^{j/p}$. On the other hand,

$$\|(\mathcal{WR})_{\alpha,\beta}\|_{L^q_v(\mathbb{R}^2_+)}$$

$$\geq \left(\int\limits_I \int\limits_{2^j}^{2^{j+1}} v(x,y) \left(\int\limits_{I_1} \int\limits_{2^{j-2}}^{2^{j-1}} (t-x)^{\alpha-1}(y-\tau)^{\beta-1} dt d\tau \right)^q dxdy \right)^{1/q}$$

$$\geq c \left(\int\limits_I \int\limits_{2^j}^{2^{j+1}} v(x,y)dxdy \right)^{1/q} |I|^\alpha 2^{j\beta}.$$

Combining these estimates, we conclude that $A_1 < \infty$. The inequality $A \leq A_1$ is obvious. The proof is complete. □

Theorem 2.4.3. *Let* $1 < p < q < \infty$, $0 < \alpha < 1/p$ *and* $\beta > 1/p$. *Then for the boundedness of the operator* $(\mathcal{IR})_{\alpha,\beta}$ *from* $L^p(\mathbb{R}^2_+)$ *to* $L^q_v(\mathbb{R}^2_+)$ *it is necessary and sufficient that* (2.4.2) *holds.*

Sufficiency of this statement follows easily by using the obvious equality

$$(\mathcal{IR})_{\alpha,\beta} f = \mathcal{R}_{\alpha,\beta} f + (\mathcal{WR})_{\alpha,\beta} f,$$

Theorems 2.4.1 and 2.4.2, and the inequality $A \leq A_1$ (see the proof of Theorem 2.4.2). Necessity is trivial if we take the functions $f_{I,j}(x,y) = \chi_{I\times(2^{j-2},2^{j-1})}(x,y)$ in the weight inequality for $(\mathcal{IR})_{\alpha,\beta}$, where $I \subset \mathbb{R}_+$ is an interval of finite length (see also proof of Theorem 2.4.2).

Finally, for the case $p = q$ we have the next result:

Theorem 2.4.4. *Let* $1 < p < \infty$. *Suppose that* $0 < \alpha < 1/p$ *and* $\beta > 1/p$. *Then the operator* $\mathcal{R}_{\alpha,\beta}$ *is bounded from* $L^p(\mathbb{R}^2_+)$ *to* $L^p_v(\mathbb{R}^2_+)$ *if and only if* $\mathcal{W}_\alpha V_j \in L^{p'}_{loc}(\mathbb{R}_+)$ *for all* $j \in \mathbb{Z}$ *and*

$$\mathcal{W}_\alpha[\mathcal{W}_\alpha V_j]^{p'}(x) \leq c\mathcal{W}_\alpha[V_j](x) \tag{2.4.3}$$

for all $x > 0$ *and* $j \in \mathbb{Z}$, *where*

$$V_j(x) := \int_{2^j}^{2^{j+1}} v(x,y) y^{\beta p - 1} dy. \tag{2.4.4}$$

Proof. *Sufficiency.* Let $f \geq 0$. Preserving the notation from the proof of Theorem 2.4.1 we have that

$$J_1 := \|\mathcal{R}^{(1)}_{\alpha,\beta} f\|^p_{L^p_v(\mathbb{R}^2_+)}$$

$$\leq c \int_0^\infty \int_0^\infty v(x,y) y^{(\beta-1)p} \Big(\int_0^x \int_0^{y/2} \frac{f(t,\tau)}{(x-t)^{1-\alpha}} dt d\tau \Big)^p dx dy$$

$$= c \sum_{j\in\mathbb{Z}} \int_0^\infty \int_{2^j}^{2^{j+1}} v(x,y) y^{(\beta-1)p} \Big(\int_0^x \int_0^{y/2} \frac{f(t,\tau)}{(x-t)^{1-\alpha}} dt d\tau \Big)^p dx dy$$

$$\leq c \sum_{j\in\mathbb{Z}} 2^{j(1-p)} \int_0^\infty V_j(x) \Big(\int_0^x (x-t)^{\alpha-1} \Big(\int_0^{2^j} f(t,\tau) d\tau \Big) dt \Big)^p dx := J_1'.$$

Proposition 2.4.2 with respect to the weights V_j, $j \in \mathbb{Z}$, and the classical Hardy inequality (see Theorem 1.1.2 with $v(x) \equiv x^{-p}$, $w(x) \equiv 1$, $p = q$) yield that

$$J_1' \leq c \sum_{j\in\mathbb{Z}} 2^{j(1-p)} \int_0^\infty \Big(\int_0^{2^j} f(t,\tau) d\tau \Big)^p dt$$

$$= c \int_0^\infty \sum_j 2^{j(1-p)} \Big(\int_0^{2^j} f(t,\tau) d\tau \Big)^p dt$$

$$\leq c \int_0^\infty \Big(\sum_j \int_{2^j}^{2^{j+1}} x^{-p} \Big(\int_0^x f(t,\tau) d\tau \Big)^p dx \Big) dt$$

$$= c \Big(\int_0^\infty \Big(\int_0^\infty x^{-p} \Big(\int_0^x f(t,\tau) d\tau \Big)^p dx \Big) dt$$

$$\leq c \int_0^\infty \int_0^\infty f^p(t,\tau) dt d\tau = c\|f\|^p_{L^p(\mathbb{R}_+)}.$$

Using again Proposition 2.4.2 we find that

$$J_2 := \|\mathcal{R}^{(2)}_{\alpha,\beta} f\|^p_{L^p_v(\mathbb{R}^2_+)} \le \int_0^\infty\int_0^\infty v(x,y)\Big(\int_0^x (x-t)^{\alpha-1}\Big(\int_{y/2}^y f^p(t,\tau)d\tau\Big)^{1/p}$$

$$\times\Big(\int_{y/2}^y (y-\tau)^{(\beta-1)p'}d\tau\Big)^{1/p'} dt\Big)^p dxdy$$

$$\le c\sum_j \int_0^\infty V_j(x)\Big(\int_0^x (x-t)^{\alpha-1}\Big(\int_{2^{j-1}}^{2^{j+1}} f^p(t,\tau)d\tau\Big)^{1/p} dt\Big)^p dxdy$$

$$\le c\sum_j \int_0^\infty \int_{2^{j-1}}^{2^{j+1}} f^p(t,\tau)dtd\tau \le c\|f\|^p_{L^p(\mathbb{R}^2_+)}.$$

Finally, the inequality

$$\|\mathcal{R}_{\alpha,\beta} f\|^p_{L^p_v(\mathbb{R}^2_+)} \le c(J_1+J_2)$$

and the estimates for J_1 and J_2 lead to the desired result.

Necessity. To prove necessity we observe that the boundedness of $\mathcal{R}_{\alpha,\beta}$ from $L^p(\mathbb{R}^2_+)$ to $L^p_v(\mathbb{R}^2_+)$ implies the inequality

$$\int_0^\infty V_j(x)\Big(\int_0^x (x-t)^{\alpha-1}g(t)dt\Big)^p dx \le c\|g\|^p_{L^p(\mathbb{R}_+)},$$

for all non-negative functions g defined on $\mathbb{R}_+$ and all $j \in \mathbb{Z}$. This is possible if we put the functions $f_j(x,y) = g(x)\chi_{(2^{j-2},2^{j-1})}(y)$ in the inequality

$$\|\mathcal{R}_{\alpha,\beta} f\|_{L^p_v(\mathbb{R}^2_+)} \le c\|f\|_{L^p(\mathbb{R}^2_+)}.$$

Further, by Proposition 2.4.2 we conclude that $\mathcal{W}_\alpha V_j \in L^{p'}_{loc}(\mathbb{R}_+)$ for all $j \in \mathbb{Z}$ and

$$\sup_{\substack{x\in\mathbb{R}_+\\ j\in\mathbb{Z}}} \left(\frac{\mathcal{W}_\alpha[\mathcal{W}_\alpha V_j]^{p'}(x)}{\mathcal{W}_\alpha V_j(x)}\right)^{1/p'} < \infty.$$

The proof is complete. □

The next statement follows similarly, therefore the details are omitted (we only need to use Proposition 2.4.3 instead of Proposition 2.4.2).

Theorem 2.4.5. *Let $1 < p < \infty$. Suppose that $0 < \alpha < 1/p$ and $\beta > 1/p$. Then the operator $(\mathcal{WR})_{\alpha,\beta}$ is bounded from $L^p(\mathbb{R}^2_+)$ to $L^p_v(\mathbb{R}^2_+)$ if and only if $\mathcal{R}_\alpha V_j \in L^{p'}_{loc}(\mathbb{R}_+)$ for all $j \in \mathbb{Z}$, and*

$$\mathcal{R}_\alpha[\mathcal{R}_\alpha V_j]^{p'}(x) \leq c\mathcal{R}_\alpha[V_j](x),\ x > 0,\ j \in \mathbb{Z}, \tag{2.4.5}$$

where V_j are defined by (2.4.4) and the positive constant c does not depend on x and j.

Theorems 2.4.4, 2.4.5 and the obvious relations between $\mathcal{R}_{\alpha,\beta}$, $(\mathcal{WR})_{\alpha,\beta}$ and $(\mathcal{IR})_{\alpha,\beta}$:

$$(\mathcal{IR})_{\alpha,\beta}f = \mathcal{R}_{\alpha,\beta}f + (\mathcal{WR})_{\alpha,\beta}f,$$

$$\mathcal{R}_{\alpha,\beta}f \leq (\mathcal{IR})_{\alpha,\beta}f; \quad (\mathcal{WR})_{\alpha,\beta}f \leq (\mathcal{IR})_{\alpha,\beta}f,\ \ f \geq 0,$$

lead us to the following statement:

Theorem 2.4.6. *Let $1 < p < \infty$. Suppose that $0 < \alpha < 1/p$ and $\beta > 1/p$. Then for the boundedness of $(\mathcal{IR})_{\alpha,\beta}$ from $L^p(\mathbb{R}^2_+)$ to $L^p_v(\mathbb{R}^2_+)$ it is necessary and sufficient that $\mathcal{R}_\alpha V_j, \mathcal{W}_\alpha V_j \in L^{p'}_{loc}(\mathbb{R}_+)$ (see (2.4.4)) for all $j \in \mathbb{Z}$ and conditions (2.4.3) and (2.4.5) hold.*

2.5 Multidimensional Hardy–Type Inequalities with General Kernels Via Convexity

The classical Hardy inequality was finally proved in 1925 by G. H. Hardy in [67] (but he announced it already in 1920 in [66]). The dramatic prehistory was recently described in [100] and the further development in several books (see, e.g., [68], [102], [100], [140] and references given there).

In this section we prove and discuss some new multidimensional Hardy–type inequalities involving arithmetic mean operators with general positive kernels. We remark that our condition is only sufficient (but see also Open Problems). Our approach is to use a convexity argument, which is completely different from the classical ones used by Hardy and others (see, e.g., [68], [102], [100] and [140]). In particular, our results further generalize and complete some results in [79] and [132]. The results and methods of proofs are mainly taken from [132] (but see also [99]).

In this section the weighted Hardy–type inequality

$$\int_0^\infty \cdots \int_0^\infty \Phi\left(A_K f(x_1,\dots,x_n)\right) v(x_1,\dots,x_n) \frac{dx_1\cdots dx_n}{x_1\cdots x_n}$$

$$\leq \int_0^\infty \cdots \int_0^\infty \Phi\left(f(x_1,\dots,x_n)\right) w(x_1,\dots,x_n) \frac{dx_1\cdots dx_n}{x_1\cdots x_n} \tag{2.5.1}$$

will be studied, where Φ is a convex function and A_K and its dual A_{K^*} are general Hardy–type (arithmetic mean) operators defined by

$$A_K f(x_1,\dots,x_n) := \frac{1}{K(x_1,\dots,x_n)} \times \int_0^{x_1}\cdots\int_0^{x_n} k(x_1,\dots,x_n,t_1,\dots,t_n)f(t_1,\dots,t_n)dt_1\cdots dt_n \tag{2.5.2}$$

and

$$A_{K^*} f(x_1,\dots,x_n) := \frac{1}{\widetilde{K}(x_1,\dots,x_n)} \times \int_{x_1}^{\infty}\cdots\int_{x_n}^{\infty} k(x_1,\dots,x_n,t_1,\dots,t_n)f(t_1,\dots,t_n)dt_1\cdots dt_n \tag{2.5.3}$$

with $K(x_1,\dots,x_n)$ and $\widetilde{K}(x_1,\dots,x_n)$ given by

$$K(x_1,\dots,x_n) := \int_0^{x_1}\cdots\int_0^{x_n} k(x_1,\dots,x_n,t_1,\dots,t_n)dt_1\cdots dt_n \tag{2.5.4}$$

and

$$\widetilde{K}(x_1,\dots,x_n) := \int_{x_1}^{\infty}\cdots\int_{x_n}^{\infty} k(x_1,\dots,x_n,t_1,\dots,t_n)dt_1\cdots dt_n, \tag{2.5.5}$$

respectively (here $k = k(x_1,\dots,x_n,t_1,\dots,t_n)$ is a kernel, i.e., a locally integrable and positive function in $\mathbb{R}^{2n}$).

All functions in this section are assumed to be measurable and expressions of the form $0.\infty$, $\frac{0}{0}$, $\frac{\infty}{\infty}$, $\frac{a}{\infty}$ $(a \in)$ are taken to be equal to zero. Finally, as usual, by a weight $v = v(x_1,\dots,x_n)$ we mean a nonnegative measurable function on a subset of $\mathbb{R}^n_+$.

Proposition 2.5.1. *Let $n \in \mathbb{N}_+$, $k(x_1,\dots,x_n,t_1,\dots,t_n)$ and $v(x_1,\dots,x_n)$ be weight functions and assume that*

$$\frac{k(x_1,\dots,x_n,t_1,\dots,t_n)v(x_1,\dots,x_n)}{x_1\cdots x_n K(x_1,\dots,x_n)}$$

is locally integrable and for each $(t_1,\dots,t_n)$, $t_i \in (0,b_i)$, define w by

$$w(x_1,\dots,x_n) := t_1\cdots t_n \int_{t_1}^{b_1}\cdots\int_{t_n}^{b_n} \frac{k(x_1,\dots,x_n,t_1,\dots,t_n)v(x_1,\dots,x_n)}{K(x_1,\dots,x_n)} \frac{dx_1\cdots dx_n}{x_1\cdots x_n} < \infty. \tag{2.5.6}$$

(i) *If* Φ *is positive and convex on* (a,c), $-\infty \le a < c \le \infty$, *then*

$$\int_0^{b_1}\cdots\int_0^{b_n} \Phi\left(A_K f(x_1,\dots,x_n)\right) v(x_1,\dots,x_n)\frac{dx_1\cdots dx_n}{x_1\cdots x_n}$$
$$\le \int_0^{b_1}\cdots\int_0^{b_n} \Phi\left(f(x_1,\dots,x_n)\right) w(x_1,\dots,x_n)\frac{dx_1\cdots dx_n}{x_1\cdots x_n} \tag{2.5.7}$$

for all f *with* $a < f(x_1,\dots,x_n) < c$, $0 \le x_i \le b_i$, $i = 1,2,\dots,n$.
(ii) *If* Φ *is positive and concave on* (a,c), $-\infty \le a < c \le \infty$, *then*

$$\int_0^{b_1}\cdots\int_0^{b_n} \Phi\left(A_K f(x_1,\dots,x_n)\right) v(x_1,\dots,x_n)\frac{dx_1\cdots dx_n}{x_1\cdots x_n}$$
$$\ge \int_0^{b_1}\cdots\int_0^{b_n} \Phi\left(f(x_1,\dots,x_n)\right) w(x_1,\dots,x_n)\frac{dx_1\cdots dx_n}{x_1\cdots x_n} \tag{2.5.8}$$

for all f *with* $a < f(x_1,\dots,x_n) < c$, $0 \le x_i \le b_i$, $i = 1,2,\dots,n$. *Here* A_K *and* $K(x_1,\dots,x_n)$ *are as defined by* (2.5.2) *and* (2.5.4), *respectively.*

Remark 2.5.1. For the case $n = 1$ Proposition 2.5.1 (i) coincides with Theorem 4.1 in [79]. Moreover, for the case $k \equiv 1$ in Proposition 2.5.1 we obtain Proposition 2.1 in [132].

Proof. (i) By applying Jensen's inequality and Fubini's theorem to the left hand side of (2.5.7) we have that

$$\int_0^{b_1}\cdots\int_0^{b_n} \Phi\left(A_K f(x_1,\dots,x_n)\right) v(x_1,\dots,x_n)\frac{dx_1\cdots dx_n}{x_1\cdots x_n}$$
$$= \int_0^{b_1}\cdots\int_0^{b_n}\Phi\left(\frac{1}{K(x_1,\dots,x_n)}\int_0^{x_1}\cdots\int_0^{x_n} k(x_1,\dots,x_n,t_1,\dots,t_n)f(t_1,\dots,t_n)dt_1\cdots dt_n\right)$$
$$\times\, v(x_1,\dots,x_n)\frac{dx_1\cdots dx_n}{x_1\cdots x_n}$$
$$\le \int_0^{b_1}\cdots\int_0^{b_n}\left(\frac{1}{K(x_1,\dots,x_n)}\int_0^{x_1}\cdots\int_0^{x_n} k(x_1,\dots,x_n,t_1,\dots,t_n)\Phi(f(t_1,\dots,t_n))dt_1\cdots dt_n\right)$$
$$\times\, v(x_1,\dots,x_n)\frac{dx_1\cdots dx_n}{x_1\cdots x_n}$$

$$= \int_0^{b_1} \cdots \int_0^{b_n} \Phi(f(t_1,\ldots,t_n)) \left(\int_{t_1}^{b_1} \cdots \int_{t_n}^{b_n} \frac{1}{K(x_1,\ldots,x_n)} k(x_1 \cdots x_n, t_1,\ldots,t_n) \right.$$

$$\left. \times v(x_1,\ldots,x_n) \frac{dx_1 \cdots dx_n}{x_1 \cdots x_n} \right) dt_1 \cdots dt_n$$

$$= \int_0^{b_1} \cdots \int_0^{b_n} \Phi\left(f(x_1,\ldots,x_n)\right) w(x_1,\ldots,x_n) \frac{dx_1 \cdot \cdots \cdot dx_n}{x_1 \cdots x_n}.$$

(ii) The proof is completely similar to that of (i). We only note that in this case, since the function Φ is concave and, hence, the inequality sign is reversed. The proof is complete. □

Proposition 2.5.2. *Let $n \in \mathbb{N}_+$, $0 \le b_i < x_i$, $t_i \le \infty$, $i = 1,2,\ldots,n$, and let $k(x_1,\ldots,x_n,t_1,\ldots,t_n)$ and $v(x_1,\ldots,x_n)$ be weight functions such that*

$$\frac{k(x_1,\ldots,x_n,t_1,\ldots,t_n)v(x_1,\ldots,x_n)}{x_1 \cdots x_n \widetilde{K}(x_1,\ldots,x_n)}$$

is locally integrable and for each $(t_1,\ldots,t_n)$, $t_i \in (b_i,\infty)$, define w by

$$w(x_1,\ldots,x_n)$$

$$= t_1 \cdots t_1 \int_{b_1}^{t_1} \cdots \int_{b_n}^{t_n} ! \frac{k(x_1,\ldots,x_n,t_1,\ldots,t_n)v(x_1,\ldots,x_n)}{\widetilde{K}(x_1,\ldots,x_n)} \frac{dx_1 \cdots dx_n}{x_1 \cdots x_n} < \infty. \tag{2.5.9}$$

(i) *If Φ is positive and convex on (a,c), $-\infty \le a < c \le \infty$ and A_{K^*} is the general dual Hardy operator defined by (2.5.3), then the inequality*

$$\int_{b_1}^{\infty} \cdots \int_{b_n}^{\infty} \Phi\left(A_{K^*} f(x_1,\ldots,x_n)\right) v(x_1,\ldots,x_n) \frac{dx_1 \cdots dx_n}{x_1 \cdots x_n}$$

$$\le \int_{b_1}^{\infty} \cdots \int_{b_n}^{\infty} \Phi\left(f(x_1,\ldots,x_n)\right) w(x_1,\ldots,x_n) \frac{dx_1 \cdots dx_n}{x_1 \cdots x_n} \tag{2.5.10}$$

holds for all f with $a < f(x_1,\ldots,x_n) < c$, $0 \le x_i \le b_i$, $i = 1,2,\ldots,n$.

(ii) *If Φ is positive and concave on (a,c), $-\infty \le a < c \le \infty$ and A_{K^*} is the general dual Hardy operator defined by (2.5.3), then the inequality*

$$\int_{b_1}^{\infty} \cdots \int_{b_n}^{\infty} \Phi\left(A_{K^*} f(x_1,\ldots,x_n)\right) v(x_1,\ldots,x_n) \frac{dx_1 \cdots dx_n}{x_1 \cdots x_n}$$

$$\ge \int_{b_1}^{\infty} \cdots \int_{b_n}^{\infty} \Phi\left(f(x_1,\ldots,x_n)\right) w(x_1,\ldots,x_n) \frac{dx_1 \cdots dx_n}{x_1 \cdots x_n} \tag{2.5.11}$$

holds for all f with $a < f(x_1, \dots, x_n) < c$, $0 \le x_i \le b_i$, $i = 1, 2, \dots, n$, and $\widetilde{K}(x_1, \dots, x_n)$ is as defined by (2.5.5).

Proof. The proof is completely similar to that of Proposition 2.5.1 so we omit the details (which can be found in [132]). □

In order to be able to prove the main result (Theorem 2.5.1) of this section we need to formulate the following Minkowski– type inequality of independent interest:

Lemma 2.5.1. *Let $p > 1$, $-\infty \le a_i < b_i \le \infty$, $k = k(x_1, \dots, x_n, t_1, \dots, t_n)$ be a locally integrable kernel and Φ and Ψ be positive and measurable functions.*
Then

$$\left(\int_{a_1}^{b_1} \cdots \int_{a_n}^{b_n} \left(\int_{a_1}^{x_1} \cdots \int_{a_n}^{x_n} k(x_1, \dots, x_n, t_1, \dots, t_n) \Psi(y_1, \dots, y_n) dy_1 \cdots dy_n\right)^p \times \Phi(x_1, \dots, x_n) dx_1 \cdots dx_n\right)^{\frac{1}{p}}$$

$$\le \int_{a_1}^{b_1} \cdots \int_{a_n}^{b_n} \left(\int_{y_1}^{b_1} \cdots \int_{y_n}^{b_n} \Phi(x_1, \dots, x_n) k^p(x_1, \dots, x_n, t_1, \dots, t_n) dx_1 \cdots dx_n\right)^{\frac{1}{p}} \times \Psi(y_1, \dots, y_n) dy_1 \cdots dy_n. \tag{2.5.12}$$

For the readers convenience we include a simple proof of this lemma in the following a little more general form:

Proposition 2.5.3. *Let $p > 1$, $-\infty \le a_i < b_i \le \infty$, and $-\infty \le c_i < d_i \le \infty$, $i = 1, 2, \dots, n$ $(n \in \mathbb{N})$. Then*

$$\left(\int_{a_1}^{b_1} \cdots \int_{a_n}^{b_n} \left(\int_{c_1}^{d_1} \cdots \int_{c_n}^{d_n} k(x_1, \dots, x_n, y_1, \dots, y_n) dy_1 \cdots dy_n\right)^p dx_1 \cdots dx_n\right)^{\frac{1}{p}}$$

$$\le \int_{c_1}^{d_1} \cdots \int_{c_n}^{d_n} \left(\int_{a_1}^{b_1} \cdots \int_{a_n}^{b_n} k^p(x_1, \dots, x_n, y_1, \dots, y_n) dx_1 \cdots dx_n\right)^{\frac{1}{p}} dy_1 \cdots dy_n. \tag{2.5.13}$$

Remark 2.5.2. We see that Lemma 2.5.1 is a special case of Proposition 2.5.3 by choosing $k = k(x_1, \dots, x_n, y_1, \dots, y_n)$ in the following way:

$$k = \begin{cases} k(x_1, \dots, x_n, y_1, \dots, y_n) \Psi(y_1, \dots, y_n) \Phi^{\frac{1}{p}}(x_1, \dots, x_n), & a_i \le y_i \le x_i \le b_i \\ 0, & \text{elsewhere}. \end{cases}$$

Proof. By using the fact that we can have equality in Hölder's inequality we find that

$$I_0 := \left(\int_{a_1}^{b_1}\cdots\int_{a_n}^{b_n}\left(\int_{c_1}^{d_1}\cdots\int_{c_n}^{d_n} k(x_1,\dots,x_n,y_1,\dots,y_n)dy_1\cdots dy_n\right)^p dx_1\cdots dx_n\right)^{\frac{1}{p}}$$

$$= \sup_{\varphi\geq 0}\int_{a_1}^{b_1}\cdots\int_{a_n}^{b_n}\varphi(x_1,\dots,x_n)$$

$$\times\int_{c_1}^{d_1}\cdots\int_{c_n}^{d_n} k(x_1,\dots,x_n,y_1,\dots,y_n)dy_1\cdots dy_n dx_1\cdots dx_n, \tag{2.5.14}$$

where the supremum is taken over all φ such that

$$\int_{a_1}^{b_1}\cdots\int_{a_n}^{b_n}\varphi(x_1,\dots,x_n)dx_1\cdots dx_n = 1.$$

Hence, by Fubini's theorem, a trivial estimate and using (2.5.14) again, we find that

$$I_0 = \sup_{\varphi\geq 0}\int_{c_1}^{d_1}\cdots\int_{c_n}^{d_n}\int_{a_1}^{b_1}\cdots\int_{a_n}^{b_n} k(x_1,\dots,x_n,y_1,\dots,y_n)$$

$$\times\varphi(x_1,\dots,x_n)dx_1\cdots dx_n dy_1\cdots dy_n$$

$$\leq \int_{c_1}^{d_1}\cdots\int_{c_n}^{d_n}\left(\sup_{\varphi\geq 0}\int_{a_1}^{b_1}\cdots\int_{a_n}^{b_n} k(x_1,\dots,x_n,y_1,\dots,y_n)\right.$$

$$\left.\times\varphi(x_1,\dots,x_n)dx_1\cdots dx_n\right)dy_1\cdots dy_n$$

$$= \int_{c_1}^{d_1}\cdots\int_{c_n}^{d_n}\left(\int_{a_1}^{b_1}\cdots\int_{a_n}^{b_n} k^p(x_1,\dots,x_n,y_1,\dots,y_n)dx_1\cdots dx_n\right)^{\frac{1}{p}} dy_1\cdots dy_n.$$

The proof is complete. □

Another proof of Proposition 2.5.3 can be found in the PhD thesis of A. Wedestig [174, Lemma 4.4].

Theorem 2.5.1. *Let* $1 < p \leq q < \infty$, $0 < b_i \leq \infty$, $s_1,\dots,s_n \in (1,p)$, $i = 1,2,\dots,n$, *and let* Φ *be a convex function on* (a,c), $-\infty \leq a < c \leq \infty$. *Let* A_K *be the general Hardy operator defined by* (2.5.2) *and let* $v(x_1,\dots,x_n)$ *and* $w(x_1,\dots,x_n)$

be weight functions, where $w(x_1,\dots,x_n)$ *is of product type, i.e.,* $w(x_1,\dots,x_n) = w_1(x_1)w_2(x_2)\cdots w_n(x_n)$. *Then the inequality*

$$\left(\int_0^{b_1}\cdots\int_0^{b_n}[\Phi(A_K f(x_1,\dots,x_n))]^q v(x_1,\dots,x_n)\frac{dx_1\cdots dx_n}{x_1\cdots x_n}\right)^{\frac{1}{q}}$$

$$\le C\left(\int_0^{b_1}\cdots\int_0^{b_n}\Phi^p(f(x_1,\dots,x_n))w(x_1,\dots,x_n)\frac{dx_1\cdots dx_n}{x_1\cdots x_n}\right)^{\frac{1}{p}} \tag{2.5.15}$$

holds for all functions $f(x_1,\dots,x_n)$, $a < f(x_1,\dots,x_n) < c$, *if*

$$A(s_1,\dots,s_n) := \sup_{0<t_1\cdots t_n<b_1\cdots b_n}\left(\int_{t_1}^{b_1}\cdots\int_{t_n}^{b_n}\left[\frac{k(x_1,\dots,x_n,y_1,\dots,y_n)}{K(x_1,\dots,x_n)}\right]^q v(x_1,\dots,x_n)\right.$$

$$\left.\cdots W_1^{\frac{q(p-s_1)}{p}}(x_1)\cdots W_n^{\frac{q(p-s_n)}{p}}(x_n)\frac{dx_1\cdots dx_n}{x_1\cdots x_n}\right)^{\frac{1}{q}}$$

$$\times W_1^{\frac{s_1-1}{p}}(t_1)\cdots W_n^{\frac{s_n-1}{p}}(t_n) < \infty \tag{2.5.16}$$

holds, where

$$W_i(x_i) := \int_0^{x_i} w^{1-p'}dt_i, \quad i = 1,2,\dots,n, \tag{2.5.17}$$

and $p' = \frac{p}{p-1}$. *Furthermore, if* C *is the best possible constant in* (2.5.15), *then*

$$C \le \inf_{1<s_1\cdots s_n<p}\left(\frac{p-1}{p-s_1}\right)^{\frac{1}{p'}}\cdots\left(\frac{p-1}{p-s_n}\right)^{\frac{1}{p'}} A(s_1,\dots,s_n). \tag{2.5.18}$$

Remark 2.5.3. For the case $n = 1$ Theorem 2.5.1 coincides with Theorem 4.4 in [79].

Proof. By applying Jensen's inequality to the left hand side of (2.5.15) we obtain that

$$\left(\int_0^{b_1}\cdots\int_0^{b_n}[\Phi(A_K f(x_1,\dots,x_n))]^q v(x_1,\dots,x_n)\frac{dx_1\cdots dx_n}{x_1\cdots x_n}\right)^{\frac{1}{q}}$$

$$=\left(\int_0^{b_1}\cdots\int_0^{b_n}\Phi\left[\frac{1}{K(x_1,\dots,x_n)}\int_0^{x_1}\cdots\int_0^{x_n}k(x_1,\dots,x_n,t_1,\dots,t_n)\right.\right.$$

$$\left.\left.\times f(t_1,\dots,t_n)dt_1\cdots dt_n\right]^q v(x_1,\dots,x_n)\frac{dx_1\cdots dx_n}{x_1\cdots x_n}\right)^{\frac{1}{q}}$$

$$\leq \left(\int_0^{b_1} \cdots \int_0^{b_n} \left[\frac{1}{K(x_1, \ldots, x_n)} \int_0^{x_1} \cdots \int_0^{x_n} k(x_1, \ldots, x_n, t_1, \ldots, t_n) \right.\right.$$

$$\left.\left. \times \Phi(f(t_1, \ldots, t_n)) dt_1 \cdots dt_n \right]^q v(x_1, \ldots, x_n) \frac{dx_1 \cdots dx_n}{x_1 \cdots x_n} \right)^{\frac{1}{q}}. \tag{2.5.19}$$

The proof will be completed if we can prove that

$$\left(\int_0^{b_1} \cdots \int_0^{b_n} \left[\frac{1}{K(x_1, \ldots, x_n)} \int_0^{x_1} \cdots \int_0^{x_n} k(x_1, \ldots, x_n, t_1, \ldots, t_n) \right.\right.$$

$$\left.\left. \Phi(f(t_1, \ldots, t_n)) dt_1 \cdots dt_n \right]^q v(x_1, \ldots, x_n) \frac{dx_1 \cdots dx_n}{x_1 \cdots x_n} \right)^{\frac{1}{q}}$$

$$\leq C \left(\int_0^{b_1} \cdots \int_0^{b_n} \Phi^p(f(x_1, \ldots, x_n)) w(x_1, \ldots, x_n) \frac{dx_1 \cdots dx_n}{x_1 \cdots x_n} \right)^{\frac{1}{p}}. \tag{2.5.20}$$

Define g so that $\Phi^p(f(x_1, \ldots, x_n)) \frac{w(x_1, \ldots, x_n)}{x_1 \cdots x_n} = \Phi(g(x_1, \ldots, x_n))$ and (2.5.20) can equivalently be rewritten as

$$L := \left(\int_0^{b_1} \cdots \int_0^{b_n} \left[\frac{1}{K(x_1, \ldots, x_n)} \int_0^{x_1} \cdots \int_0^{x_n} k(x_1, \ldots, x_n, t_1, \ldots, t_n) \Phi^{\frac{1}{p}}(g(t_1, \ldots, t_n)) \right.\right.$$

$$\left.\left. \times \left(\frac{t_1 \cdots t_n}{w(t_1, \ldots, t_n)} \right)^{\frac{1}{p}} dt_1 \cdots dt_n \right]^q v(x_1, \ldots, x_n) \frac{dx_1 \cdots dx_n}{x_1 \cdots x_n} \right)^{\frac{1}{q}}$$

$$\leq C \left(\int_0^{b_1} \cdots \int_0^{b_n} \Phi(g(x_1, \ldots, x_n)) \frac{dx_1 \cdots dx_n}{x_1 \cdots x_n} \right)^{\frac{1}{p}}. \tag{2.5.21}$$

By applying Hölder's inequality and Lemma 2.5.1 with p replaced by $\frac{q}{p}$ to the left hand side of (2.5.21) and using the fact that $w(x_1, \ldots, x_n) = w_1(x_1) \cdots w_n(x_n)$ and $\frac{-p'}{p} = 1 - p'$ we find that

$$L = \left(\int_0^{b_1} \cdots \int_0^{b_n} \left[\frac{1}{K(x_1, \ldots, x_n)} \int_0^{x_1} \cdots \int_0^{x_n} k(x_1, \ldots, x_n, t_1, \ldots, t_n) \Phi^{\frac{1}{p}}(g(t_1, \ldots, t_n)) \right.\right.$$

$$\times W_1^{\frac{s_1-1}{p}}(t_1)\cdots W_n^{\frac{s_n-1}{p}}(t_n)W_1^{\frac{-(s_1-1)}{p}}(t_1)\cdots W_n^{\frac{-(s_n-1)}{p}}(t_n)w_1^{-\frac{1}{p}}(t_1)\cdots w_n^{-\frac{1}{p}}(t_n)$$

$$\times t_1^{\frac{1}{p}}\cdots t_n^{\frac{1}{p}}dt_1\cdots dt_n\Big]^q v(x_1,\dots,x_n)\frac{dx_1\cdots dx_n}{x_1\cdots x_n}\Big)^{\frac{1}{q}}$$

$$\le \Big(\int_0^{b_1}\cdots\int_0^{b_n}\Big[\int_0^{x_1}\cdots\int_0^{x_n}k^p(x_1,\dots,x_n,t_1,\dots,t_n)$$

$$\times\Phi(g(t_1,\dots,t_n))W_1^{s_1-1}(t_1)\cdots W_n^{s_n-1}(t_n)dt_1\cdots dt_n\Big]^{\frac{q}{p}}$$

$$\times\Big[\int_0^{x_1}\cdots\int_0^{x_n}W_1^{\frac{-p'(s_1-1)}{p}}(t_1)\cdots W_n^{\frac{-p'(s_n-1)}{p}}(t_n)w_1^{-\frac{p'}{p}}(t_1)\cdots w_n^{-\frac{p'}{p}}(t_n)$$

$$\times t_1^{\frac{p'}{p}}\cdots t_n^{\frac{p'}{p}}dt_1\cdots dt_n\Big]^{\frac{q}{p'}}\frac{w(x_1,\dots,x_n)}{K^q(x_1,\dots,x_n)}\frac{dx_1\cdots dx_n}{x_1\cdots x_n}\Big)^{\frac{1}{q}}$$

$$=\Big(\frac{p-1}{p-s_1}\Big)^{\frac{1}{p'}}\cdots\Big(\frac{p-1}{p-s_n}\Big)^{\frac{1}{p'}}\Big(\int_0^{b_1}\cdots\int_0^{b_n}\Big[\int_0^{x_1}\cdots\int_0^{x_n}k^p(x_1,\dots,x_n,t_1,\dots,t_n)$$

$$\times\Phi(g(t_1,\dots,t_n))W_1^{s_1-1}(t_1)\cdots W_n^{s_n-1}(t_n)dt_1\cdots dt_n\Big]^{\frac{q}{p}}W_1^{\frac{q(p-s_1)}{p}}(x_1)\cdots W_n^{\frac{q(p-s_n)}{p}}(x_n)$$

$$\times\frac{v(x_1,\dots,x_n)}{K^q(x_1,\dots,x_n)}\frac{dx_1\cdots dx_n}{x_1\cdots x_n}\Big)^{\frac{1}{q}}$$

$$\le\Big(\frac{p-1}{p-s_1}\Big)^{\frac{1}{p'}}\cdots\Big(\frac{p-1}{p-s_n}\Big)^{\frac{1}{p'}}\Big(\int_0^{b_1}\cdots\int_0^{b_n}\Phi(g(t_1,\dots,t_n))W_1^{s_1-1}(t_1)\cdots W_n^{s_n-1}(t_n)$$

$$\times\Big[\int_{t_1}^{b_1}\cdots\int_{t_n}^{b_n}W_1^{\frac{q(p-s_1)}{p}}(x_1)\cdots W_n^{\frac{q(p-s_n)}{p}}(x_n)v(x_1,\dots,x_n)\Big(\frac{k(x_1,\dots,x_n,t_1,\dots,t_n)}{K(x_1,\dots,x_n)}\Big)^q$$

$$\times\frac{dx_1\cdots dx_n}{x_1\cdots x_n}\Big]^{\frac{p}{q}}dt_1\cdots dt_n\Big)^{\frac{1}{p}}\le\Big(\frac{p-1}{p-s_1}\Big)^{\frac{1}{p'}}\cdots\Big(\frac{p-1}{p-s_n}\Big)^{\frac{1}{p'}}A(s_1,\dots,s_n)$$

$$\times\Big(\int_0^{b_1}\cdots\int_0^{b_n}\Phi(g(t_1,\dots,t_n))dt_1\cdots dt_n\Big)^{\frac{1}{p}}.$$

Hence, (2.5.21) and, thus, (2.5.20) holds and the proof is complete. □

By using our results in special cases and making suitable variable transformation we obtain directly some multidimensional Hardy–type inequalities. For example if the kernel $k \equiv 1$, then Proposition 2.5.1 with $\Phi(v) = v^p$ implies the following result:

Example 2.5.1. (c.f. [79]) Let $0 \le d_i \le \infty$, $i = 1, 2, \ldots, n$ $(n \in \mathbb{N})$.
(i) If $p > 1$ or $p < 0$, then,

$$\int_0^{d_1} \cdots \int_0^{d_n} \left(\frac{1}{y_1 \cdots y_n} \int_0^{y_1} \cdots \int_0^{y_n} g(s_1, \ldots, s_n) ds_1 \cdots ds_n \right)^p dy_1 \cdots dy_n$$

$$\le \left(\frac{p}{p-1} \right)^{pn} \int_0^{d_1} \cdots \int_0^{d_n} g^p(y_1, \ldots, y_n)$$

$$\times \left(1 - \left(\frac{y_1}{d_1} \right)^{\frac{p-1}{p}} \right) \cdots \left(1 - \left(\frac{y_n}{d_n} \right)^{\frac{p-1}{p}} \right) dy_1 \cdots dy_n$$

for each positive measurable function g.
(ii) If $0 < p < 1$, then

$$\int_0^{d_1} \cdots \int_0^{d_n} \left(\frac{1}{y_1 \cdots y_n} \int_0^{y_1} \cdots \int_0^{y_n} g(s_1, \ldots, s_n) ds_1 \cdots ds_n \right)^p dy_1 \cdots dy_n$$

$$\ge \left(\frac{p}{1-p} \right)^{pn} \int_0^{d_1} \cdots \int_0^{d_n} g^p(y_1, \ldots, y_n)$$

$$\times \left(1 - \left(\frac{y_1}{d_1} \right)^{\frac{p-1}{p}} \right) \cdots \left(1 - \left(\frac{y_n}{d_n} \right)^{\frac{p-1}{p}} \right) dy_1 \cdots dy_n$$

for every positive measurable function g.

Moreover, by choosing $\Phi(v) = \exp v$ and replacing f by $\ln f^p$ for any $p \in \mathbb{R}$ we obtain the following multidimensional version of Pólya-Knopp's inequality:

Example 2.5.2. The assumptions of Proposition 2.5.1 yields that

$$\int_0^{b_1} \cdots \int_0^{b_n} v(x_1, \ldots, x_n) \exp \left(\frac{1}{K(x_1, \ldots, x_n)} \int_0^{x_1} \cdots \int_0^{x_n} k(x_1, \ldots, x_n, t_1, \ldots, t_n) \right.$$

$$\left. \times \ln f(t_1, \ldots, t_n) dt_1 \cdots dt_n \right)^p \frac{dx_1 \cdots dx_n}{x_1 \cdots x_n}$$

$$\le \int_0^{b_1} \cdots \int_0^{b_n} w(x_1, \ldots, x_n) f^p(x_1, \ldots, x_n) \frac{dx_1 \cdots dx_n}{x_1 \cdots x_n}. \tag{2.5.22}$$

Remark 2.5.4. By choosing $p = 1$, $n = 1$, $b_1 = \infty$, $v(x) \equiv 1$ (so that $w(x) \equiv 1$) and replacing $f(t)/t$ by $f(t)$ in (2.5.22) (and that $\int_0^x \ln t dt = x/e$) we obtain the classical Pólya–Knopp inequality

$$\int_0^\infty \exp\left(\frac{1}{x}\int_0^x \ln f(t)dt\right)dx \le e\int_0^\infty f(x)dx.$$

2.6 Weighted Integral Inequalities for Monotonic Functions, the Case $p \le q$

Let $n \in \mathbb{N}$ and consider measurable, non–negative, and decreasing function in each variable $f : \mathbb{R}^n_+ \to \mathbb{R}_+$. Let $(\mathcal{M}_j, \mu_j), j = 1, 2$, denote two σ -finite measure spaces. Moreover, for every $x \in \mathcal{M}_j$, we let $d\sigma_j^x(y)$ denote a positive measure on $\mathbb{R}^n_+$, and

$$T_j f(x) = \int_{\mathbb{R}^n_+} f(y) d\sigma_j^x(y), \tag{2.6.1}$$

$j = 1, 2$. In particular, if $\mathcal{M}_1 = \mathcal{M}_2 = \mathbb{R}^n_+$, and $d\sigma_j^x(y) = k_j(x, y)dy$, $j = 1, 2$, then T_j is a linear operator with kernel and if $d\sigma_j^x(y) = \delta_x(y)$ (Dirac's delta function), then $T_j f(x) = f(x)$ for $j = 1$ and/or $j = 2$. We need the following definition:

Definition 2.6.1. We say that a set $D \subset \mathbb{R}^n_+$ is decreasing (and write $D \in \Delta_d$) if the function χ_D is decreasing (separately in each variable). Similarly, we say that a set $I \subset \mathbb{R}^n_+$ is increasing (and write $I \in \Delta_i$) if the function χ_I is increasing (in each variable).

Theorem 2.6.1. *Let*

$$C_n = C_n(p, q, T_1, T_2) = \sup_{D \in \Delta_d} \frac{\left(\int_{\mathcal{M}_1} (T_1 \chi_D)^q d\mu_1\right)^{1/q}}{\left(\int_{\mathcal{M}_2} (T_2 \chi_D)^p d\mu_2\right)^{1/p}}$$

and f be decreasing.

1. *If $0 < p \le 1 \le q \le \infty$, then*

$$\left(\int_{\mathcal{M}_1} (T_1 f(x))^q d\mu_1(x)\right)^{1/q} \le C_n \left(\int_{\mathcal{M}_2} (T_2 f(x))^p d\mu_2(x)\right)^{1/p}.$$

2. *If $T_1 = Id$, $0 < \max(1, p) \le q \le \infty$, then*

$$\left(\int_{\mathbb{R}^n_+} f^q(x) d\mu_1(x)\right)^{1/q} \le C_n \left(\int_{\mathcal{M}_2} (T_2 f(x))^p d\mu_2(x)\right)^{1/p}.$$

3. *If* $T_2 = Id$, $0 < p \le \min(1,q) < \infty$, *then*

$$\left(\int_{\mathcal{M}_1} (T_1 f(x))^q d\mu_1(x)\right)^{1/q} \le C_n \left(\int_{\mathbb{R}^n_+} f^p(x) d\mu_2(x)\right)^{1/p}.$$

4. *If* $T_1 = T_2 = Id$, $0 < p \le q < \infty$, *then*

$$\left(\int_{\mathbb{R}^n_+} f^q(x) d\mu_1(x)\right)^{1/q} \le C_n \left(\int_{\mathbb{R}^n_+} f^p(x) d\mu_2(x)\right)^{1/p}. \tag{2.6.2}$$

In all cases C_n *is the best constant.*

Proof. The main property we are going to use is the following identity for the operator T_j, which is easily proved using Fubini's theorem:

$$T_j f(x) = \int_0^\infty \left(\int_{\{y: f(y)>t\}} d\sigma_j^x(y) \right) dt = \int_0^\infty T_j \chi_{D_t}(x) dt, \tag{2.6.3}$$

where $D_t = \{y : f(y) > t\}$ (which is a decreasing set).
(1) In the proof we use Minkowski's integral inequality twice, for $q \le 1$ and for $1/p \ge 1$:

$$\left(\int_{\mathcal{M}_1} (T_1 f(x))^q d\mu_1(x)\right)^{1/q} = \left(\int_{\mathcal{M}_1} \left(\int_0^\infty T_1 \chi_{D_t}(x) dt\right)^q d\mu_1(x)\right)^{1/q}$$

$$\le \int_0^\infty \left(\int_{\mathcal{M}_1} (T_1 \chi_{D_t}(x) dt)^q \, d\mu_1(x)\right)^{1/q} dt \le C_n \int_0^\infty \left(\int_{\mathcal{M}_2} (T_2 \chi_{D_t}(x) dt)^p \, d\mu_2(x)\right)^{1/p} dt$$

$$= C_n \left(\int_{\mathcal{M}_2} (T_2 f(x) dt)^p \, d\mu_2(x)\right)^{1/p}.$$

(2) In view of (1), we only have to prove the case $1 \le p \le q < \infty$. We use now [22, Theorem 2(b)] (i.e., the embedding $L^{p/q} \subset L^1$ with the best constant) applied to the decreasing function $\mu_1(D_t)$:

$$\left(\int_{\mathbb{R}^n_+} f^q(x) d\mu_1(x)\right)^{1/q} = q^{1/q} \left(\int_0^\infty t^q \mu_1(D_t) \frac{dt}{t}\right)^{1/q}$$

$$\le q^{1/q}\left[\left(\frac{p}{q}\right)^{q/p} q^{q/p-1}\left(\int_0^\infty (t^q\mu_1(D_t))^{p/q}\frac{dt}{t}\right)^{q/p}\right]^{1/q}$$

$$= p^{1/p}\left(\int_0^\infty \left(t^{p-1}\mu_1(D_t)\right)^{p/q} dt\right)^{1/p} \le C_n p^{1/p}\left(\int_0^\infty t^{p-1}\int_{\mathcal{M}_2}(T_2\chi_{D_t}(x)dt)^p\, d\mu_2(x)dt\right)^{1/p}$$

$$= C_n p^{1/p}\left(\int_{\mathcal{M}_2}\left(\int_0^\infty t^{p-1}\left(T_2\chi_{D_t}(x)dt\right)^p dt\right) d\mu_2(x)\right)^{1/p}.$$

Using again [22, Theorem 2(b)] (i.e., the embedding $L^1 \subset L^p$, $p \ge 1$) for the decreasing function $T_2\chi_{D_t}(x)$ we get that

$$\left(\int_{\mathbb{R}^n_+} f^q(x)d\mu_1(x)\right)^{1/q} \le C_n p^{1/p}\left(\int_{\mathcal{M}_2}\left(p^{-1/p}\int_0^\infty tT_2\chi_{D_t}(x)\frac{dt}{t}\right)^p d\mu_2(x)\right)^{1/p}$$

$$= C_n\left(\int_{\mathcal{M}_2}\left(\int_0^\infty T_2\chi_{D_t}(x)dt\right)^p d\mu_2(x)\right)^{1/p} = C_n\left(\int_{\mathcal{M}_2}(T_2f(x))^p d\mu_2(x)\right)^{1/p}.$$

(3) In this case, again, according to (1), it suffices to prove the case $0 < p \le q \le 1$. We use now twice [22, Theorem 2(b)]; first for the embedding $L^q \subset L^1$ and then for the embedding $L^1 \subset L^{q/p}$:

$$\left(\int_{\mathcal{M}_1}(T_1f(x))^q d\mu_1(x)\right)^{1/q} = \left(\int_{\mathcal{M}_1}\left(\int_0^\infty (T_1\chi_{D_t}(x))dt\right)^q d\mu_1(x)\right)^{1/q}$$

$$\le \left(\int_{\mathcal{M}_1}\left(q\int_0^\infty t^{q-1}(T_1\chi_{D_t}(x))^q dt\right) d\mu_1(x)\right)^{1/q}$$

$$= \left(q\int_0^\infty t^{q-1}\left(\int_{\mathcal{M}_1}(T_1\chi_{D_t}(x))^q d\mu_1(x)\right) dt\right)^{1/q} \le \left(q\int_0^\infty t^{q-1}C_n^q\left(\mu_2(D_t)\right)^{q/p} dt\right)^{1/q}$$

$$\le C_n q^{1/q}\left(\frac{q}{p}\right)^{-p/q} p^{1-p/q}\int_0^\infty t^p\left(\mu_2(D_t)\frac{dt}{t}\right)^{1/p} = C_n\left(p\int_0^\infty t^{p-1}\mu_2(D_t)dt\right)^{1/p}$$

$$= C_n \left(\int_{\mathbb{R}^n_+} f^p(x) d\mu_2(x) \right)^{1/p}.$$

(4) As before, we use [22, Theorem 2(b)] for the embedding $L^1 \subset L^{q/p}$

$$\left(\int_{\mathbb{R}^n_+} f^q(x) d\mu_1(x) \right)^{1/q} = \left(q \int_0^\infty t^{q-1} \mu_1(D_t) dt \right)^{1/q}$$

$$\leq \left(qC_n^q \int_0^\infty t^{q-1} \mu_2^{q/p}(D_t) dt \right)^{1/q} \leq C_n q^{1/q} \left(\left(\frac{q}{p} \right)^{-p/q} p^{1-p/q} \int_0^\infty t^{p-1} \mu_2(D_t) dt \right)^{1/q}$$

$$= C_n \left(p \int_0^\infty t^{p-1} \mu_2(D_t) dt \right)^{1/p} = C_n \left(\int_{\mathbb{R}^n_+} f^p(x) d\mu_2(x) \right)^{1/p}.$$

To finish the proof, we observe that C_n is the best constant in all cases (1)-(4) simply by evaluating the operators for the decreasing function $f = \chi_D$. □

Remark 2.6.1. A corresponding theorem for increasing functions separately in each variable is given in [16, Theorem 2.4] (c.f. also [9, Theorem 2.3]).

The constant C_n defined in Theorem 2.6.1 is not always easy to be calculated. However in [16, Theorem 2.5] it was shown that if the weights are of product type, then the supremum can be reduced to a more simpler class of decreasing sets (rectangles) and this is now something much simpler to compute.

Theorem 2.6.2. *If* $0 < p \leq q < \infty$, $v(x_1, \ldots, x_N) = v_1(x_1) \ldots v_N(x_N)$ *and* $w(x_1, \ldots x_N) = w_1(x_1) \ldots w_N(x_N)$, *then*

$$\sup_{D \in \Delta_d} \frac{\left(\int_D v(x)dx\right)^{1/q}}{\left(\int_D w(x)dx\right)^{1/p}} = \sup_{a_i > 0} \frac{\left(\int_0^{a_1} \cdots \int_0^{a_n} (v(x)dx\right)^{1/q}}{\left(\int_0^{a_1} \ldots \int_0^{a_n} w(x)dx\right)^{1/p}}. \tag{2.6.4}$$

Proof. By induction, it suffices to prove (2.6.4) for $n = 2$. Without loss of generality we may also only consider the case $q = 1$. Note that

$$A = \sup_{a,b>0} \frac{\int_0^a \int_0^b v(x_1, x_2) dx_1 dx_2}{\left(\int_0^a \int_0^b w(x_1, x_2) dx_1 dx_2\right)^{1/p}}$$

$$= \sup_{a>0} \frac{\int_0^a v_1(x_1) dx_1}{\left(\int_0^b w_1(x_1) dx_1\right)^{1/p}} \sup_{b>0} \frac{\int_0^b v_2(x_2) dx_2}{\left(\int_0^b w_2(x_2) dx_2\right)^{1/p}} < \infty,$$

because the weights are of product type. Consequently,

$$\left(\int_0^b w_2(x_2)dx_2\right)^{1/p} \geq A^{-1}\left(\int_0^b v_2(x_2)dx_2\right)\sup_{a>0}\frac{\int_0^a v_1(x_1)dx_1}{\left(\int_0^b w_1(x_1)dx_1\right)^{1/p}},$$

and, hence, since any decreasing set D can be written as

$$D = \{(x,y) : 0 < x < r,\ \ 0 < y < f(x)\},$$

for some $r > 0$ and some decreasing function f, we have that

$$\left(\int_0^r w_1(x_1)\int_0^{f(x_1)} w_2(x_2)dx_2\right)^{1/p}$$

$$\geq A^{-1}\sup_{a>0}\frac{\int_0^a v_1(x_1)dx_1}{\left(\int_0^b w_1(x_1)dx_1\right)^{1/p}}\left(\int_0^r w_1(x_1)\left(\int_0^{f(x_1)} v_2(x_2)dx_2\right)^p dx_1\right)^{1/p}.$$

Thus, using [26, Theorem 2.12] (with the trivial observation that the equivalence in that theorem is in fact an equality), we obtain the following chain of inequalities:

$$C_2 = \sup_{D\in\Delta_d}\frac{\int_D v(x_1,x_2)dx_1dx_2}{\left(\int_D w(x_1,x_2)dx_1dx_2\right)^{1/p}}$$

$$= \sup_{r>0}\left(\sup_{f\downarrow}\frac{\int_0^r v_1(x_1)\left(\int_0^{f(x_1)} v_2(x_2)dx_2\right)dx_1}{\left(\int_0^r w_1(x_1)\left(\int_0^{f(x_1)} w_2(x_2)dx_2\right)dx_1\right)^{1/p}}\right)$$

$$\leq A\left(\sup_{a>0}\frac{\int_0^a v_1(x_1)dx_1}{\left(\int_0^a w_1(x_1)dx_1\right)^{1/p}}\right)^{-1}$$

$$\times\sup_{r>0}\left(\sup_{f\downarrow}\frac{\int_0^r v_1(x_1)\left(\int_0^{f(x_1)} v_2(x_2)dx_2\right)dx_1}{\left(\int_0^r w_1(x_1)\left(\int_0^{f(x_1)} v_2(x_2)dx_2\right)dx_1\right)^{1/p}}\right)$$

$$\leq A\left(\sup_{a>0}\frac{\int_0^a v_1(x_1)dx_1}{\left(\int_0^a w_1(x_1)dx_1\right)^{1/p}}\right)^{-1}\sup_{r>0}\left(\sup_{g\downarrow}\frac{\int_0^r v_1(x_1)g(x_1)dx_1}{\left(\int_0^r w_1(x_1)g^p(x_1)dx_1\right)^{1/p}}\right)$$

$$= A\left(\sup_{a>0}\frac{\int_0^a v_1(x_1)dx_1}{\left(\int_0^a w_1(x_1)dx_1\right)^{1/p}}\right)^{-1}\sup_{r,s>0}\frac{\int_0^{\min(r,s)} v_1(x_1)dx_1}{\left(\int_0^{\min(r,s)} w_1(x_1)dx_1\right)^{1/p}} = A.$$

The equality of the two constants now follows since, clearly $A \leq C_2$. □

Remark 2.6.2. The same kind of proof can be used to show that in the case of increasing sets the intervals are of the form $I = [a_1, \infty) \times \cdots \times [a_n, \infty)$.

Remark 2.6.3. As observed in [16, Remark 3.1] the equality of Theorem 2.6.2 is not always true. We take $p = 1$, $n = 2$, $D_0 = \{(x, y) \in \mathbb{R}^2_+ : 0 \leq x \leq 1, 0 \leq y \leq 1 - x\}$, $v(x, y) = x + y$ on D_0, $v(x, y) = 0$ elsewhere, and $w(x, y) = 1$ on $\mathbb{R}^2_+$. Then a simple calculation shows that the following supremum

$$\sup_{a,b>0} \frac{\int_a^b \int_a^b v(x, y)dxdy}{\int_a^b \int_a^b w(x, y)dxdy}$$

is approximately equal to 0.5407895304, and

$$\sup_{D \in \Delta_d} \frac{\int_D v(x, y)dxdy}{\int_D w(x, y)dxdy} \geq \frac{\int_{D_0} v(x, y)dxdy}{\int_{D_0} w(x, y)dxdy} = \frac{2}{3}.$$

Moreover, it was proved in [19, Example 2.3] that the two constants of Theorem 2.6.2 are not even comparable in general. For a particular choice of weights it was proved that the left–hand side is always 1 whereas the right hand side tends to 0.

A complete characterization of the constant C_n was given in [19, Theorem 3.1] in the case $n = 2$ and weights of the type $w(x, y) = v(xy)$, namely

Theorem 2.6.3. *Let $0 < p \leq q < \infty$ and $v, w \geq 0$ be two measurable functions on $\mathbb{R}_+$ such that $V(t) = \int_0^t v < \infty$ and $W(t) = \int_0^t w < \infty$ for all $t > 0$.*
Then the inequality

$$\left(\int_{\mathbb{R}^2_+} f^q(x, y)v(xy)dxdy \right)^{1/q} \leq C \left(\int_{\mathbb{R}^2_+} f^p(x, y)w(xy)dxdy \right)^{1/p} \tag{2.6.5}$$

holds for all decreasing f, with a finite constant $C > 0$ independent of f if and only if

$$A = A_{p,q} := \sup \left(\frac{V(t)}{W(t)} \right)^{1/q} \left(\int_0^\infty W(x) \frac{dx}{x} \right)^{1/q - 1/p} < \infty. \tag{2.6.6}$$

Moreover,

$$C = A, \quad \text{if} \quad p = q$$

and

$$2^{-1/p} A \leq C \leq \left(\frac{p}{q} \right)^{1/q} A, \quad \text{if} \quad p < q.$$

Proof. The best constant in (2.6.5) is $C = I$, where

$$I = \sup_{t>0,h\downarrow} I_h(t) := \sup_{t>0,h\downarrow} \frac{\left(\int_0^t dx \int_0^{h(x)} v(xy)dy\right)^{1/q}}{\left(\int_0^t dx \int_0^{h(x)} w(xy)dy\right)^{1/p}} \tag{2.6.7}$$

and thus, by changing variables, we find that

$$I_h(t) = \frac{\left(\int_0^t V(xh(x))\frac{dx}{x}\right)^{1/q}}{\left(\int_0^t W(xh(x))\frac{dx}{x}\right)^{1/p}}. \tag{2.6.8}$$

We begin with the upper bound. By using (2.6.6) we obtain that

$$\int_0^t V(xh(x))\frac{dx}{x} \le A^q \int_0^t \left(\int_0^{xh(x)} W(s)\frac{ds}{s}\right)^{q/p-1} W(xh(x))\frac{dx}{x}$$

$$\left[\text{changing the variables}: \ s = h(x)\xi\,\right]$$

$$= A^q \int_0^t \left(\int_0^x W(h(x)\xi)\frac{d\xi}{\xi}\right)^{q/p-1} W(xh(x))\frac{dx}{x}$$

$$\left[\, h(x) \le h(\xi) \text{ if } \xi \le x\,\right]$$

$$\le A^q \int_0^t \left(\int_0^x W(h(\xi)\xi)\frac{d\xi}{\xi}\right)^{q/p-1} W(xh(x))\frac{dx}{x}$$

$$= \frac{p}{q} A^q \left(\int_0^t W(\xi h(\xi))\frac{d\xi}{\xi}\right)^{q/p}.$$

This implies that

$$I_h(t) \le \left(\frac{p}{q}\right)^{1/q} A$$

for all $t > 0$ and $h \downarrow$. Thus, (2.6.7) brings the upper bound $\left(\frac{p}{q}\right)^{1/q} A$ and, in particular, $C \le A$ when $p = q$.
For the lower bound, let $0 < \delta < t < \infty$ and $h_\delta(s)$ be defined as follows:

$$h_\delta(s) = \begin{cases} 1 & if \quad 0 \le s < \delta \\ \frac{\delta}{s} & if \quad \delta \le s < t \\ 0 & if \quad\quad s \ge t. \end{cases} \tag{2.6.9}$$

Then, by using (2.6.8), we find in the case $p<q$ that

$$I_\delta^q(t)=I_{h_\delta}^q(t)=\frac{\int_0^\delta V(x)\frac{dx}{x}+V(\delta)\log\frac{t}{\delta}}{\left(\int_0^\delta W(x)\frac{dx}{x}+W(\delta)\log\frac{t}{\delta}\right)^{q/p}}. \tag{2.6.10}$$

Since $\log\frac{t}{\delta}$ takes all the values of $(0,\infty)$, when $t>\delta$, we can choose such a t_δ so that

$$\log\frac{t_\delta}{\delta}=\frac{1}{W(\delta)}\int_0^\delta W(x)\frac{dx}{x}.$$

With this t_δ, (2.6.10) gives

$$\begin{aligned} I_\delta^q(t_\delta) &= \frac{\int_0^\delta V(x)\frac{dx}{x}+\frac{V(\delta)}{W(\delta)}\int_0^\delta W(x)\frac{dx}{x}}{2^{q/p}\left(\int_0^\delta W(x)\frac{dx}{x}\right)^{q/p}} \\ &\geq 2^{-q/p}\frac{V(\delta)}{W(\delta)}\left(\int_0^\delta W(x)\frac{dx}{x}\right)^{1-q/p}. \end{aligned}$$

Since $\delta>0$ is arbitrary this implies that

$$C\geq 2^{-1/p}A,\quad p<q.$$

In the case $p=q$ we find from (2.6.10) that

$$I_\delta^q(t)=\frac{\int_0^\delta V(x)\frac{dx}{x}+V(\delta)\log\frac{t}{\delta}}{\int_0^\delta W(x)\frac{dx}{x}+W(\delta)\log\frac{t}{\delta}}$$

and observe that the right hand side tends to $\frac{V(\delta)}{W(\delta)}$, when $t\to\infty$, so that

$$C\geq A,\quad p=q$$

and the proof is complete. □

In the case $n=2$ and homogeneous weights the characterization of the best constant C_n was given in [10, Theorem 3], namely

Theorem 2.6.4. *Let $0<p\leq q<\infty$, $\alpha,\beta>-2$ and $v,w\geq 0$ be two homogeneous weight functions on $\mathbb{R}_+^2$, i.e., $v(\varepsilon x,\varepsilon y)=\varepsilon^\alpha v(x,y)$ and $w(\varepsilon x,\varepsilon y)=\varepsilon^\beta w(x,y)$, $\varepsilon>0$. Let $V(t)=\int_0^t v(1,s)ds<\infty$ and $W(t)=\int_0^t w(1,s)ds<\infty$ for all $t>0$. Then inequality (2.6.2) holds for all decreasing f, with a finite constant $C>0$ independent of f if and only if*

$$A=A_{p,q}:=\sup_{t>0}\left(t^{\beta-\alpha}\frac{V(t)}{W(t)}\right)^{1/q}\left(\int_t^\infty W(s)\frac{ds}{s^{\beta+3}}\right)^{1/q-1/p}<\infty.$$

Moreover,

$$C_2 = A, \quad \textit{if} \quad p = q$$

and

$$A \le C_2 \le \left(\frac{p}{q}\right)^{1/q} A, \quad \textit{if} \quad p < q.$$

Proof. By homogeneity it suffices to suppose that $(\alpha+2)/q = (\beta+2)/p$. We start with the upper bound. The best constant in inequality (2.6.2) is

$$C_2 = \sup_{t>0,h\downarrow} I_h(t) = \sup_{t>0,h\downarrow} \frac{\left(\int_0^\infty dx \int_0^{h(x)} v(x,y)dx\right)^{1/q}}{\left(\int_0^\infty dx \int_0^{h(x)} w(x,y)dx\right)^{1/p}}.$$

By using homogeneity and change of variables we find that

$$I_h(t) = \frac{\left(\int_0^\infty x^{\alpha+1} V(\frac{h(x)}{x})dx\right)^{1/q}}{\left(\int_0^\infty x^{\beta+1} W(\frac{h(x)}{x})dx\right)^{1/p}}.$$

By hypothesis we have that

$$V(t) \le A^q \left(\int_t^\infty W(s)\frac{ds}{s^{\beta+3}}\right)^{q/p-1} W(t) t^{(\beta+2)(q/p-1)}, \quad t > 0.$$

If we substitute now t by $\frac{h(x)}{x}$ and multiply by $x^{\alpha+1}$ we get that

$$x^{\alpha+1} V\left(\frac{h(x)}{x}\right)$$

$$\le A^q \left(\int_{\frac{h(x)}{x}}^\infty W(s)\frac{ds}{s^{\beta+3}}\right)^{q/p-1} W\left(\frac{h(x)}{x}\right) x^{\beta+1} h^{(\beta+2)(q/p-1)}(x),$$

i.e.,

$$x^{\alpha+1} V\left(\frac{h(x)}{x}\right) \le A^q \left(\int_{\frac{h(x)}{x}}^\infty W(s)\left(\frac{h(x)}{s}\right)^{(\beta+2)} \frac{ds}{s}\right)^{q/p-1} W\left(\frac{h(x)}{x}\right) x^{\beta+1}.$$

Integrating now with respect to x, from 0 to t, changing the variables $\xi = \frac{h(x)}{s}$ and using the facts that h and W are decreasing, respectively increasing we obtain that

$$\int_0^t x^{\alpha+1} V\left(\frac{h(x)}{x}\right) \le A^q \int_0^t \left(\int_0^x W(\frac{h(x)}{\xi})\xi^{\beta+1} d\xi\right)^{q/p-1} W\left(\frac{h(x)}{x}\right) x^{\beta+1} dx$$

$$\le A^q \int_0^t \left(\int_0^x W(\frac{h(\xi)}{\xi})\xi^{\beta+1} d\xi\right)^{q/p-1} W\left(\frac{h(x)}{x}\right) x^{\beta+1} dx$$

$$= \frac{p}{q} A^q \int_0^t \left(W(\frac{h(x)}{x})x^{\beta+1}\right)^{q/p}.$$

This implies that

$$I_h \le \left(\frac{p}{q}\right)^{1/q} A$$

for all $t > 0$ and $h \downarrow$ and we get the upper bound.
For the lower bound, let $(a_n)_{n\ge 1}$, $a_1 = 0$ be a strictly increasing sequence, convergent to 1 and let $h(x) \equiv 1$. Let $t > 0$. Since V is increasing we get that

$$\int_0^{a_n t} x^{\alpha+1} V\left(\frac{1}{x}\right) dx = \sum_{k=2}^n \int_{a_{k-1}t}^{a_k t} x^{\alpha+1} V\left(\frac{1}{x}\right) dx$$

$$= \sum_{k=2}^n \int_{\frac{1}{a_k t}}^{\frac{1}{a_{k-1}t}} y^{-(\alpha+3)} V(y) dy \ge \sum_{k=2}^n V\left(\frac{1}{a_k t}\right) (a_{k-1}t)^{\alpha+3}$$

$$\ge t^{\alpha+3} V\left(\frac{1}{t}\right) \sum_{k=2}^n a_{k-1}^{\alpha+3}. \tag{2.6.11}$$

In the same way, since W is also increasing, we find that

$$\int_0^{a_n t} x^{\beta+1} W\left(\frac{1}{x}\right) dx = \int_0^{a_2 t} x^{\beta+1} W\left(\frac{1}{x}\right) dx + \sum_{k=3}^n \int_{\frac{1}{a_k t}}^{\frac{1}{a_{k-1}t}} y^{-(\beta+3)} W(y) dy$$

$$\le \int_0^{a_2 t} x^{\beta+1} W\left(\frac{1}{x}\right) dx + t^{\beta+3} \sum_{k=3}^n W\left(\frac{1}{a_{k-1}t}\right) a_k^{\beta+3}. \tag{2.6.12}$$

By (2.6.11) and (2.6.12) we have that

$$\frac{\int_0^{a_n t} x^{\alpha+1} V\left(\frac{1}{x}\right) dx}{\int_0^{a_n t} x^{\beta+1} W\left(\frac{1}{x}\right) dx} \geq \frac{t^{\alpha+3} V\left(\frac{1}{t}\right) \sum_{k=2}^{n} a_{k-1}{}^{\alpha+3}}{\int_0^{a_2 t} x^{\beta+1} W\left(\frac{1}{x}\right) dx + t^{\beta+3} \sum_{k=3}^{n} W\left(\frac{1}{a_{k-1}t}\right) a_k{}^{\beta+3}}.$$

Dividing now by $t^{\alpha+3} \sum_{k=2}^{n} a_{k-1}^{\alpha+3}$ we get that

$$\frac{\int_0^{a_n t} x^{\alpha+1} V\left(\frac{1}{x}\right) dx}{\int_0^{a_n t} x^{\beta+1} W\left(\frac{1}{x}\right) dx} \geq \frac{V\left(\frac{1}{t}\right)}{\frac{\int_0^{a_2 t} x^{\beta+1} W\left(\frac{1}{x}\right) dx}{t^{\alpha+3} \sum_{k=2}^{n} a_{k-1}^{\alpha+3}} + \frac{t^{\beta-\alpha} \sum_{k=3}^{n} W\left(\frac{1}{a_{k-1}t}\right) a_k^{\alpha+3}}{\sum_{k=2}^{n} a_{k-1}^{\alpha+3}}}. \tag{2.6.13}$$

Since W is continuous, we get by Stolz–Cesàro's theorem that

$$\lim_{n\to\infty} \frac{\sum_{k=3}^{n} W\left(\frac{1}{a_{k-1}t}\right) a_k^{\alpha+3}}{\sum_{k=2}^{n} a_{k-1}^{\alpha+3}} = W\left(\frac{1}{t}\right).$$

By letting now $n \to \infty$ in (2.6.13) we finally conclude that

$$\frac{\left(\int_0^t x^{\alpha+1} V\left(\frac{1}{x}\right) dx\right)^{1/q}}{\left(\int_0^t x^{\beta+1} W\left(\frac{1}{x}\right) dx\right)^{1/q}} \geq \left(t^{\alpha-\beta} \frac{V(\frac{1}{t})}{W(\frac{1}{t})}\right)^{1/q}.$$

Hence

$$C_2 \geq \frac{\left(\int_0^t x^{\alpha+1} V\left(\frac{1}{x}\right) dx\right)^{1/q}}{\left(\int_0^t x^{\beta+1} W\left(\frac{1}{x}\right) dx\right)^{1/p}}$$

$$= \frac{\left(\int_0^t x^{\alpha+1} V\left(\frac{1}{x}\right) dx\right)^{1/q}}{\left(\int_0^t x^{\beta+1} W\left(\frac{1}{x}\right) dx\right)^{1/q}} \left(\int_0^t x^{\beta+1} W\left(\frac{1}{x}\right) dx\right)^{1/p-1/q}$$

$$\geq \left(t^{\alpha-\beta} \frac{V(\frac{1}{t})}{W(\frac{1}{t})}\right)^{1/q} \left(\int_{\frac{1}{t}}^{\infty} W(s) \frac{ds}{s^{\beta+3}}\right)^{1/p-1/q}.$$

Since $t > 0$ was arbitrary, the above inequality implies the lower bound for the constant and this completes the proof. □

We give now an important corollary of Theorem 2.6.1.

Corollary 2.6.1. *If $f \downarrow$ and $-1 < a_i, b_i < \infty$, $i = 1, \ldots, n$, then*

$$\left(\int_{\mathbb{R}_+^n} f^q(x) x_1^{a_1} \ldots x_n^{a_n} dx \right)^{1/q} \leq \prod_{i=1}^n \frac{(b_i+1)^{1/p}}{(a_i+1)^{1/q}} \left(\int_{\mathbb{R}_+^n} f^p(x) x_1^{b_1} \ldots x_n^{b_n} dx \right)^{1/p}$$

if and only if,

$$\frac{n + \sum_{i=1}^n a_i}{q} = \frac{n + \sum_{i=1}^n b_i}{p}.$$

In particular this is an improvement of the result proved in [15, Theorem 1.5].

Another important corollary of Theorem 2.6.1 was given in [141, Theorem 1.1]. Some related results, in particular Hardy–type inequalities for decreasing functions and weights of product type can be found in [11]. A complete characterization of the Hardy inequality for decreasing functions of several variables and general weights was given in [8]. The two–dimensional version reads:

Theorem 2.6.5. *Let $0 < p < \infty$, $H_1 f(x,y) = \frac{1}{x}\int_0^x f(s,y)ds$, $H_2 f(x,y) = \frac{1}{y}\int_0^y f(x,t)dt$, $Hf(x,y) = \frac{1}{xy}\int_0^x \int_0^y f(s,t)dsdt$ and D be a decreasing set. We denote by $D_x^y = D \bigcap([0,x] \times [0,y])$. The following conditions are equivalent:*

- *The inequality*

$$\left(\int_{\mathbb{R}_+^2} (Hf(x,y))^p v(x,y) dxdy \right)^{1/p} \leq C \left(\int_{\mathbb{R}_+^2} f^p(x,y) v(x,y) dxdy \right)^{1/p}.$$

  holds for all $f \downarrow$.

- *There exists a constant $C > 0$ such that for every decreasing set D:*

$$\int_{\mathbb{R}_+^2 \setminus D} \frac{|D_x^y|^p}{(xy)^p} v(x,y) dxdy < C \int_D v(x,y) dxdy.$$

- *The inequalities*

$$\left(\int_{\mathbb{R}_+^2} (H_1 f(x,y))^p v(x,y) dxdy \right)^{1/p} \leq C \left(\int_{\mathbb{R}_+^2} f^p(x,y) v(x,y) dxdy \right)^{1/p}$$

  and

$$\left(\int_{\mathbb{R}_+^2} (H_2 f(x,y))^p v(x,y) dxdy \right)^{1/p} \leq C \left(\int_{\mathbb{R}_+^2} f^p(x,y) v(x,y) dxdy \right)^{1/p}.$$

  hold for all $f \downarrow$.

To extend the theorem to arbitrary dimension is a trivial fact.

2.7 Weighted Integral Inequalities for Monotonic Functions, the Case $0 < q < p < \infty$

Throughout this section we suppose that $h(x) \geq 0$, $h \neq 0$ a.e., denotes a decreasing function on $\mathbb{R}^n_+$. For $t > 0$, we use the following notation:

$$D_{h,t} := \{x \in \mathbb{R}^n_+ ; h(x) > t\}$$

and for an increasing sequence $\{t_k\} \subset \mathbb{R}_+$ we set

$$D_k = D_{h,k} := \{x \in \mathbb{R}^n_+ ; h(x) > t_k\}, \quad k \in \mathbb{Z}. \tag{2.7.1}$$

Obviously, $D_k \supset D_{k+1}$ and we define

$$\Delta_k = \Delta_{h,k} := D_k \setminus D_{k+1}. \tag{2.7.2}$$

Hence, $\Delta_k \cap \Delta_j = \Phi$, $\quad k \neq j$ and $\mathbb{R}^n_+ = \bigcup_k \Delta_k$. Let $0 < q < p < \infty$ and let $r \in \mathbb{R}_+$ be defined by $\frac{1}{r} = \frac{1}{q} - \frac{1}{p}$.

If v and w are non–negative, measurable functions on $\mathbb{R}^n_+$ we define the following quantities:

$$B^r = \sup_{0 \leq h \downarrow} \int_0^\infty \Bigg(\int_{D_{h,t}} w \Bigg)^{-r/p} d\left(- \Bigg(\int_{D_{h,t}} v \Bigg)^{r/q} \right),$$

and

$$\mathcal{B}^r := \sup_{0 \leq h \downarrow} \sup_{\{t_k\}} \sum_k \Bigg(\int_{\Delta_k} v \Bigg)^{r/q} \Bigg(\int_{\Delta_k} w \Bigg)^{-r/p}.$$

The following result was proved in [19, Theorem 4.1] (see also [10]):

Theorem 2.7.1. *Let $0 < q < p < \infty$.*

1. *The inequality*

$$\Bigg(\int_{\mathbb{R}^n_+} f^q(x) v(x) dx \Bigg)^{1/q} \leq C \Bigg(\int_{\mathbb{R}^n_+} f^p(x) w(x) dx \Bigg)^{1/p} \tag{2.7.3}$$

   holds for all decreasing f, with a finite constant $C > 0$ independent of f if and only if $\mathcal{B} < \infty$. Moreover,

$$\mathcal{B} \leq C \leq 4^{1/q} \mathcal{B}. \tag{2.7.4}$$

2. *The following inequalities hold:*

$$\mathcal{B} \leq B \leq 2^{1/q} (2^{r/q} + 2^{r/p})^{1/r} \mathcal{B}. \tag{2.7.5}$$

3. *The following representation takes place:*

$$B^r = \frac{\left(\int_{\mathbb{R}^n_+} v\right)^{r/q}}{\left(\int_{\mathbb{R}^n_+} w\right)^{r/p}} + \sup_{0\le h\downarrow} \int_0^\infty \left(\int_{D_{h,t}} v\right)^{r/q} d\left(\left(\int_{D_{h,t}} w\right)^{-r/p}\right). \tag{2.7.6}$$

Proof. For a fixed $0 \le h \downarrow$ and an increasing sequence $\{t_k\}$ we define the function $f_h(x)$ by

$$f_h(x) = \sum_k \left(\sum_{j\le k}\left(\int_{\Delta_j} v\right)^{r/q}\left(\int_{D_j} w\right)^{-r/q}\right)^{1/p} \chi_{\Delta_k}(x).$$

Then $f_h(x) \ge 0$ is a decreasing function and

$$\int_{\mathbb{R}^n_+} f_h^p w = \sum_k \left(\sum_{j\le k}\left(\int_{\Delta_j} v\right)^{r/q}\left(\int_{D_j} w\right)^{-r/q}\right)\int_{\Delta_k} w$$

$$\left[\text{ changing the order of sums }\right]$$

$$= \sum_j \left(\int_{\Delta_j} v\right)^{r/q}\left(\int_{D_j} w\right)^{-r/q} \sum_{k\ge j}\int_{\Delta_k} w$$

$$\left[\text{ using } \sum_{k\ge j}\int_{\Delta_k} w = \int_{D_j} w, \quad -r/q+1 = -r/p\,\right]$$

$$= \sum_j \left(\int_{\Delta_j} v\right)^{r/q}\left(\int_{D_j} w\right)^{-r/p} =: \mathcal{B}^r_{h,\{t_k\}}. \tag{2.7.7}$$

Suppose now that (2.7.3) is valid with a finite constant $C > 0$, and assume temporarily that $\mathcal{B} \in (0,\infty)$. Then, for any $h\downarrow$ and $\{t_k\}$ such that $\mathcal{B}^r_{h,\{t_k\}} > 0$, we obtain, by using the representation formula (2.7.7)

$$C^q\left(\mathcal{B}_{h,\{t_k\}}\right)^{qr/p} = C^q\left(\int_{\mathbb{R}^n_+} f_h^p w\right)^{q/p} \ge \int_{\mathbb{R}^n_+} f_h^q v$$

$$= \sum_k \int_{\Delta_k} v \left(\sum_{j\le k}\left(\int_{\Delta_j} v\right)^{r/q}\left(\int_{D_j} w\right)^{-r/q}\right)^{q/p}.$$

$$\Big[\text{reducing the interior sum to one term with } k=j\Big]$$

$$\geq \sum_k \left(\int_{\Delta_k} v\right)^{r/q} \left(\int_{D_k} w\right)^{-r/p} = \mathcal{B}^r_{h,\{t_k\}}.$$

Hence,

$$C \geq \mathcal{B}_{h,\{t_k\}}$$

and the lower bound of (2.7.7) follows. The temporary assumption $\mathcal{B} \in (0,\infty)$ can be removed in the usual way (see [164, p. 178]).
Next we consider the upper bound. Given $f\downarrow$ we define

$$V(t) = \int_{D_{f,t}} v; \quad W(t) = \int_{D_{f,t}} w.$$

Obviously, $V(t)$ and $W(t)$ are decreasing functions.
Now we construct a special increasing sequence $\{\tau_k\} \in \mathcal{R}_+$ as follows: Put

$$\begin{aligned} \tau_0 &= 1, \\ \tau_{k+1} &= \inf\left\{t : \min\left(\frac{W(\tau_k)}{W(t)}, \frac{V(\tau_k)}{V(t)}\right) = 2\right\}, \quad k \geq 0, \\ \tau_{k-1} &= \sup\left\{t : \min\left(\frac{W(t)}{W(\tau_k)}, \frac{V(t)}{V(\tau_k)}\right) = 2\right\}, \quad k \leq 0, \end{aligned}$$

and let

$$\begin{aligned} Z_1 &= \left\{k \in \mathbb{Z} : W(\tau_{k+1}) = \frac{1}{2}W(\tau_k)\right\}, \\ Z_2 &= \left\{k \in \mathbb{Z} : V(\tau_{k+1}) = \frac{1}{2}V(\tau_k)\right\}. \end{aligned} \tag{2.7.8}$$

We assume without loss of generality that

$$\mathbb{Z} = Z_1 \bigcup Z_2 \tag{2.7.9}$$

and note that $Z_1 \bigcap Z_2 \neq \phi$. Now, we define

$$I := \int_{\mathbb{R}^n_+} f^q v = \sum_k \int_{\Delta_{k-1}} f^q v,$$

where

$$\Delta_k := \Delta_{f,k} = D_{f,k} \setminus D_{f,k+1} := D_k \setminus D_{k+1}.$$

Since

$$\tau_k < f(x) \leq \tau_{k+1}, \quad x \in \Delta_k,$$

we find that

$$I \le \sum_k \tau_k{}^q \int\limits_{\Delta_{k-1}} v = \sum_k \frac{\tau_k{}^q(\int_{\Delta_{k-1}} v)^{q/p}}{\left(\sum_{j\le k}(\int_{\Delta_{j-1}} v)^{r/p} W^{-r/p}(\tau_j)\right)^{q/r}}$$

$$\times \left(\int\limits_{\Delta_{k-1}} v\right)^{1-q/p} \left(\sum_{j\le k}\left(\int\limits_{\Delta_{j-1}} v\right)^{r/p} W^{-r/p}(\tau_j)\right)^{q/r}.$$

[applying Hölder's inequality with $\frac{p}{q}$ and $\frac{r}{q}$]

$$\le \left(\sum_k \frac{\tau_k{}^p(\int_{\Delta_{k-1}} v)}{\left(\sum_{j\le k}(\int_{\Delta_{j-1}} v)^{r/p} W^{-r/p}(\tau_j)\right)^{p/r}}\right)^{q/p}$$

$$\times \left(\sum_k \left(\int\limits_{\Delta_{k-1}} v\right) \sum_{j\le k}\left(\int\limits_{\Delta_{j-1}} v\right)^{r/q} W^{-r/p}(\tau_j)\right)^{q/r} =: I_1^{q/p} I_2^{q/r}.$$

We have that

$$\sum_{j\le k}\left(\int\limits_{\Delta_{j-1}} v\right)^{r/q} W^{-r/p}(\tau_j) \ge \left(\int\limits_{\Delta_{k-1}} v\right)^{r/p} W^{-r/p}(\tau_j).$$

Thus,

$$I_1 \le \sum_k \tau_k^p W(\tau_k).$$

We note also that the sequence $\{\tau_k\}$ is constructed in such a way that

$$W(\tau_k) \ge 2W(\tau_{k+1}), \quad V(\tau_k) \ge 2V(\tau_{k+1}), \quad \text{for all} \quad k \in \mathbb{Z}. \tag{2.7.10}$$

Therefore, in particular,

$$W(\tau_k) = \int\limits_{\Delta_k} w + W(\tau_{k+1}) \ge W(\tau_{k+1}).$$

Hence,

$$W(\tau_{k+1}) \le \int\limits_{\Delta_k} w$$

and, consequently,

$$W(\tau_k) \le 2 \int\limits_{\Delta_k} w. \tag{2.7.11}$$

This implies that

$$I_1 \le 2 \sum_k {\tau_k}^p \int\limits_{\Delta_k} w \le 2 \sum_k \int\limits_{\Delta_k} f^p w \le \int\limits_{\mathcal{R}^n_+} f^p w.$$

Now we return to the estimate of I_2. Write

$$I_2 = \sum_j \Bigg(\int\limits_{\Delta_{j-1}} v \Bigg)^{r/p} W^{-r/p}(\tau_j) \sum_{k \ge j} \int\limits_{\Delta_{j-1}} v$$

$$= \sum_j \Bigg(\int\limits_{\Delta_{j-1}} v \Bigg)^{r/p} W^{-r/p}(\tau_j) V(\tau_{j-1}) := I_{2,1} + I_{2,2},$$

where using (2.7.9), we put

$$I_{2,1} = \sum_{\{j: j-1 \in Z_1\}} \Bigg(\int\limits_{\Delta_{j-1}} v \Bigg)^{r/p} W^{-r/p}(\tau_j) V(\tau_{j-1}),$$

and

$$I_{2,2} = \sum_{\{j: j-1 \in Z_2\}} \Bigg(\int\limits_{\Delta_{j-1}} v \Bigg)^{r/p} W^{-r/p}(\tau_j) V(\tau_{j-1}).$$

Similar to the derivation of (2.7.11) we obtain

$$V(\tau_k) \le 2 \int\limits_{\Delta_k} v. \tag{2.7.12}$$

Hence, by using (2.7.8) and (2.7.12), we find that

$$I_{2,1} \le 2 \sum_{\{j: j-1 \in Z_1\}} \Bigg(\int\limits_{\Delta_{j-1}} v \Bigg)^{r/q} \left(\frac{1}{2} W(\tau_{j-1}) \right)^{-r/p}$$

$$\le 2^{1+r/p} \sum_j \Bigg(\int\limits_{\Delta_j} v \Bigg)^{r/q} W^{-r/p}(\tau_j) \le 2^{1+r/p} \mathcal{B}^r.$$

For the second term we use again (2.7.8) and (2.7.12). We have that

$$V(\tau_{j-1}) = 2V(\tau_j), \quad j-1 \in Z_2,$$

$$\int\limits_{\Delta_{j-1}} v = V(\tau_{j-1}) - V(\tau_j) = V(\tau_j) \le 2 \int\limits_{\Delta_j} v.$$

Thus,

$$I_{2,2} \le 2^{1+r/p} \sum_{\{j:j-1\in Z_2\}} \left(\int\limits_{\Delta_j} v \right)^{r/q} W^{-r/p}(\tau_j) \le 2^{1+r/p}\mathcal{B}^r.$$

Summarizing the above estimates we obtain the upper bound

$$\left(\int\limits_{\mathbb{R}^n_+} f^q u \right)^{1/q} \le 4^{1/q}\mathcal{B} \left(\int\limits_{\mathbb{R}^n_+} f^p w \right)^{1/p}$$

and part (2.7.1) of the theorem has been proved.

For the proof of the lower bound (2.7.5) we fix $0 \le h(x) \downarrow$ and define

$$\Delta_{k,t} = \{x : t < h(x) \le t_{k+1}\}.$$

Then

$$\left(\int\limits_{\Delta_k} v \right)^{r/q} = \int\limits_{t_k}^{t_{k+1}} d\left(- \left(\int\limits_{\Delta_{k,t}} v \right)^{r/q} \right) = \frac{r}{q} \int\limits_{t_k}^{t_{k+1}} \left(\int\limits_{\Delta_{k,t}} v \right)^{r/p} d\left(- \int\limits_{\Delta_{k,t}} v \right).$$

Since

$$\int\limits_{\Delta_{k,t}} v \le V(t); \quad d\left(- \int\limits_{\Delta_{k,t}} v \right) = d\left(-V(t) + \int\limits_{D_{k+1}} v \right) = d(-V(t))$$

we obtain that

$$\left(\int\limits_{\Delta_k} v \right)^{r/q} \le \frac{r}{q} \int\limits_{t_k}^{t_{k+1}} V^{r/p}(t) d(-V(t)) = \int\limits_{t_k}^{t_{k+1}} d(-V^{r/q}(t)).$$

Applying this estimate and the fact

$$\left(\int\limits_{D_k} w \right)^{-r/p} = W^{-r/p}(t_k) \le W^{-r/p}(t), \quad t \in [t_k, t_{k+1}],$$

we find that

$$\sum_k \left(\int_{\Delta_k} v\right)^{r/q} \left(\int_{D_k} w\right)^{-r/p} \le \sum_k \int_{t_k}^{t_{k+1}} W^{-r/p}(t)d(-V^{r/q}(t))$$

$$\le \int_0^\infty W^{-r/p}(t)d(-V^{r/q}(t)) \le B^r.$$

Thus,

$$\mathcal{B} \le B.$$

For the proof of the upper bound of (2.7.5) we observe that for $0 \le h(x) \downarrow$ and an increasing sequence $\{t_k\} \subset \mathcal{R}_+$ we have that

$$B_h^r := \int_0^\infty W^{-r/p}(t)d(-V^{r/q}(t)) = \sum_k \int_{t_k}^{t_{k+1}} W^{-r/p}(t)d(-V^{r/q}(t))$$

$$\le \sum_k W^{-r/p}(t_{k+1})(V^{r/q}(t_k)) =: \mathcal{I}.$$

Now suppose that $\{t_k\}$ is taken in the same way as the sequence τ_k was taken in the proof of part (2.7.1), that is $t_k = \tau_k$, $k \in \mathbb{Z}$. Then

$$\mathcal{I} = \sum_{k \in Z_1} + \sum_{k \in Z_2} =: \mathcal{I}_1 + \mathcal{I}_2.$$

Therefore, by using (2.7.10), (2.7.11) and (2.7.12), we find that

$$\mathcal{I}_1 \le 2^{r/q+r/p} \sum_{k \in Z_1} W^{-r/p}(\tau_k) \left(\int_{\Delta_k} v\right)^{r/q},$$

$$\mathcal{I}_2 \le 2^{2r/q} \sum_{k \in Z_2} W^{-r/p}(\tau_{k+1}) \left(\int_{\Delta_{k+1}} v\right)^{r/q}.$$

Thus,

$$\mathcal{I} \le 2^{r/q}(2^{r/q+r/p}) \sum_{k \in \mathbb{Z}} W^{-r/p}(\tau_k) \left(\int_{\Delta_k} v\right)^{r/q} \le 2^{r/q}(2^{r/q+r/p})\mathcal{B}^r.$$

This implies that

$$B \le 2^{r/q}(2^{r/q+r/p})^{1/r}\mathcal{B}$$

and hence, the upper bound (2.7.5) is proved.

For the proof of part (2.7.1) we suppose first that $B < \infty$. Then by putting, for a fixed $0 \le h(x) \downarrow$,

$$W(t) = \int\limits_{D_{h,t}} w, \quad V(t) = \int\limits_{D_{h,t}} v,$$

we see that

$$\infty > B^r \ge \int\limits_{\tau}^{\infty} W^{-r/q}(t)d(-V^{r/q}(t)) \to 0, \quad \tau \to \infty.$$

Hence,

$$\int\limits_{\tau}^{\infty} W^{-r/p}(t)d(-V^{r/q}(t)) = W^{-r/q}(\tau)V^{r/q}(\tau) \to 0, \quad \tau \to \infty.$$

This implies, by integration by parts, that

$$\int\limits_{0}^{\infty} W^{-r/p}(t)d(-V^{r/q}(t)) = \frac{V^{r/q}(0)}{W^{r/p}(0)} + \int\limits_{0}^{\infty} V^{r/q}(t)dW^{-r/p}(t)$$

and the inequality

$$\infty > B^r \ge \frac{(\int_{\mathbb{R}^n_+} v)^{r/q}}{(\int_{\mathbb{R}^n_+} w)^{r/p}} + \sup_{0 \le h} \int\limits_{0}^{\infty} \left(\int\limits_{D_{h,t}} v \right)^{r/q} d \left(\int\limits_{D_{h,t}} w \right)^{-r/p} \tag{2.7.13}$$

follows. Now suppose that the right hand side of (2.7.6) is finite. Then, for a fixed $h \downarrow$, integration by parts gives

$$\int\limits_{0}^{\infty} V^{r/p}(t)d(-W^{r/q}(t)) \ge -\frac{V^{r/q}(0)}{W^{r/p}(0)} + \int\limits_{0}^{\infty} W^{-r/q}(t)d(-V^{-r/p}(t))$$

and we obtain the reversed inequality to (2.7.13). Thus also (2.7.6) is proved and the proof is complete. □

Example 2.7.1. Let $w = v \in L^1(\mathbb{R}^n_+)$. Then $B^r = \frac{r}{q}\int_{\mathbb{R}^n_+} w$.

Other equivalent characterizations in this case and in the more general context of partially ordered measure spaces were given in [50].

2.8 Further Results and Applications

Theorem 2.6.1 can be applied and generalized in different ways. For example, in [17, Theorem 2.2] a multidimensional Chebyshev–type inequality with different weights (not necessarily of product type) and different parameters was proved. This result generalizes and unifies also several previous results concerning Chebyshev's inequality. Denote

$$C_{m,n} = \sup_{D_j \in \Delta_d} \frac{\prod_{j=1}^m \left(\int_{\mathcal{M}_j} (T_j \chi_{D_j})^{q_j} d\mu_j \right)^{1/q_j}}{\left(\int_{\bigcap_{j=1}^m D_j} d\mu \right)^{1/p}}$$

where $(\mathcal{M}, \mu)$, $(\mathcal{M}_j, \mu_j)$, are σ-finite measure spaces and $T_j f(x) = \int_{\mathbb{R}_+^n} f(y) d\sigma_j^x(y)$, $j = 1, \ldots, m$ are linear operators.

Theorem 2.8.1. *Let $C_{m,n}$ be defined as above and $0 \le f_j \downarrow$, $j = 1, \ldots, m$.*

1. *If $0 < p \le \min(1, q_j) < \infty$, $j = 1, \ldots, m$, then*

$$\prod_{j=1}^m \left(\int_{\mathcal{M}_j} (T_j f_j(x))^{q_j} d\mu_j(x) \right)^{1/q_j} \le C_{m,n} \left(\int_{\mathbb{R}_+^n} \left(\prod_{j=1}^m f_j(x) \right)^p d\mu(x) \right)^{1/p}.$$

2. *If $T_j = Id$, $0 < p \le q_j < \infty$, $\forall j = 1, \ldots, m$, then*

$$\prod_{j=1}^m \left(\int_{\mathbb{R}_+^n} f_j^{q_j}(x) d\mu_j(x) \right)^{1/q_j} \le C_{m,n} \left(\int_{\mathbb{R}_+^n} \left(\prod_{j=1}^m f_j(x) \right)^p d\mu(x) \right)^{1/p}.$$

In both cases $C_{m,n}$ is the best constant.

Proof. 1. By induction it suffices to prove the statement for $m = 2$. According to Theorem 2.6.1 (3) we have that

$$\left(\int_{\mathcal{M}_1} (T_1 f_1(x))^{q_1} d\mu_1(x) \right)^{1/q_1}$$

$$\le \sup_{D \in \Delta_d} \frac{\left(\int_{\mathcal{M}_1} (T_1 \chi_D(x))^{q_1} d\mu_1(x) \right)^{1/q_1}}{\left(\int_D (f_2^p(x)) d\mu(x) \right)^{1/p}} \left(\int_{\mathbb{R}_+^n} (f_1(x) f_2(x))^p d\mu(x) \right)^{1/p}$$

for all decreasing functions f_1. Multiplying by

$$\left(\int_{\mathcal{M}_2} (T_2 f_2(x))^{q_2} d\mu_2(x) \right)^{1/q_2}$$

we get that

$$\left(\int_{\mathcal{M}_1}(T_1f_1(x))^{q_1}d\mu_1(x)\right)^{1/q_1}\left(\int_{\mathcal{M}_2}(T_2f_2(x))^{q_2}d\mu_2(x)\right)^{1/q_2}$$

$$\leq \sup_{D\in\Delta_d}\frac{\left(\int_{\mathcal{M}_1}(T_1\chi_D(x))^{q_1}d\mu_1(x)\right)^{1/q_1}\left(\int_{\mathcal{M}_2}(T_2f_2(x))^{q_2}d\mu_2(x)\right)^{1/q_2}}{\left(\int_D(f_2^p(x))d\mu(x)\right)^{1/p}}$$

$$\times\left(\int_{\mathbb{R}^n_+}(f_1(x)f_2(x))^pd\mu(x)\right)^{1/p}.$$

By using Theorem 2.6.1 (3) again, we find that

$$C_{2,n}=\sup_{f_2\downarrow}\sup_{D\in\Delta_d}\frac{\left(\int_{\mathcal{M}_1}(T_1\chi_D(x))^{q_1}d\mu_1(x)\right)^{1/q_1}\left(\int_{\mathcal{M}_2}(T_2f_2(x))^{q_2}d\mu_2(x)\right)^{1/q_2}}{\left(\int_D(f_2^p(x))d\mu(x)\right)^{1/p}}$$

$$=\sup_{f_2\downarrow}\sup_{D\in\Delta_d}\frac{\left(\int_{\mathcal{M}_2}(T_2f_2(x))^{q_2}d\mu_2(x)\right)^{1/q_2}}{\left(\int_D(f_2^p(x))d\mu(x)\right)^{1/p}}\left(\int_{\mathcal{M}_1}(T_1\chi_D(x))^{q_1}d\mu_1(x)\right)^{1/q_1}.$$

Hence

$$C_{2,n}=\sup_{D\in\Delta_d}\left(\sup_{\widetilde{D}\in\Delta_d}\frac{\left(\int_{\mathcal{M}_2}(T_2\chi_{\widetilde{D}}(x))^{q_2}d\mu_2(x)\right)^{1/q_2}}{\left(\int_{\widetilde{D}}(\chi_D(x))d\mu(x)\right)^{1/p}}\right)$$

$$\times\left(\int_{\mathcal{M}_1}(T_1\chi_D(x))^{q_1}d\mu_1(x)\right)^{1/q_1}$$

$$=\sup_{D,\widetilde{D}\in\Delta_d}\frac{\left(\int_{\mathcal{M}_1}(T_1\chi_D(x))^{q_1}d\mu_1(x)\right)^{1/q_1}\left(\int_{\mathcal{M}_2}(T_2\chi_{\widetilde{D}}(x))^{q_2}d\mu_2(x)\right)^{1/q_2}}{\left(\int_{D\cap\widetilde{D}}d\mu(x)\right)^{1/p}}.$$

2. The proof is essentially the same as in (1). Here we just use Theorem 2.6.1 (4).

To finish the proof, we observe that $C_{m,n}$ is the best constant both in cases 1 and 2 simply by evaluating the operators for the decreasing functions $f_j=\chi_{D_j}$, $j=1,\ldots,m$. □

Remark 2.8.1. The above weighted inequalities have a lot of applications in the theory of multidimensional Lorentz spaces which in many cases can give refinements of the classical Lorentz spaces. It is very easy to see that embedding results for the multidimensional Lorentz spaces defined in [18] are equivalent to embeddings for the cone of decreasing functions on $L^p_{\mathbb{R}^n_+}(w)$ characterized in Theorems 2.6.1 and 2.7.1. Moreover, in [12, Theorem 2.4, 2.10] the normability of the two–dimensional Lorentz space was characterized by using Theorems 2.6.1 and 2.7.1. These theorems were also used to prove functional properties of these spaces. Other related inequalities for radially decreasing functions can be found in [13].

Finally we present and discuss some modular inequalities. The inequality 2.7.3 was extended to modular inequalities of the form

$$\Phi_2^{-1}\left(\int\limits_{\mathbb{R}^n_+}\Phi_2(\omega(x)f(x))v(x)dx\right)\le\Phi_1^{-1}\left(\int\limits_{\mathbb{R}^n_+}\Phi_1(Cf(x))w(x)dx\right). \tag{2.8.1}$$

More exactly, the following characterization was given in [17, Theorem 2.1]:

Theorem 2.8.2. *Let Φ_1 and Φ_2 be two N-functions satisfying*

$$\sum_k\Phi_2\circ\Phi_1^{-1}(a_k)\le K\Phi_2\circ\Phi_1^{-1}\left(\sum_k a_k\right) \tag{2.8.2}$$

for all $a_k\ge 0$ with a constant $K\ge 1$ independent of $\{a_k\}$. Given $\omega(x),v(x),w(x)\ge 0$ inequality (2.8.1) holds for all $0\le f\downarrow$ if and only if there exists a constant $A=A(\Phi_1,\Phi_2,v,w,\omega)$ such that for all $\varepsilon>0$ and decreasing D sets the inequality

$$\Phi_2^{-1}\left(\int\limits_D\Phi_2(\varepsilon\omega(x))v(x)dx\right)\le\Phi_1^{-1}\left(\Phi_1(A\varepsilon)\int\limits_D w(x)dx\right). \tag{2.8.3}$$

holds.

Proof. Necessity follows, if we replace f in (2.8.1) by $f=\varepsilon\chi_{D_{h,t}}$.

For sufficiency we define for a fixed $f\downarrow$,

$$\begin{aligned}\Delta_k&=\{x\in\mathbb{R}^n_+;2^k<f(x)\le 2^{k+1}\},\quad k\in\mathbb{Z},\\ D_k&=\{x\in\mathbb{R}^n_+;f(x)>2^k\},\end{aligned}$$

and note that

$$D_k\supset D_{k+1},\quad D_k=\bigcup_{j\le k}\Delta_j,\quad ,\mathbb{R}^n_+=\bigcup_k\Delta_k.$$

Obviously, $\Delta_j \bigcap \Delta_k = \emptyset$ for $j \neq k$. We have, using the fact that Φ_2 is an N-function, that

$$\int\limits_{\mathbb{R}^n_+} \Phi_2(\omega(x)f(x))v(x)dx \le \frac{1}{K}\int\limits_{\mathbb{R}^n_+} \Phi_2(K\omega(x)f(x))v(x)dx$$

$$= \frac{1}{K}\sum_k \int\limits_{\Delta_k} \Phi_2\big(K\omega(x)f(x)\big)v(x)dx \le \frac{1}{K}\sum_k \int\limits_{\Delta_k} \Phi_2(2^{k+1}K\omega(x))v(x)dx$$

$$\le \frac{1}{K}\sum_k \int\limits_{D_k} \Phi_2(2^{k+1}K\omega(x))v(x)dx$$

$\big[$ applying (2.8.3) with $\varepsilon = 2^{k+1}K$ $\big]$

$$\le \frac{1}{K}\sum_k \Phi_2\circ\Phi_1^{-1}\left(\Phi_1(AK2^{k+1})\int\limits_{D_k} w\right)$$

$\big[$ applying (2.8.2) $\big]$

$$\le \Phi_2\circ\Phi_1^{-1}\left(\sum_k \Phi_1(AK2^{k+1})\sum_{j\ge k}\int\limits_{\Delta_j} w\right)$$

$$= \Phi_2\circ\Phi_1^{-1}\left(\sum_j \left(\int\limits_{\Delta_j} w\right)\sum_{k\le j}\Phi_1(AK2^{k+1})\right)$$

$\big[$ using the convexity of Φ_1 $\big]$

$$\le \Phi_2\circ\Phi_1^{-1}\left(\sum_j \left(\int\limits_{\Delta_j} w\right)\Phi_1(4AK2^j)\right)$$

$$\le \Phi_2\circ\Phi_1^{-1}\left(\sum_j \int\limits_{\Delta_j} \Phi_1(4AKf(x))w(x)dx\right)$$

$$= \Phi_2\circ\Phi_1^{-1}\left(\int\limits_{\mathbb{R}^n_+} \Phi_1(4AKf(x))w(x)dx\right).$$

Thus, the least possible constant C in (2.8.1) satisfies $A \le C \le 4AK$ and the theorem is proved. □

A function Φ on $[0,\infty)$ is called modular if it is strictly increasing and $\lim_{x\to 0}\Phi(x)=0$ and $\lim_{x\to\infty}\Phi(x)=\infty$. We say that a modular function is weakly convex if $2\Phi(t)\le\Phi(Mt)$ for all $t>0$ and some constant $M>1$. All convex modular functions are weakly convex. The functions $\Phi(t)=t^p$, $0<p<1$ and $\Phi(t)=e^{\sqrt{t}}-1$ are weakly convex but not convex. Suppose that D_k and Δ_k are defined by (2.7.1) and (2.7.2) respectively. In [14, Theorem 3.1] the characterization of inequality (2.8.1) was generalized in the following way:

Theorem 2.8.3. *Let Φ_1,Φ_2 be two modular functions and Φ_1 be weakly convex. Then (2.8.1) holds for all $0\le f\downarrow$ if and only if there exists a constant $A=A(\Phi_1,\Phi_2,v,w,\omega)>0$ such that for all decreasing sequences $(\varepsilon_k)_k$, $k\in\mathbb{Z}$ and all increasing sequences of decreasing sets $(D_k)_{k\in\mathbb{Z}}$, $\int_{D_k}w=2^k$, the inequality*

$$\Phi_2^{-1}\left(\sum_{k\in\mathbb{Z}}\int_{\Delta_k}\Phi_2\Big(\frac{\varepsilon_k}{A}\omega(x)\Big)v(x)dx\right)\le\Phi_1^{-1}\left(\sum_{k\in\mathbb{Z}}\Phi_1(\varepsilon_k)\int_{\Delta_k}w(x)dx\right).$$

holds.

Further, in [50] the hypothesis was weakened. More precisely, the restriction of Φ_1 to be weakly convex was not anymore necessary. The following theorem was proved:

Theorem 2.8.4. *Let Φ_1 and Φ_2 be two modular functions. Let A be the infimum of the constants $C>0$ such that inequality (2.8.1) (with $\omega\equiv 1$) holds for all $0\le f\downarrow$ and let B be the infimum of the constants C such that the inequality*

$$\Phi_2^{-1}\left(\sum_{k\in\mathbb{Z}}\Phi_2(\varepsilon_k)\int_{\Delta_k}v(x)dx\right)\le\Phi_1^{-1}\left(\sum_{k\in\mathbb{Z}}\Phi_1(C\varepsilon_k)\int_{\Delta_k}w(x)dx\right).$$

holds for all $0\le(\varepsilon)_k\downarrow$ and for all $\{D_k\}$ decreasing sets. Then $A=B$.

2.9 Notes and Comments on Chapter 2.

Two–weight criteria for one–sided potentials in the non–diagonal case were obtained in [53, Chapter 3], [40] and [39, Chapter 2] (see also [109] for Sawyer–type two–weight criteria).

Necessary and sufficient conditions for the boundedness/ compactness of the operator $\mathcal{R}_\alpha$ from $L^p(\mathbb{R}_+)$ to $L^q_v(\mathbb{R})$, where $1<p\le q<\infty$, $1/p\le\alpha<1$, were established in [117] and independently in [146]. The same problem in the case $0<\alpha<1/p$ was studied in [89] (see also [147]).

The boundedness problem of the operator

$$Tf(x) = \int_0^x k(x,y)f(y)dy$$

from $L^p(\mathbb{R}_+)$ to $L^q_v(\mathbb{R}_+)$, $k \in V \cap V_p$, was investigated in [119] (see also [39, Chapter 2]).

Necessary and sufficient condition on a weight v defined $\mathbb{R}^n$ guaranteeing the trace inequality

$$\|J_\alpha f\|_{L^q(\mathbb{R}^n)} \leq c\|f\|_{L^p(\mathbb{R})}, \quad 1 < p \leq q < \infty,$$

for the truncated potentials with single kernel

$$(J_\alpha f)(x) := \int_{\{y\in\mathbb{R}^n:|y|<2|x|\}} \frac{f(y)}{|x-y|^{n-\alpha}}dy$$

can be found in [152] for $p = q$, and by different proof in [90] for $p \leq q$ (see also [39, Chapter 5]). The same problem for the ball fractional integrals

$$(B_\alpha f)(x) := \int_{\{y\in\mathbb{R}^n:|y|<2|x|\}} \frac{f(y)(|x|^2-|y|^2)^\alpha}{|x-y|^n}dy$$

was solved in [118] (see also [39, Chapter 4]).

Sections 2.1– 2.4 are mainly based on the results of the papers [91], [92] (see also the survey papers [94], [95]).

The results and proofs of Section 2.5 are mainly taken from the paper [135] (see also [99]). We remark that our conditions (or, more exactly, scale of conditions) (2.5.16) to guarantee that (2.5.15) holds is only sufficient with both general weights and kernels. We remark also that necessary and sufficient conditions are also not known in this situation even in the one–dimensional case (see, e.g., [102] and [100]). However, we conjecture that our condition (2.5.15) can be both necessary and sufficient at least with some mild additional conditions on weights or kernels (see also Open Problems).

The main reference for Sections 2.6–2.8 is the PhD thesis [9] by Sorina Barza but also the closely related papers [17], [18] and [19]. Moreover, the results and ideas in [8], [10], [11], [12], [13], [14], [15], [18], [22], [26], [50] and [141] has influenced the final presented results of Sections 2.6–2.8 in various ways.

Chapter 3

One–sided Fractional Multiple Operators

This chapter is devoted to two–weight criteria for one–sided strong fractional maximal functions and potentials with product kernels. In particular, necessary and sufficient conditions on a weight governing the trace inequality for these operators are established. Applications of some derived results in two–weight Strichartz estimates for certain non–linear wave equations are presented.

3.1 One–dimensional Operators

Let $D(\mathbb{R})$ be the set of all dyadic intervals in $\mathbb{R}$ (sometimes we will use the symbol D for $D(\mathbb{R})$). By dyadic interval we mean an interval of the form $\left[2^k n, 2^k(n+1)\right)$, where k and n are integers. The main property of the dyadic intervals is that if $|I'| \leq |I|$, then either $I' \subset I$ or $I' \cap I = \oslash$. Let us denote $\Lambda_k = 2^{-k}\mathbb{Z}$ for $k \in \mathbb{Z}$. Suppose that $D^{(k)}(\mathbb{R})$ is the collection of the intervals determined by Λ_k. It is clear that $D(\mathbb{R}) = \cup_{k\in\mathbb{Z}} D^{(k)}(\mathbb{R})$. Each $I \in D^{(k)}(\mathbb{R})$ is the union of 2 non– overlapping intervals belonging to $D_{k+1}(\mathbb{R})$ (for some properties of the dyadic intervals see, for instance, [49], p. 136).

In the sequel we will use the following notation:

$$\alpha I := (x - \alpha r, x + \alpha r), \text{ where } \alpha > 0 \text{ and } I = (x - r, x + r);$$

$$I^+ := [2a - b, a), \quad I^- := [b, 2b - a) \text{ for } I = [a, b);$$

$$v(E) := \int_E v(x)dx,$$

where v is a weight function on $E \subseteq \mathbb{R}$;

$$\rho_x(E) := \int_E \rho(x,y)dy, \quad \rho_y(E) := \int_E \rho(x,y)dx, \quad E \subseteq \mathbb{R},$$

where ρ is a weight on $\mathbb{R}^2_+$.

Definition 3.1.1. A measure μ on $\mathbb{R}$ satisfies the dyadic doubling condition ($\mu \in DC^{(d)}(\mathbb{R})$) if there exists a constant $a > 1$ such that

$$\mu(I) \leq a\mu(I')$$

for all I', $I \in D$ with $I' \subset I$ and $|I| = 2|I'|$. Further, μ satisfies the doubling condition on $\mathbb{R}$ ($\mu \in DC(\mathbb{R})$) if there is a constant $b > 1$ such that

$$\mu(2I) \leq b\mu(I)$$

for all intervals $I \subset \mathbb{R}$.

Definition 3.1.2. We say that a measure μ defined on $\mathbb{R}$ satisfies the dyadic reverse doubling condition $\big(\mu \in RD^{(d)}(\mathbb{R})\big)$ if there exists a constant $c > 1$ such that

$$\mu(I) \geq c\mu(I'),$$

for all I', $I \in D$ with $I' \subset I$ and $|I| = 2|I'|$.

Remark 3.1.1. It is easy to see that if $\mu \in DC(\mathbb{R})$, then $\mu \in DC^{(d)}(\mathbb{R})$.

Remark 3.1.2. If $\mu \in DC^{(d)}(\mathbb{R})$, then $\mu \in RD^{(d)}(\mathbb{R})$.

Indeed, let I, I_1, I_2 be dyadic intervals such that $I_1 \cup I_2 = I$. Arguing as in the proof of Lemma 1.6.1 we have

$$\mu(I) = \mu(I_1) + \mu(I_2) \geq \mu(I_1) + (1/a)\mu(I),$$

where $a > 1$. Hence

$$\mu(I) \geq \frac{a}{a-1}\mu(I_1)$$

and the constant $\frac{a}{a-1}$ is greater than 1.

Definition 3.1.3. We say that a weight function $\rho(x, y)$ on $\mathbb{R}^2$ satisfies the dyadic doubling condition with respect to the second variable uniformly to x ($\rho \in RD^{(d)}(y)$) if there is a constant $a > 1$ such that for all dyadic intervals I, I' with $I' \subset I$, $|I'| = |I|/2$, and almost every $x \in \mathbb{R}$, it holds that

$$\rho_x(I) \leq a\rho_x(I').$$

The class $DC^{(d)}(x)$ is defined analogously. Further, a weight function $\rho(x, y)$ on $\mathbb{R}^2$ belongs to the class $RD^{(d)}(y)$ (resp. $RD^{(d)}(x)$) uniformly to x (resp. y) if there is a constant $c > 1$ such that for all dyadic intervals I, I' with $I' \subset I$, $|I'| = |I|/2$, and almost every $x \in \mathbb{R}$ (resp. $y \in \mathbb{R}$) the inequality

$$\rho_x(I) \geq c\rho_x(I')$$

$$\left(\text{resp.} \quad \rho_y(I) \geq c\rho_y(I') \right)$$

holds.

Remark 3.1.3. Taking into account Remark 3.1.2 we can conclude that if $\mu \in DC^{(d)}(y)$ (resp. $\mu \in DC^{(d)}(x)$), then $\mu \in RD^{(d)}(y)$ (resp. $\mu \in RD^{(d)}(y)$).

The next statement is a corollary of the well–known Carleson–Hörmander ([25], [71]) embedding theorem regarding dyadic intervals (see, e.g., [159], [169]).

Theorem 3.1.1. *Let $1 < p < q < \infty$ and let w be a weight function on $\mathbb{R}$ such that $w^{1-p'} \in RD^{(d)}(\mathbb{R})$. Then the inequality*

$$\sum_{I \in D} \left(\int_I w^{1-p'}(x)dx \right)^{-\frac{q}{p'}} \left(\int_I f(x)dx \right)^q \leq c \left(\int_{\mathbb{R}} f^p(x)w(x)dx \right)^{\frac{q}{p}}$$

holds for all non-negative $f \in L^p_w(\mathbb{R})$.

Let us formulate some results regarding one–dimensional one–sided potentials:

$$(R_\alpha f)(x) := \int_{-\infty}^{x} \frac{f(y)}{(x-y)^{1-\alpha}}dy, \quad (W_\alpha f)(x) := \int_{x}^{\infty} \frac{f(t)}{(t-x)^{1-\alpha}}dt, \ x \subset \mathbb{R};$$

$$(\mathcal{R}_\alpha f)(x) := \int_{0}^{x} \frac{f(y)}{(x-y)^{1-\alpha}}dy, \quad (\mathcal{W}_\alpha f)(x) := \int_{x}^{\infty} \frac{f(t)}{(t-x)^{1-\alpha}}dt, \ x \in \mathbb{R}_+,$$

where $0 < \alpha < 1$.

Recall that the symbol $L^p_\rho(E)$ (where $1 \leq p < \infty$, E is a measurable subset of $\mathbb{R}^n$ and ρ is a weight function) denotes the weighted Lebesgue space. If $\rho \equiv 1$, then we denote $L^p_\rho(E)$ by $L^p(E)$.

The next statement can be found in the Hardy–Littlewood–Pólya book [68].

Theorem 3.1.2. *Let $1 < p < \infty$, $0 < \alpha < 1/p$ and $p^* = \frac{p}{1-\alpha p}$. Then*

(i) *$\mathcal{R}_\alpha$ and $\mathcal{W}_\alpha$ are bounded from $L^p(\mathbb{R}_+)$ to $L^{p^*}(\mathbb{R}_+)$;*
(ii) *R_α and W_α are bounded from $L^p(\mathbb{R})$ to $L^{p^*}(\mathbb{R})$.*

Now we discuss the classes of weights A^+_∞ and A^-_∞ introduced in [112].

Definition 3.1.4. We say that a weight w is in the class $A^-_\infty(\mathbb{R})$ if there exist positive constants k and δ such that, for all numbers $a < b < c$ and measurable sets $E \subset (a, b)$,

$$\frac{|E|}{c-a} \leq k \left(\frac{w(E)}{w([b,c])} \right)^\delta .$$

Further, $w \in A_\infty^+(\mathbb{R})$ if there exists positive constants k and δ such that for all numbers $a < b < c$ and measurable sets $E \subset (b, c)$,

$$\frac{|E|}{c-a} \le k \left(\frac{w(E)}{w([a,b])} \right)^{\delta}.$$

For the next three statements we refer to [112].

Theorem 3.1.3. *The following conditions are equivalent:*

(a) $w \in A_\infty^-(\mathbb{R})$;

(b) *there exist positive numbers k and δ such that*

$$\frac{w(E)}{w([a,c])} \le k \left(\frac{|E|}{b-a} \right)^{\delta}$$

for all numbers $a < b < c$ and measurable sets $E \subset (b, c)$;

(c) *there exists p, $1 \le p < \infty$, such that $w \in A_p^-(\mathbb{R})$, that is,*

$$\sup_{\substack{x \in \mathbb{R} \\ h>0}} \left(\frac{1}{h} \int_x^{x+h} w(y)dy \right) \left(\frac{1}{h} \int_{x-h}^{x} w^{1-p'}(y)dy \right)^{p-1} < \infty \quad \textit{for} \quad 1 < p < \infty,$$

$$(M^+ w)(x) := \sup_{r>0} \frac{1}{r} \int_x^{x+r} w(t)dt \le cw(x), \quad \textit{a.e.,} \quad \textit{when } p = 1.$$

Theorem 3.1.4. *The following statements are equivalent:*

(a) $w \in A_\infty^+(\mathbb{R})$;

(b) *there exist positive numbers k and δ such that*

$$\frac{w(E)}{w([a,c])} \le k \left(\frac{|E|}{c-b} \right)^{\delta}$$

for all numbers $a < b < c$ and measurable sets $E \subset (a, b)$;

(c) *there exists p, $1 \le p < \infty$, such that $w \in A_p^+(\mathbb{R})$, i.e.,*

$$\sup_{\substack{x \in \mathbb{R} \\ h>0}} \left(\frac{1}{h} \int_{x-h}^{x} w(y)dy \right) \left(\frac{1}{h} \int_x^{x+h} w^{1-p'}(y)dy \right)^{p-1} < \infty \ \textit{for}\ 1 < p < \infty,$$

$$(M^- w)(x) := \sup_{r>0} \frac{1}{r} \int_{x-r}^{x} w(t)dt \le cw(x), \quad \textit{a.e.,} \quad \textit{when } p = 1.$$

The following result was proved in [112] (for the Riesz potentials it was derived in [126]).

Theorem 3.1.5. *Let $1 < q < \infty$. If $\rho \in A^+_\infty(\mathbb{R})$, then*

$$\left(\int_{\mathbb{R}} |(W_\alpha f)(x)|^q \rho(x)dx \right)^{\frac{1}{q}} \le c \left(\int_{\mathbb{R}} \left(M^+_\alpha f\right)^q (x)\rho(x)dx \right)^{\frac{1}{q}}.$$

Further, the condition $\rho \in A^-_\infty(\mathbb{R})$ implies the inequality

$$\left(\int_{\mathbb{R}} |(R_\alpha f)(x)|^q \rho(x)dx \right)^{\frac{1}{q}} \le c \left(\int_{\mathbb{R}} \left(M^-_\alpha f\right)^q (x)\rho(x)dx \right)^{\frac{1}{q}}.$$

Definition 3.1.5. We say that a weight $\rho(x, y)$ on $\mathbb{R}^2$ belongs to the class $A^+_{\infty,x}(\mathbb{R})$ if there exist positive numbers k and δ such that for all numbers $a < b < c$, measurable sets $E \subset (b, c)$ and $y \in \mathbb{R}$ the inequality

$$\frac{|E|}{c-a} \le k \left(\frac{\rho_y(E)}{\rho_y([a,b])} \right)^\delta$$

holds. Analogously, it can be defined the classes $A^+_{\infty,y}(\mathbb{R})$, $A^-_{\infty,x}(\mathbb{R})$, $A^-_{\infty,y}(\mathbb{R})$.

The next statement was established in [40] (see also [39, Chapter 2]) which is a consequence of the solution of the two–weight problem for positive kernel operators given in [52] (see also [53, Chapter 3]) and the Sawyer–type two–weight criteria in [109].

Theorem 3.1.6. *Let $1 < p < q < \infty$ and let $0 < \alpha < 1$. Then W_α is bounded from $L^p_w(\mathbb{R})$ to $L^q_v(\mathbb{R})$ if and only if*

(i)

$$A_1 := \sup_{\substack{a \in \mathbb{R} \\ h>0}} \left(\int_{a-h}^{a+h} w^{1-p'}(y)dy \right)^{\frac{1}{p'}} \left(\int_{-\infty}^{a-h} \frac{v(y)}{(a-y)^{(1-\alpha)q}}dy \right)^{\frac{1}{q}} < \infty;$$

(ii)

$$A_2 := \sup_{\substack{a \in \mathbb{R} \\ h>0}} \left(\int_{a-h}^{a+h} v(y)dy \right)^{\frac{1}{p'}} \left(\int_{a+h}^{\infty} \frac{w^{1-p'}(y)}{(y-a)^{(1-\alpha)p'}}dy \right)^{\frac{1}{p'}} < \infty.$$

Moreover, $\|W_\alpha\| \approx \max\{A_1, A_2\}$.

Theorem 3.1.7. *Let $1 < p < q < \infty$ and let $\alpha \in (0,1)$. Then R_α is bounded from $L^p_w(\mathbb{R})$ to $L^q_v(\mathbb{R})$ if and only if*

(i)

$$B_1 := \sup_{\substack{a\in\mathbb{R}\\ h>0}} \left(\int\limits_{a-h}^{a+h} v(y)dy \right)^{\frac{1}{q}} \left(\int\limits_{-\infty}^{a-h} \frac{w^{1-p'}(y)}{(a-y)^{(1-\alpha)p'}} dy \right)^{\frac{1}{p'}} < \infty;$$

(ii)

$$B_2 := \sup_{\substack{a\in\mathbb{R}\\ h>0}} \left(\int\limits_{a-h}^{a+h} w^{1-p'}(y)dy \right)^{\frac{1}{p'}} \left(\int\limits_{a+h}^{+\infty} \frac{v(y)}{(y-a)^{(1-\alpha)q}} dy \right)^{\frac{1}{q}} < \infty.$$

Moreover, $\|R_\alpha\| \approx \max\{B_1, B_2\}$.

The $L^p \to L^q_v$ $(1 < p < q < \infty)$ boundedness criterion for the potential operator

$$(I_\alpha f)(x) := \int\limits_{\mathbb{R}} \frac{f(y)}{|x-y|} dy \quad x \in \mathbb{R}, \ \ 0 < \alpha < 1,$$

has been established by D. Adams [4]. Here we formulate the statement regarding the trace inequality for one–sided potentials (see [39], p. 133).

Theorem 3.1.8. *Let* $1 < p < q < \infty$ *and let* $0 < \alpha < \frac{1}{p}$. *Then the following conditions are equivalent:*

(i) I_α *is bounded from* $L^p(\mathbb{R})$ *to* $L^q_v(\mathbb{R})$;

(ii) R_α *is bounded from* $L^p(\mathbb{R})$ *to* $L^q_v(\mathbb{R})$;

(iii) W_α *is bounded from* $L^p(\mathbb{R})$ *to* $L^q_v(\mathbb{R})$;

(iv) $D := \sup\limits_I v(I)|I|^{(\alpha-1/p)q} < \infty$, *where the supremum is taken over all one-dimensional intervals* I. *Moreover, the norms of these operators are equivalent to* D.

The next two statements by K. Andersen and E. Sawyer [7] give one-weight criteria for one-dimensional one–sided potentials (for I_α see [126]).

Theorem 3.1.9. *Let* $0 < \alpha < 1/p$, *where* $1 < p < \infty$. *Let* $p^* = \frac{p}{1-\alpha p}$. *Then* W_α *is bounded from* $L^p_{w^p}(\mathbb{R})$ *to* $L^{p^*}_{w^{p^*}}(\mathbb{R})$ *if and only if*

$$\overline{A}^{+}_\alpha := \sup_{\substack{a\in\mathbb{R}\\ h>0}} h^{\alpha-1} \left(\int\limits_{a-h}^{a} w^{p^*}(x)dx \right)^{1/p^*} \left(\int\limits_{a}^{a+h} w^{-p'}(x)dx \right)^{1/p'} < \infty.$$

Moreover, $\|W_\alpha\| \approx \overline{A}^{+}_\alpha$.

Theorem 3.1.10. *Let* $1 < p < \infty$, $0 < \alpha < 1/p$. *Then* R_α *is bounded from* $L^p_{w^p}(\mathbb{R})$ *to* $L^{p^*}_{w^{p^*}}(\mathbb{R})$ *if and only if*

$$\overline{A}^{-}_\alpha := \sup_{\substack{a\in\mathbb{R}\\ h>0}} h^{\alpha-1} \left(\int\limits_{a}^{a-h} w^{p^*}(x)dx \right)^{1/p^*} \left(\int\limits_{a-h}^{a} w^{p}(x)dx \right)^{1/p'} < \infty.$$

Moreover, $\|R_\alpha\| \approx \overline{A_\alpha^-}$.

We will also need two–weight Muckenhoupt–type criteria for the operator

$$H'f(x) := \int_x^\infty f(t)dt$$

(see, e.g., [102] and the references given there).

Theorem 3.1.11. *Let* $1 < p \le q < \infty$. *Then the operator* H' *is bounded from* $L^p_w(\mathbb{R}_+)$ *to* $L^q_v(\mathbb{R}_+)$ *if and only if*

$$A' := \sup_{t>0} \Big(\int_0^t v(x)dx \Big)^{1/q} \Big(\int_t^\infty w^{1-p'}(x)dx \Big)^{1/p'} < \infty.$$

Moreover, $\|H'\| \approx A'$.

The next statement regrading the boundedness of the Riemann–Liouville transform

$$\mathcal{R}_\alpha f(x) := \int_0^x \frac{f(y)}{(x-t)^{1-\alpha}}dt, \quad x \in \mathbb{R},$$

will also be useful for us.

Proposition 3.1.1 *Let* $0 < \alpha < 1$, $1 < p \le q < \infty$, $\alpha \in [1/p - 1/q, 1/p) \cup (1/p, 1)$. *Suppose that one of the following conditions holds:*

(a) $\alpha \in [1/p - 1/q, 1/p)$, $\overline{v}$ *and* $\overline{w}$ *are monotonic weights on* $\mathbb{R}_+$;

(b) $\alpha \in (1/p, 1)$ *and* $\overline{w}$ *is increasing weight on* $\mathbb{R}_+$;

(c) $\alpha \in (1/p, 1)$, $\overline{w}$ *is decreasing weight on* $\mathbb{R}_+$ *and* $\overline{w}^{1-p'} \in DC(\mathbb{R})$.

Then $\mathcal{R}_\alpha$ *is bounded from* $L^p_{\overline{w}}(\mathbb{R}_+)$ *to* $L^p_{\overline{v}}(\mathbb{R}_+)$ *if and only if*

$$G := \sup_{a>0} \left(\int_a^\infty \overline{v}(t)t^{(\alpha-1)q}dt \right)^{1/q} \left(\int_0^a \overline{w}^{1-p'}(t)dt \right)^{1/p'} < \infty. \tag{3.1.1}$$

Moreover, $c_1 G \le \|\mathcal{R}_\alpha\| \le c_2 G$, *where the positive constants* c_1 *and* c_2 *depend only on* p, q *and* α.

Proof. *Sufficiency.* Let $f \ge 0$. We have

$$\|\mathcal{R}_\alpha f\|_{L^q_{\overline{v}}(\mathbb{R}_+)} \le c\Big(\|\mathcal{R}^{(1)}_\alpha f\|_{L^q_{\overline{v}}(\mathbb{R}_+)} + \|\mathcal{R}^{(2)}_\alpha f\|_{L^q_{\overline{v}}(\mathbb{R}_+)} \Big) := c(I_1 + I_2),$$

where

$$\mathcal{R}^{(1)}_\alpha f(x) := \int_0^{x/2} \frac{f(t)}{(x-t)^{1-\alpha}}dt; \quad \mathcal{R}^{(2)}_\alpha f(x) := \int_{x/2}^x \frac{f(t)}{(x-t)^{1-\alpha}}dt.$$

Using Theorem 1.1.2 and the inequality $\frac{x}{2} \le x - y$ which holds for $0 < y < \frac{x}{2}$, we find that (3.1.1) yields

$$I_1^q \le c\|\mathcal{R}_\alpha^{(1)} f\|^q_{L^q_{\overline{v}}(\mathbb{R}_+)} \le cG^q \|f\|^q_{L^p_{\overline{w}}(\mathbb{R}_+)},$$

where c depends only on α, p and q.

Suppose that $p < q$, $\alpha \in (1/p - 1/q, 1/p)$ and $\overline{v}$ and $\overline{w}$ are increasing (other cases can be proved in a similar way). We set $p^* = \frac{p}{1-\alpha p}$. Then the condition $\frac{1}{p} - \frac{1}{q} < \alpha$ guarantees that $q < p^*$. Further, $\left(\frac{p^*}{q}\right)' = \frac{p}{p-q+\alpha pq}$. Besides this, it is easy to see that

$$G \ge \left(\int_a^\infty \overline{v}(x) x^{(\alpha-1)q} dx \right)^{1/q} \left(\int_0^a \overline{w}^{1-p'}(x) dx \right)^{1/p'} \ge c\overline{v}^{1/q}(4a)\overline{w}^{-1/p}(a)\, a^{\alpha+\frac{1}{q}-\frac{1}{p}},$$

where c depends only on α, p and q. Hence,

$$\overline{v}(4a) \le cG^q \overline{w}^{\frac{q}{p}}(a)\, a^{(\frac{1}{p}-\frac{1}{q}-\alpha)q} \tag{3.1.2}$$

for all $a > 0$.

Due to Theorem 3.1.2, Hölder's inequality and (3.1.2) we have that

$$I_2^q = \sum_k \int_{2^k}^{2^{k+1}} \overline{v}(x) \left(\int_{x/2}^x \frac{f(t)dt}{(x-t)^{1-\alpha}} \right)^q dx$$

$$\le \sum_k \left(\int_{2^k}^{2^{k+1}} [\overline{v}(x)]^{\left(\frac{p^*}{q}\right)'} dx \right)^{\frac{1}{\left(\frac{p^*}{q}\right)'}} \left(\int_{2^k}^{2^{k+1}} \left(\int_{x/2}^x \frac{f(t)dt}{(x-t)^{1-\alpha}} \right)^{p^*} dx \right)^{\frac{q}{p^*}}$$

$$\le \sum_k \overline{v}(2^{k+1}) 2^{k(1-\frac{q}{p}+\alpha q)} \left(\int_0^\infty \left(\mathcal{R}_\alpha \left(f \cdot \chi_{[2^{k-1}, 2^{k+1}]} \right)(x) \right)^{p^*} dx \right)^{\frac{q}{p^*}}$$

$$\le cG^q \sum_k \overline{w}^{\frac{q}{p}} \left(2^{k-1}\right) \left(\int_{2^{k-1}}^{2^{k+1}} (f(x))^p dx \right)^{q/p} \le cG^q \left(\int_0^\infty (f(x))^p \overline{w}(x) dx \right)^{q/p},$$

where the constant c depends only on p, q and α.

The proof of the case $1/p - 1/q = \alpha$ is similar to that of $1/p - 1/q < \alpha$; the details are omitted.

Let now $\alpha \in (1/p, 1)$ and let $\overline{w}$ is increasing. In this case we estimate I_2 as follows:

$$I_2^q \le \int_0^\infty \overline{v}(x)\Big(\int_x^{2x} f^p(t)\overline{w}(t)dt\Big)^{q/p}\Big(\int_{x/2}^{x}(x-t)^{(\alpha-1)p'}(\overline{w})^{1-p'}(t)dt\Big)^{q/p'}dx$$

$$\le c\sum_k \int_{2^k}^{2^{k+1}} \overline{v}(x)\Big(\int_{x/2}^{x} f^p(t)\overline{w}(t)dt\Big)^{q/p}(\overline{w})^{-q/p}(x/2)x^{(\alpha-1)q+q/p'}dx$$

$$\le c\sum_k \Big(\int_{2^{k-1}}^{2^{k+1}}(f(t))^p w(t)dt\Big)^{q/p}\Big(\int_{2^k}^{2^{k+1}}\overline{v}(x)x^{(\alpha-1)q+q/p'}dx\Big)$$

$$\times\Big(\int_0^{2^k-1} w^{1-p'}(t)dt\Big)^{q/p'} \le c\|f\|^q_{L^p_w(\mathbb{R})}.$$

Necessity follows easily if we substitute the test function

$$f(t) = \chi_{(0,a)}(t)w^{1-p'}(t)$$

in the two–weight inequality for $\mathcal{R}_\alpha$ and observe that $\int_0^a w^{1-p'}(t)dt < \infty$ for all positive a. The proof is complete. □

3.2 One–sided Strong Fractional Maximal Functions

In this section we establish two–weight criteria for one–sided strong fractional maximal functions :

$$(M^{+,+}_{\alpha,\beta}f)(x,y) := \sup_{h,s>0}\frac{1}{h^{1-\alpha}s^{1-\beta}}\int_x^{x+h}\int_y^{y+s}|f(t,\tau)|dtd\tau,$$

$$(M^{-,-}_{\alpha,\beta}f)(x,y) := \sup_{h,s>0}\frac{1}{h^{1-\alpha}s^{1-\beta}}\int_{x-h}^{x}\int_{y-s}^{y}|f(t,\tau)|dtd\tau,$$

$$(M^{+,-}_{\alpha,\beta}f)(x,y) := \sup_{h,s>0}\frac{1}{h^{1-\alpha}s^{1-\beta}}\int_x^{x+h}\int_{y-s}^{y}|f(t,\tau)|dtd\tau,$$

$$(M^{-,+}_{\alpha,\beta}f)(x,y) := \sup_{h,s>0}\frac{1}{h^{1-\alpha}s^{1-\beta}}\int_{x-h}^{x}\int_y^{y+s}|f(t,\tau)|dtd\tau$$

provided that the right–hand side weight satisfies some additional conditions.

To prove the main results of this section we need some auxiliary results regarding one–sided dyadic maximal function given by

$$(\overline{M}^{+,+}_{\alpha,\beta} f)(x,y) := \sup_{\substack{I\ni x,\ J\ni y\\ I,J\in D(\mathbb{R})}} \frac{1}{|I|^{1-\alpha}|J|^{1-\beta}} \iint\limits_{I^+\times J^+} |f(t,\tau)|dtd\tau,$$

Together with this operator the maximal function

$$(N^{+,+}_{\alpha,\beta} f)(x,y) := \sup_{h,s>0} \frac{1}{(h/2)^{1-\alpha}(s/2)^{1-\beta}} \int\limits_{x+h/2}^{x+h} \int\limits_{y+s/2}^{y+s} |f(t,\tau)|dtd\tau$$

will be useful for us.

For one–sided dyadic fractional maximal functions we refer to [113].

Lemma 3.2.1. *There are positive constants b_1 and b_2 such that for all locally integrable f and $x,y\in\mathbb{R}$, the inequality*

$$b_1(N^{+,+}_{\alpha,\beta} f)(x,y) \le (M^{+,+}_{\alpha,\beta} f)(x,y) \le b_2(N^{+,+}_{\alpha,\beta} f)(x,y) \tag{3.2.1}$$

holds.

Proof. To show the left–hand side of inequality (3.2.1) we observe that

$$\left(\frac{h}{2}\right)^{\alpha-1}\left(\frac{s}{2}\right)^{\beta-1} \int\limits_{x+h/2}^{x+h} \int\limits_{y+s/2}^{y+s} |f(t,\tau)|dtd\tau \le 2^{2-\alpha-\beta}\frac{1}{h^{1-\alpha}s^{1-\beta}} \int\limits_{x}^{x+h}\int\limits_{y}^{y+s} |f(t,\tau)|dtd\tau$$

$$\le 2^{2-\alpha-\beta}(M^{+,+}_{\alpha,\beta} f)(x,y).$$

Hence,

$$(N^{+,+}_{\alpha,\beta} f)(x,y) \le 2^{2-\alpha-\beta}(M^{+,+}_{\alpha,\beta} f)(x,y).$$

To prove the right–hand side of (3.2.1), notice that

$$h^{\alpha-1}\sup_{s>0} s^{\beta-1} \int\limits_{x}^{x+h}\int\limits_{y}^{y+s} |f(t,\tau)|dtd\tau \le h^{\alpha-1}\sup_{s>0} s^{\beta-1} \int\limits_{x}^{x+h/2}\int\limits_{y}^{y+s} |f(t,\tau)|dtd\tau$$

$$+h^{\alpha-1}\sup_{s>0} s^{\beta-1} \int\limits_{x+h/2}^{x+h}\int\limits_{y}^{y+s} |f(t,\tau)|dtd\tau$$

$$\le 2^{\alpha-1}\Big[(M^{+,+}_{\alpha,\beta} f)(x,y) + (N^{+}_{\alpha} M^{+}_{\beta} f)(x,y)\Big],$$

where

$$(N_\alpha^+ M_\beta^+ f)(x,y) = \sup_{h,s>0} \frac{1}{(h/2)^{1-\alpha} s^{1-\beta}} \int\limits_{x+h/2}^{x+h} \int\limits_{y}^{y+s} |f(t,\tau)|dtd\tau.$$

Consequently,

$$(1-2^{\alpha-1})(M_{\alpha,\beta}^{+,+} f)(x,y) \le 2^{\alpha-1}(N_\alpha^+ M_\beta^+ f)(x,y).$$

Moreover, it is easy to see that

$$s^{\beta-1} \sup_{h>0} \frac{1}{(h/2)^{1-\alpha}} \int\limits_{x+h/2}^{x+h} \int\limits_{y}^{y+s} |f(t,\tau)|dtd\tau$$

$$\le s^{\beta-1} \sup_{h>0} \frac{1}{(h/2)^{1-\alpha}} \int\limits_{x+h/2}^{x+h} \int\limits_{y}^{y+s/2} |f(t,\tau)|dtd\tau$$

$$+ s^{\beta-1} \sup_{h>0} \frac{1}{(h/2)^{1-\alpha}} \int\limits_{x+h/2}^{x+h} \int\limits_{y+s/2}^{y+s} |f(t,\tau)|dtd\tau$$

$$\le 2^{\beta-1}\Big[(N_\alpha^+ M_\beta^+ f)(x,y) + (N_{\alpha,\beta}^{+,+} f)(x,y)\Big].$$

From this we have that

$$(N_\alpha^+ M_\beta^+ f)(x,y) \le 2^{\beta-1}\Big[(N_\alpha^+ M_\beta^+ f)(x,y) + (N_{\alpha,\beta}^{+,+} f)(x,y)\Big].$$

Summarizing the above–derived inequalities we conclude that

$$(M_{\alpha,\beta}^{+,+} f)(x,y) \le \frac{1}{2^{1-\alpha}-1} \frac{1}{2^{1-\beta}-1} (N_{\alpha,\beta}^{+,+} f)(x,y).$$

The proof is complete. □

Lemma 3.2.2. *There exist positive constants b_1 and b_2 such that for all locally integrable f and $x, y \in \mathbb{R}$, the following two–sided inequality holds:*

$$(\overline{M}_{\alpha,\beta}^{+,+} f)(x,y) \le b_1 (M_{\alpha,\beta}^{+,+} f)(x,y) \le b_2 (\overline{M}_{\alpha,\beta}^{+,+} f)(x,y)$$

Proof. Let us fix $f \in L_{\mathrm{loc}}(\mathbb{R})$ and $x, y \in \mathbb{R}$. Suppose that $I = [a,b)$, $J = [c,d)$ are dyadic intervals such that $x < a$, $y < c$ and $0 \le a - x < b - a$, $0 \le c - y < d - c$. Then

$$|I^-|^{\alpha-1} |J^-|^{\beta-1} \iint\limits_{I \times J} |f(t,\tau)|dtd\tau \le \Big(\frac{b-x}{|I|}\Big)^{1-\alpha} \Big(\frac{d-y}{|J|}\Big)^{1-\beta} (M_{\alpha,\beta}^{+,+} f)(x,y)$$

$$\leq 2^{2-\alpha-\beta}(M^{+,+}_{\alpha,\beta} f)(x,y).$$

Consequently,

$$(\overline{M}^{+,+}_{\alpha,\beta} f)(x,y) \leq 2^{2-\alpha-\beta}(M^{+,+}_{\alpha,\beta} f)(x,y).$$

To prove the reverse inequality with some constant we first observe that

$$N^{+,+}_{\alpha,\beta} f \approx \widetilde{N}^{+,+}_{\alpha,\beta} f, \tag{3.2.2}$$

where

$$(\widetilde{N}^{+,+}_{\alpha,\beta} f)(x,y) = \sup_{i,j\in\mathbb{Z}} 2^{(\alpha-1)(j-1)} 2^{(\beta-1)(i-1)} \int\limits_{x+2^{j-1}}^{x+2^j} \int\limits_{y+2^{i-1}}^{y+2^i} |f(t,\tau)| dt d\tau.$$

Indeed, let us take $h, s > 0$. Then $h \in [2^{j-1}, 2^j)$, $s \in [2^{i-1}, 2^i)$ for some integers j, i. Then

$$\frac{1}{(h/2)^{1-\alpha}(s/2)^{1-\beta}} \int\limits_{x+h/2}^{x+h} \int\limits_{y+s/2}^{y+s} |f(t,\tau)| dt d\tau$$

$$\leq 2^{(\alpha-1)(j-2)} 2^{(\beta-1)(i-2)} \int\limits_{x+2^{j-2}}^{x+2^j} \int\limits_{y+2^{i-2}}^{y+2^i} |f(t,\tau)| dt d\tau$$

$$= 2^{(\alpha-1)(j-2)} 2^{(\beta-1)(i-2)} \Bigg[\int\limits_{x+2^{j-2}}^{x+2^{j-1}} \int\limits_{y+2^{i-2}}^{y+2^{i-1}} |f(t,\tau)| dt d\tau$$

$$+ \int\limits_{x+2^{j-1}}^{x+2^j} \int\limits_{y+2^{i-1}}^{y+2^i} |f(t,\tau)| dt d\tau$$

$$+ \int\limits_{x+2^{j-2}}^{x+2^{j-1}} \int\limits_{y+2^{i-1}}^{y+2^i} |f(t,\tau)| dt d\tau + \int\limits_{x+2^{j-1}}^{x+2^j} \int\limits_{y+2^{i-2}}^{y+2^{i-1}} |f(t,\tau)| dt d\tau \Bigg]$$

$$= 2^{(\alpha-1)(j-2)} 2^{(\beta-1)(i-2)} \int\limits_{x+2^{j-2}}^{x+2^{j-1}} \int\limits_{y+2^{i-2}}^{y+2^{i-1}} |f(t,\tau)| dt d\tau$$

$$+ 2^{2-\alpha-\beta} 2^{(\alpha-1)(j-1)} 2^{(\beta-1)(i-1)} \int\limits_{x+2^{j-1}}^{x+2^j} \int\limits_{y+2^{i-1}}^{y+2^i} |f(t,\tau)| dt d\tau$$

$$+2^{1-\beta}2^{(\alpha-1)(j-2)}2^{(\beta-1)(i-1)}\int\limits_{x+2^{j-2}}^{x+2^{j-1}}\int\limits_{y+2^{i-1}}^{y+2^{i}}|f(t,\tau)|dtd\tau$$

$$+2^{1-\alpha}2^{(\alpha-1)(j-1)}2^{(\beta-1)(i-2)}\int\limits_{x+2^{j-1}}^{x+2^{j}}\int\limits_{y+2^{i-2}}^{y+2^{i-1}}|f(t,\tau)|dtd\tau$$

$$\leq (1+2^{2-\alpha-\beta}+2^{1-\beta}+2^{1-\alpha})(\widetilde{N}^{+,+}_{\alpha,\beta}f)(x,y).$$

Consequently,

$$(N^{+,+}_{\alpha,\beta}f)(x,y)\leq c_{\alpha,\beta}(\widetilde{N}_{\alpha,\beta}f)(x,y).$$

The reverse inequality of the latter one is obvious and thus (3.2.2) is proved. By Lemma 3.2.1 and (3.2.2) it is enough to prove the inequality

$$(\widetilde{N}_{\alpha,\beta}f)(x,y)\leq c(\overline{M}^{+,+}_{\alpha,\beta}f)(x,y). \tag{3.2.3}$$

Suppose that $h=2^j$ for some $j\in\mathbb{Z}$. Let us take dyadic intervals I and I' with the length 2^{j-1} such that $[x+h/2,x+h)\subset I\cup I'$. Analogously, for $s=2^i$, we take two dyadic intervals J and J' with the length 2^{j-1} so that $[y+s/2,y+s)\subset J\cup J'$.

If $I\cup I'$ and $J\cup J'$ are dyadic, then

$$\int\limits_{x+h/2}^{x+h}\int\limits_{y+s/2}^{y+s}|f(t,\tau)|dtd\tau\leq\int\limits_{I\cup I'}\int\limits_{J\cup J'}|f(t,\tau)|dtd\tau$$

$$\leq 2^{j(1-\alpha)}2^{i(1-\beta)}(\overline{M}^{+,+}_{\alpha,\beta}f)(x,y). \tag{3.2.4}$$

If $I\cup I'$ and $J\cup J'$ are not dyadic, then we take the dyadic interval I_1 (resp. J_1) with the length 2^j (resp. 2^i) containing I' (resp. J'). In this case we have that

$$\int\limits_{x+h/2}^{x+h}\int\limits_{y+s/2}^{y+s}|f(t,\tau)|dtd\tau\leq\iint\limits_{(I\cup I_1)\times(J\cup J_1)}|f(t,\tau)|dtd\tau$$

$$=\iint\limits_{I\times J}|f(t,\tau)|dtd\tau+\iint\limits_{I\times J_1}|f(t,\tau)|dtd\tau+\iint\limits_{I_1\times J}|f(t,\tau)|dtd\tau+\iint\limits_{I_1\times J_1}|f(t,\tau)|dtd\tau$$

$$\leq 2^{(j-1)(1-\alpha)}2^{(i-1)(1-\beta)}(\overline{M}^{+,+}_{\alpha,\beta}f)(x,y)$$

$$+2^{(j-1)(1-\alpha)}2^{i(1-\beta)}(\overline{M}^{+,+}_{\alpha,\beta}f)(x,y)$$

$$+2^{j(1-\alpha)}2^{(i-1)(1-\beta)}(\overline{M}^{+,+}_{\alpha,\beta}f)(x,y)+2^{j(1-\alpha)}2^{i(1-\beta)}(\overline{M}^{+,+}_{\alpha,\beta}f)(x,y)$$

$$=2^{(j-1)(1-\alpha)}2^{(i-1)(1-\beta)}(\overline{M}^{+,+}_{\alpha,\beta}f)(x,y).$$

According to this estimate and (3.2.4) we find that (3.2.3) holds and the proof is complete. □

Now we formulate and prove the main results concerning one–sided strong maximal operators.

Theorem 3.2.1. *Let $0 < \alpha,\ \beta < 1$, $1 < p < q < \infty$. Suppose that $w(x,y) = w_1(x)w_2(y)$ with $w_i^{1-p'} \in RD^{(d)}(\mathbb{R})$, $i = 1,2$. Then the following conditions are equivalent:*

(i) $M^{+,+}_{\alpha,\beta}$ *is bounded from $L^p_w(\mathbb{R})$ to $L^q_v(\mathbb{R})$;*

(ii) $\overline{M}^{+,+}_{\alpha,\beta}$ *is bounded from $L^p_w(\mathbb{R})$ to $L^q_v(\mathbb{R})$;*

(iii)

$$D_{+,+} =: \sup_{\substack{a,b\in\mathbb{R}\\ h,s>0}} \frac{1}{h^{1-\alpha}s^{1-\beta}} \left[\int\limits_{a-h}^{a}\int\limits_{b-s}^{b} v(x,y)dxdy\right]^{\frac{1}{q}} \left[\int\limits_{a}^{a+h}\int\limits_{b}^{b+s} w^{1-p'}(x,y)dxdy\right]^{\frac{1}{p'}} < \infty.$$

Proof. First we show that (iii) $\Longrightarrow$ (ii). Let $f \geq 0$. Without loss of generality we can assume that f is bounded and has a compact support. In this case $\left(M^{+,+}_{\alpha,\beta} f\right)(x,y) < \infty$ almost everywhere. If we prove the statement for such f, then we can pass to any $f \in L^p_w(\mathbb{R}^n)$. Indeed, let $f \in L^p(\mathbb{R}^2)$. We take the sequence $\{f_n\}$, $f_n = f\chi_{B(a,k_n)}\chi_{\{f<j_n\}}$, where $a \in \mathbb{R}^2$ and integers k_n, j_n tending to ∞ as $n \to \infty$. It is obvious that $f_n \to f$ in $L^p_w(\mathbb{R}^2)$ and also pointwise. Moreover, $f_n(x,y) \leq f(x,y)$. Further, $M^{+,+}_{\alpha,\beta} f_n$ is a Cauchy sequence, because

$$\left\|M^{+,+}_{\alpha,\beta} f_n - M^{+,+}_{\alpha,\beta} f_m\right\|_{L^q_v(\mathbb{R}^2)} \leq \left\|M^{+,+}_{\alpha,\beta} (f_n - f_m)\right\|_{L^q_v(\mathbb{R}^2)}$$
$$\leq c\|f_n - f_m\|_{L^p_w(\mathbb{R}^2)}.$$

Consequently, there exists $g \in L^q_v(\mathbb{R}^2)$ such that

$$\left\|M^{+,+}_{\alpha,\beta} f_n - g\right\|_{L^q_v(\mathbb{R}^2)} \to 0.$$

Hence, there exists a subsequence $M^{+,+}_{\alpha,\beta} f_{n_m}$ converging to g in norm and almost everywhere. But f_{n_m} converges to f in L^p_w norm and almost everywhere. Therefore we have that

$$\|g\|_{L^q_v(\mathbb{R}^2)} \leq c\|f\|_{L^p_w(\mathbb{R}^2)}, \tag{3.2.5}$$

where the positive constant c does not depend on f. Further, for all $x,y \in \mathbb{R}$ and $h,s > 0$, it follows that

$$h^{\alpha-1}s^{\beta-1}\int\limits_{x}^{x+h}\int\limits_{y}^{y+s} f(t,\tau)dtd\tau = \lim_{m\to\infty} h^{\alpha-1}s^{\beta-1}\int\limits_{x}^{x+h}\int\limits_{y}^{y+s} f_{n_m}(t,\tau)dtd\tau$$

$$\leq \lim_{m\to\infty}\sup_{h,s>0} s^{\alpha-1}s^{\beta-1}\int\limits_{x}^{x+h}\int\limits_{y}^{y+s} f_{m_k}(t,\tau)dtd\tau = \lim_{m\to\infty}\left(M^{+,+}_{\alpha,\beta} f_{n_m}\right)(x,y).$$

The latter limit exists because it converges to g a.e.

Hence,

$$\left(M^{+,+}_{\alpha,\beta} f\right)(x,y) \le \lim_{m\to\infty} \left(M^{+,+}_{\alpha,\beta} f_{nm}\right)(x,y) = g(x,y), \text{ a. e.,}$$

Now (3.2.5) implies that

$$\left\|M^{+,+}_{\alpha,\beta} f\right\|_{L^q_v(\mathbb{R}^2)} \le c\|f\|_{L^p_w(\mathbb{R}^2)}.$$

Let us now take $(x,y)\in\mathbb{R}^2$ so that $(M_{\alpha,\beta}f)(x,y)\neq 0$. Then there exist dyadic intervals I, J of type $I_x=[a_1,b_1)$, $I_y=[a_2,b_2)$, $0\le a_1-x<b_1-a_1$, $0\le a_2-y<b_2-a_2$ such that

$$\frac{1}{|(I_x)^-|^{1-\alpha}|(I_y)^-|^{1-\beta}} \iint\limits_{I_x\times I_y} f(t,\tau)dtd\tau > \frac{1}{2}\left(\overline{M}^{+,+}_{\alpha,\beta} f\right)(x,y). \tag{3.2.6}$$

Since f is bounded, we have that

$$\frac{1}{|I_x|^{1-\alpha}|I_y|^{1-\beta}} \iint\limits_{I_x\times I_y} f(t,\tau)dtd\tau \le c|I_x|^{1-\alpha}|I_y|^{1-\beta},$$

where the positive constant c depends on the bound of f. From the latter inequality we conclude that there exists a "minimal" rectangle $I_x\times I_y$ for which (3.2.6) holds. Note that the "minimal" rectangle might be non–unique.

Let $I, J\in D$ and let us introduce the set

$$F_{I,J} := \Big\{(x,y)\in\mathbb{R}^2 : \left(\overline{M}^{+,+}_{\alpha,\beta} f\right)(x,y)\neq 0,\ x\in I^-,\ y\in I^+,$$
$$I\times J \text{ is "minimal" for which (3.2.6) holds}\Big\}.$$

Let

$$S_m := \{(I,J) :\ F_{I,J}=\emptyset\}.$$

We have that

$$\iint\limits_{\mathbb{R}^2} \left(\overline{M}^{+,+}_{\alpha,\beta} f\right)^q (x,y)v(x,y)dxdy$$

$$\le 2^q \sum_{(I,J)\in S_m} \iint\limits_{F_{I,J}} \left(|I|^{\alpha-1}|J|^{\beta-1} \iint\limits_{I\times J} f(t,\tau)dtd\tau\right)^q v(x,y)dxdy$$

$$\le C\sum_{(I,J)\in S_m} \left(\iint\limits_{I^-\times J^-} v(x,y)dxdy\right)\left(|I|^{\alpha-1}|J|^{\beta-1}\right)^q \left(\iint\limits_{I\times J} f(t,\tau)dtd\tau\right)^q$$

$$\le CD^q_{+,+}\sum_{I\in D}\Bigg(\int_I w_1^{1-p'}(t)dt\Bigg)^{-\frac{q}{p'}}\sum_J\Bigg(\int_J w_2^{1-p'}(\tau)d\tau\Bigg)^{-\frac{q}{p'}}$$

$$\times\Bigg(\int_J\Bigg(\int_I f(x,y)dx\Bigg)dy\Bigg)^q$$

[by Theorem 3.1.1]

$$\le CD^q_{+,+}\sum_{I\in D}\Bigg(\int_I w_1^{1-p'}(t)dt\Bigg)^{-\frac{q}{p'}}\Bigg(\int_{\mathbb{R}}\Bigg(\int_I f(x,y)dx\Bigg)^p w_2(y)dy\Bigg)^{\frac{q}{p}}$$

[by Minkowski's inequality]

$$CD^q_{+,+}\sum_{I\in D}\Bigg(\int_I w_1^{1-p'}(t)dt\Bigg)^{-\frac{q}{p'}}\Bigg(\int_I\Bigg(\int_{\mathbb{R}} f^p(x,y)w_2(y)dy\Bigg)^{\frac{1}{p}}dx\Bigg)^q$$

[by Theorem 3.1.1 again]

$$\le CD^q_{+,+}\Bigg(\iint_{\mathbb{R}^2} f^p(x,y)w_2(y)w_1(x)dxdy\Bigg)^{q/p}.$$

Hence the implication (iii) $\Longrightarrow$ (ii) is proved. The implication (ii) $\Longrightarrow$ (i) follows from Lemma 3.2.2. Finally we will show that (i) $\Longrightarrow$ (iii). Let $a,b\in\mathbb{R}$; $h,s>0$. Then by taking the test function

$$f(x,y)=\chi_{(a,a+h)\times(b,b+h)}(x,y)w^{1-p'}(x,y),$$

we find that

$$\left\|M^{+,+}_{\alpha,\beta}f\right\|^q_{L^q_v(\mathbb{R}^2)}\ge c\int\limits_{a-h}^{a}\int\limits_{b-s}^{b}v(x,y)\Bigg(\frac{1}{h^{1-\alpha}s^{1-\beta}}\int\limits_x^{x+2h}\int\limits_y^{y+2s}f(t,\tau)dtd\tau\Bigg)^q dxdy$$

$$\ge c\Bigg(\int\limits_{a-h}^{a}\int\limits_{b-s}^{b}v(x,y)dxdy\Bigg)h^{(\alpha-1)q}s^{(\beta-1)q}\Bigg(\int\limits_a^{a+h}\int\limits_b^{b+s}w^{1-p'}(t,\tau)dtd\tau\Bigg)^q.$$

On the other hand, it is easy to check that

$$\|f\|^q_{L^p_w(\mathbb{R}^2)}=\Bigg(\int\limits_a^{a+h}\int\limits_b^{b+s}w^{1-p'}(t,\tau)dtd\tau\Bigg)^{q/p}<\infty.$$

By using these facts together with (i) we obtain that $D_{+,+} < \infty$, and the proof is complete. □

The following three statements follow analogously; therefore we omit the proofs.

Theorem 3.2.2. *Let $1 < p < q < \infty$ and let $0 < \alpha,\ \beta < 1$. Then $M^{-,-}_{\alpha,\beta}$ is bounded from $L^p_w(\mathbb{R}^2)$ to $L^q_v(\mathbb{R}^2)$ if and only if*

$$D_{-,-} := \sup_{\substack{a,b\in\mathbb{R}\\ h,s>0}} h^{\alpha-1}s^{\beta-1}\left(\int_a^{a+h}\int_b^{b+s} v(x,y)dxdy\right)^{1/q}\left(\int_{a-h}^{a}\int_{b-s}^{b} w^{1-p'}(x,y)dxdy\right)^{1/p'} < \infty$$

provided that $w(x,y) = w_1(x)w_2(y)$, where $w_i^{1-p'} \in RD^{(d)}(\mathbb{R})$, $i = 1,2$.

Theorem 3.2.3. *Let $1 < p < q < \infty$ and let $0 < \alpha,\ \beta < 1$. Then $M^{+,-}_{\alpha,\beta}$ is bounded from $L^p_w(\mathbb{R}^2)$ to $L^q_v(\mathbb{R}^2)$ if and only if*

$$D_{+,-} := \sup_{\substack{a,b\in\mathbb{R}\\ h,s>0}} h^{\alpha-1}s^{\beta-1}\left(\int_{a-h}^{a}\int_b^{b+s} v(x,y)dxdy\right)^{1/q}\left(\int_{a}^{a+h}\int_{b-s}^{b} w^{1-p'}(x,y)dxdy\right)^{1/p'} < \infty$$

provided that $w(x,y) = w_1(x)w_2(y)$, $w_i^{1-p'} \in RD^{(d)}(\mathbb{R})$, $i = 1,2$.

Theorem 3.2.4. *Let $1 < p < q < \infty$ and let $0 < \alpha,\ \beta < 1$. Suppose that $w(x,y) = w_1(x)w_2(y)$ with $w_i^{1-p'} \in RD^{(d)}(\mathbb{R})$, $i = 1,2$. Then the operator $M^{-,+}_{\alpha,\beta}$ is bounded from $L^p_w(\mathbb{R}^2)$ to $L^q_v(\mathbb{R}^2)$ if and only if*

$$D_{-,+} := \sup_{\substack{a,b\in\mathbb{R}\\ h,s>0}} h^{\alpha-1}s^{\beta-1}\left(\int_a^{a+h}\int_{b-s}^{b} v(x,y)dxdy\right)^{1/q}\left(\int_{a-h}^{a}\int_{b}^{b+s} w^{1-p'}(x,y)dxdy\right)^{1/p'} < \infty.$$

Next we establish a special type of the two–weight inequality for one–sided strong fractional maximal functions.

Theorem 3.2.5. *Let $1 < p < q < \infty$ and let $1/p - 1/q < \alpha, \beta < 1/p$. Then the following inequality holds*

$$\left(\iint_{\mathbb{R}^2} (M^{+,+}_{\alpha,\beta} f)^q(x,y)v(x,y)dxdy\right)^{1/q} \le c\left(\iint_{\mathbb{R}^2} |f(x,y)|^p(\mathcal{M}^{-,-}_{\alpha,\beta} v)^{p/q}(x,y)dxdy\right)^{1/p},$$

where

$$(\mathcal{M}^{-,-}_{\alpha,\beta} v)(x,y) := \sup_{\substack{h>0\\ s>0}} h^{(\alpha-1/p)q}s^{(\beta-1/p)q}\int_{x-h}^{x}\int_{y-s}^{y} v(t,\tau)dtd\tau$$

and the positive constant c does not depend on f and v.

Proof. Due to Lemma 3.2.2 it is enough to prove the inequality

$$E := \Big(\iint_{\mathbb{R}^2} (\overline{M}^{+,+}_{\alpha,\beta} f)^q(x,y) v(x,y)dxdy \Big)^{1/q}$$

$$\leq c \Big(\iint_{\mathbb{R}^2} |f(x,y)|^p (\overline{\mathcal{M}}^{-,-}_{\alpha,\beta} v)^{p/q}(x,y)dxdy \Big)^{1/p},$$

where

$$(\overline{\mathcal{M}}^{-,-}_{\alpha,\beta} v)(x,y) := \sup_{\substack{I\ni x, J\ni y \\ I\in D, J\in D}} \iint_{I^-\times J^-} |f(t,\tau)|dtd\tau.$$

Suppose that f is non–negative, is bounded and has a compact support. Keeping the notation of the proof of Theorem 3.2.1, using Theorem 3.1.1 (for $w \equiv 1$) twice and Minkowski's inequality we find that

$$E^q \leq 2^q \sum_{(I,J)\in S_m} \iint_{F_{I,J}} \Big(|I|^{\alpha-1}|J|^{\alpha-1} \iint_{I\times J} f(t,\tau)dtd\tau \Big)^q v(x,y)dxdy$$

$$\leq c \sum_{(I,J)\in S_m} \Big(\iint_{I^-\times J^-} v(x,y)dxdy \Big) \Big(|I|^{\alpha-1}|J|^{\beta-1} \iint_{I\times J} f(t,\tau)dtd\tau \Big)^q$$

$$\leq c \sum_{I,J\in D} |I|^{-q/p'}|J|^{-q/p'} \Big(\iint_{I\times J} f(x,y) \big(\overline{\mathcal{M}}^{-,-}_{\alpha,\beta} v\big)^{1/q}(x,y)dxdy \Big)^q$$

$$\leq c \sum_{I\in D} |I|^{-q/p'} \Big(\int_{\mathbb{R}} \Big(\int_I f(x,y) \big(\overline{\mathcal{M}}^{-,-}_{\alpha,\beta} v\big)^{1/q}(x,y)dx \Big)^p dy \Big)^{q/p}$$

$$\leq c \sum_{I\in D} |I|^{-q/p'} \Big(\int_I \Big(\int_{\mathbb{R}} f^p(x,y) \big(\overline{\mathcal{M}}^{-,-}_{\alpha,\beta} v\big)^{p/q}(x,y)dy \Big)^{1/p} dx \Big)^q$$

$$\leq c \Big(\iint_{\mathbb{R}^2} f^p(x,y) \big(\overline{\mathcal{M}}^{-,-}_{\alpha,\beta} v\big)^{p/q}(x,y)dxdy \Big)^{q/p}.$$

The proof is complete. □

The proofs of the following three statements are similar to the proof of Theorem 3.2.5; the details are omitted.

Theorem 3.2.6. *Let $1 < p < q < \infty$ and let $1/p - 1/q < \alpha, \beta < 1/p$. Then the following inequality holds*

$$\Big(\iint_{\mathbb{R}^2} (M^{-,-}_{\alpha,\beta} f)^q(x,y) v(x,y)dxdy \Big)^{1/q} \leq c \Big(\iint_{\mathbb{R}^2} |f(x,y)|^p (\mathcal{M}^{+,+}_{\alpha,\beta} v)^{p/q}(x,y)dxdy \Big)^{1/p},$$

where

$$(\mathcal{M}^{+,+}_{\alpha,\beta}v)(x,y) := \sup_{\substack{h>0\\ s>0}} h^{(\alpha-1/p)q}s^{(\beta-1/p)q} \int\limits_{x}^{x+h}\int\limits_{y}^{y+s} v(t,\tau)dtd\tau$$

and the positive constant c does not depend on f and v.

Theorem 3.2.7. *Let $1 < p < q < \infty$ and let $1/p - 1/q < \alpha, \beta < 1/p$. Then the following inequality holds*

$$\Big(\iint\limits_{\mathbb{R}^2} (M^{+,-}_{\alpha,\beta} f)^q(x,y)v(x,y)dxdy\Big)^{1/q} \le c\Big(\iint\limits_{\mathbb{R}^2} |f(x,y)|^p(\mathcal{M}^{-,+}_{\alpha,\beta}v)^{p/q}(x,y)dxdy\Big)^{1/p},$$

where

$$(\mathcal{M}^{-,+}_{\alpha,\beta}v)(x,y) := \sup_{\substack{h>0\\ s>0}} h^{(\alpha-1/p)q}s^{(\beta-1/p)q} \int\limits_{x-h}^{x}\int\limits_{y}^{y+s} v(t,\tau)dtd\tau$$

and the positive constant c does not depend on f and v.

Theorem 3.2.8. *Let $1 < p < q < \infty$ and let $1/p - 1/q < \alpha, \beta < 1/p$. Then the following inequality holds*

$$\Big(\iint\limits_{\mathbb{R}^2} (M^{-,+}_{\alpha,\beta} f)^q(x,y)v(x,y)dxdy\Big)^{1/q} \le c\Big(\iint\limits_{\mathbb{R}^2} |f(x,y)|^p(\mathcal{M}^{+,-}_{\alpha,\beta}v)^{p/q}(x,y)dxdy\Big)^{1/p},$$

where

$$(\mathcal{M}^{+,-}_{\alpha,\beta}v)(x,y) := \sup_{\substack{h>0\\ s>0}} h^{(\alpha-1/p)q}s^{(\beta-1/p)q} \int\limits_{x}^{x+h}\int\limits_{y-s}^{y} v(t,\tau)dtd\tau$$

and the positive constant c does not depend on f and v.

3.3 Mixed–type Operators

This section deals with the two–weight estimates for the following operators:

$$(M^{+}_{\alpha}W_{\beta})f(x,y) := \sup_{h>0} h^{\alpha-1} \int\limits_{x}^{x+h}\Big|\int\limits_{y}^{\infty} \frac{f(t,\tau)}{(\tau-y)^{1-\beta}}d\tau\Big|dt;$$

$$(M^{-}_{\alpha}R_{\alpha})f(x,y) := \sup_{h>0} h^{\alpha-1} \int\limits_{x-h}^{x}\Big|\int\limits_{-\infty}^{y} \frac{f(t,\tau)}{(y-\tau)^{1-\beta}}d\tau\Big|dt;$$

$$(M_\alpha^+ R_\beta)f(x,y) := \sup_{h>0} h^{\alpha-1} \int\limits_x^{x+h} \left| \int\limits_{-\infty}^{y} \frac{f(t,\tau)}{(y-\tau)^{1-\beta}} d\tau \right| dt;$$

$$(M_\alpha^- W_\beta)f(x,y) := \sup_{h>0} h^{\alpha-1} \int\limits_{x-h}^{x} \left| \int\limits_{y}^{\infty} \frac{f(t,\tau)}{(\tau-y)^{1-\beta}} d\tau \right| dt,$$

where $x, y \in \mathbb{R}$ and $0 < \alpha, \beta < 1$.

Suppose that

$$(\overline{M}_\alpha^+ W_\beta)f(x,y) := \sup_{I} |I|^{\alpha-1} \int\limits_I \left| \int\limits_{y}^{\infty} \frac{f(t,\tau)}{(\tau-y)^{1-\beta}} d\tau \right| dt,$$

where the supremum is taken over all dyadic intervals $I = [a,b)$, $x < a$, $a - x < b - a$. We will also need the following operator:

$$(N_\alpha^+ W_\beta)f(x,y) := \sup_{h>0} \frac{1}{(h/2)^{1-\alpha}} \int\limits_{x+h/2}^{x+h} \left| \int\limits_{y}^{\infty} \frac{f(t,\tau)}{(\tau-y)^{1-\beta}} d\tau \right| dt, \ x, y \in \mathbb{R}.$$

Lemma 3.3.1. *There exist positive constants b_1 and b_2 such that for all $f \in L_{loc}(\mathbb{R})$ and $x, y \in \mathbb{R}$ the following relation holds:*

$$b_1(M_\alpha^+ W_\beta f)(x,y) \le b_2(N_\alpha^+ W_\beta f)(x,y) \le (M_\alpha^+ W_\beta f)(x,y).$$

Proof. It is clear that

$$\left(\frac{h}{2}\right)^{\alpha-1} \int\limits_{x+h/2}^{x+h} \left| \int\limits_{y}^{\infty} \frac{f(t,\tau)}{(\tau-y)^{1-\beta}} d\tau \right| dt \le 2^{1-\alpha} \left(M_\alpha^+ W_\beta\right) f(x,y),$$

which implies that

$$\left(N_\alpha^+ W_\beta\right) f(x,y) \le 2^{1-\alpha} \left(M_\alpha^+ W_\beta\right) f(x,y).$$

To prove the reverse inequality with different constant we observe that

$$h^{\alpha-1} \int\limits_{x}^{x+h} \left| \int\limits_{y}^{\infty} \frac{f(t,\tau)}{(\tau-y)^{1-\beta}} d\tau \right| dt = h^{\alpha-1} \int\limits_{x}^{x+h/2} \left| \int\limits_{y}^{\infty} \frac{f(t,\tau)}{(\tau-y)^{1-\beta}} d\tau \right| dt$$

$$+ h^{\alpha-1} \int\limits_{x+h/2}^{x+h} \left| \int\limits_{y}^{\infty} \frac{f(t,\tau)}{(\tau-y)^{1-\beta}} d\tau \right| dt \le 2^{\alpha-1} \left[\left(M_\alpha^+ W_\beta\right) f(x,y) + \left(N_\alpha^+ W_\beta\right) f(x,y) \right].$$

Hence

$$\left(M_\alpha^+ W_\beta\right) f(x,y) \le \frac{2^{\alpha-1}}{1-2^{\alpha-1}} \left(N_\alpha^+ W_\beta\right) f(x,y),$$

and the proof is complete. □

Lemma 3.3.2. *There are positive constants b_1 and b_2 such that for all $f \in L_{loc}(\mathbb{R})$ and $(x,y) \in \mathbb{R}^2$,*

$$(M_\alpha^+ W_\beta) f(x,y) \le b_1 (\overline{M}_\alpha^+ W_\beta) f(x,y) \le b_2 (M_\alpha^+ W_\beta) f(x,y).$$

Proof. We take $(x,y) \in \mathbb{R}^2$. Assume that $I = [a,b)$ is a dyadic interval with $x < a$, $0 \le a - x < b - a$.

Then

$$|I|^{\alpha-1} \int_I \left| \int_y^\infty \frac{f(t,\tau)}{(\tau-y)^{1-\beta}} d\tau \right| dt$$

$$\le \left(\frac{b-x}{|I|}\right)^{1-\alpha} \left(M_\alpha^+ W_\beta\right) f(x,y) \le c_{\alpha,\beta} \left(M_\alpha^+ W_\beta\right) f(x,y).$$

Hence

$$\left(\overline{M}_\alpha^+ W_\beta\right) f(x,y) \le c_{\alpha,\beta} (M_\alpha^+ W_\beta) f(x,y).$$

To prove the reverse inequality we show that the relation

$$(N_\alpha^+ W_\beta) f \approx (\widetilde{N}_\alpha^+ W_\beta) f \tag{3.3.1}$$

holds, where

$$\left(\widetilde{N}_\alpha^+ W_\beta\right) f(x,y) = \sup_j \frac{1}{2^{(j-1)(1-\alpha)}} \int_{x+2^{j-1}}^{x+2^j} \left| \int_y^\infty \frac{f(t,\tau)}{(\tau-y)^{1-\beta}} d\tau \right| dt.$$

Let $h > 0$. Then $h \in \left[2^{j-1}, 2^j\right)$ for some $j \in \mathbb{Z}$. Consequently,

$$\frac{1}{(h/2)^{1-\alpha}} \int_{x+h/2}^{x+h} \left| \int_y^\infty \frac{f(t,\tau)}{(\tau-y)^{1-\beta}} d\tau \right| dt \le \frac{1}{2^{(j-2)(1-\alpha)}} \int_{x+2^{j-2}}^{x+2^j} \left| \int_y^\infty \frac{f(t,\tau)}{(\tau-y)^{1-\beta}} d\tau \right| dt$$

$$= 2^{(2-j)(1-\alpha)} \int_{x+2^{j-1}}^{x+2^j} \left| \int_y^\infty \frac{f(t,\tau)}{(\tau-y)^{1-\beta}} d\tau \right| dt + 2^{(2-j)(1-\alpha)} \int_{x+2^{j-2}}^{x+2^{j-1}} \left| \int_y^\infty \frac{f(t,\tau)}{(\tau-y)^{1-\beta}} d\tau \right| dt$$

$$\le 2^{1-\alpha} \left(\widetilde{N}_\alpha^+ W_\beta\right) f(x,y) + \left(\widetilde{N}_\alpha^+ W_\beta\right) f(x,y) = \left(2^{1-\alpha}+1\right) \left(\widetilde{N}_\alpha^+ W_\beta\right) f(x,y).$$

Hence

$$\left(N_\alpha^+ W_\beta\right) f(x,y) \le \left(2^{1-\alpha}+1\right)\left(\widetilde{N}_\alpha^+ W_\beta\right) f(x,y).$$

The reverse inequality of the latter one is obvious and (3.3.1) is proved .

Due to (3.3.1) and Lemma 3.3.1 it suffices to prove that

$$\left(\widetilde{N}_\alpha^+ W_\beta\right) f(x,y) \le c\left(\overline{M}_\alpha^+ W_\beta\right) f(x,y). \tag{3.3.2}$$

Assume that $h = 2^j$ for some $j \in \mathbb{Z}$. Let us take dyadic intervals I, I' of the length 2^{j-1} so that $[x+h/2, x+h) \subset I \cup I'$.

If $I \cup I'$ is a dyadic interval, then

$$\int_{x+h/2}^{x+h}\left|\int_y^\infty \frac{f(t,\tau)}{(\tau-y)^{1-\beta}}d\tau\right|dt \le \int_{I\cup I'}\left|\int_y^\infty \frac{f(t,\tau)}{(\tau-y)^{1-\beta}}d\tau\right|dt$$

$$\le 2^{j(1-\alpha)}\left(\overline{M}_\alpha^+ W_\beta\right) f(x,y) = \left(\frac{h}{2}\right)^{1-\alpha} 2^{1-\alpha}(\overline{M}_\alpha^+ W_\beta) f(x,y).$$

If $I \cup I'$ is not dyadic, then we have that

$$\int_{x+h/2}^{x+h}\left|\int_y^\infty \frac{f(t,\tau)}{(\tau-y)^{1-\beta}}d\tau\right|dt \le \int_{I\cup I_1}\left|\int_y^\infty \frac{f(t,\tau)}{(\tau-y)^{1-\beta}}d\tau\right|dt$$

$$\le \int_I\left|\int_y^\infty \frac{f(t,\tau)}{(\tau-y)^{1-\beta}}d\tau\right|dt + \int_{I_1}\left|\int_y^\infty \frac{f(t,\tau)}{(\tau-y)^{1-\beta}}d\tau\right|dt \le 2^{(j-1)(1-\alpha)}\left(\overline{M}_\alpha^+ W_\beta\right) f(x,y)$$

$$+2^{j(1-\alpha)}\left(\overline{M}_\alpha^+ W_\beta\right) f(x,y) \le 2^{(j-1)(1-\alpha)}\left(1+2^{1-\alpha}\right)\left(\overline{M}_\alpha^+ W_\beta\right) f(x,y)$$

$$= \left(\frac{h}{2}\right)^{1-\alpha}\left(1+2^{1-\alpha}\right)\left(\overline{M}_\alpha^+ W_\beta\right) f(x,y).$$

Summarizing these estimates we find that (3.3.2) holds with $c = 1 + 2^{1-\alpha}$ and the proof is complete. □

We now formulate and prove the main results if this section.

Theorem 3.3.1. *Let $1 < p < q < \infty$ and let $0 < \alpha,\ \beta < 1$. Suppose that the weight w on $\mathbb{R}^2$ has the form $w(x,y) = w_1(x)w_2(y)$ where $w_1^{1-p'} \in RD^{(d)}(\mathbb{R})$. Then $M_\alpha^+ W_\beta$ is bounded from $L_w^p(\mathbb{R}^2)$ to $L_v^q(\mathbb{R}^2)$ if and only if*

(i)

$$E_{+,+}^{(1)} := \sup_{\substack{a,b\in\mathbb{R}\\ h,s>0}} h^{\alpha-1}\left(\int_{a-h}^{a}\int_{-\infty}^{b-s}\frac{v(x,y)}{(b-y)^{(1-\beta)q}}dxdy\right)^{1/q}\left(\int_a^{a+h}\int_{b-s}^{b+s} w^{1-p'}(x,y)dxdy\right)^{1/p'} < \infty;$$

(ii)

$$E_{+,+}^{(2)} := \sup_{\substack{a,b\in\mathbb{R}\\ h,s>0}} h^{\alpha-1}\left(\int_{a-h}^{a}\int_{b-s}^{b+s} v(x,y)dxdy\right)^{1/q}\left(\int_{a}^{a+h}\int_{b+s}^{\infty} \frac{w^{1-p'}(x,y)}{(y-b)^{(1-\beta)p'}}dxdy\right)^{1/p'} < \infty.$$

Proof. *Sufficiency.* By the argument used in the proof of Theorem 3.2.1 we can assume that f is non-negative, bounded and has a compact support. It is obvious that in this case $(M_\alpha^+ W_\beta f)(x,y) < \infty$ for any $x, y \in \mathbb{R}$. By Lemma 3.3.2 it suffices to show that the condition $\max\left\{E_{+,+}^{(1)}, E_{+,+}^{(2)}\right\} < \infty$ implies the boundedness of $\overline{M}_\alpha^+ W_\beta$ from $L_w^p(\mathbb{R}^2)$ to $L_v^q(\mathbb{R}^2)$. We take $(x,y) \in \mathbb{R}^2$ so that $\left(\overline{M}_\alpha^+ W_\beta\right) f(x,y) \neq 0$ (note that if x is outside of some sufficiently large interval, then $\left(\overline{M}_\alpha^+ W_\beta\right) f(x,y) = 0$). Further, there exists an interval $I_y(x) \subset \mathbb{R}$ of the type $I_y(x) = [a,b)$ with $0 \le a - x < b - a$ such that

$$\frac{1}{|I_y(x)|^{1-\alpha}} \int_{I_y(x)} \left(\int_{y}^{\infty} \frac{f(t,\tau)}{(\tau-y)^{1-\beta}}d\tau\right) dt > \frac{1}{2}\left(\overline{M}_\alpha^+ W_\beta\right) f(x,y). \tag{3.3.3}$$

For these x, y and f, there exists a "minimal" dyadic interval $I_y(x)$ for which (3.3.3) holds. Due to the main property of dyadic intervals such an interval is unique.

Recall that by the symbol I^- we denote an interval $[2a-b, a)$, where $I := [a,b)$. Moreover, for an interval I we introduce the set:

$$F_I := \left\{(x,y) : (\overline{M}_\alpha^+ W_\beta) f(x,y) \neq 0,\ x \in I^-, I \text{ is "minimal" for which (3.3.3) holds.}\right\}$$

It is obvious that $F_I \subset I^- \cup \mathbb{R}$. Denote by D_m a subset of D for which $F_I \neq 0$. In the sequel we will use the following notation:

$$\mathbf{V}_J(y) := |J|^{(\alpha-1)q} \int_J v(x,y)dx; \quad \overline{F}_I(\tau) := \int_I f(t,\tau)dt,$$

where J and I are intervals in $\mathbb{R}$.

We have

$$\iint_{\mathbb{R}^2} (\overline{M}_\alpha^+ W_\beta)^q f(x,y) v(x,y) dxdy$$

$$\le c \sum_{I\in D_m} \iint_{F_I} \left[\frac{1}{|I|^{1-\alpha}} \int_I W_\beta(f(t,\cdot))(y)dt\right]^q v(x,y)dxdy$$

$$\le c \sum_{I\in D_m} \iint\limits_{I^-\times\mathbb{R}} \left[\frac{1}{|I|^{1-\alpha}} \int\limits_I W_\beta(f(t,\cdot))(y)dt\right]^q v(x,y)dxdy$$

$$= c \sum_{I\in D_m} \int\limits_{\mathbb{R}} \mathbf{V}_{I^-}(y) \left[\int\limits_I \left(\int\limits_y^\infty \frac{f(t,\tau)d\tau}{(\tau-y)^{1-\beta}}\right) dt\right]^q dy$$

[change the order of integration in the inner integral]

$$= c \sum_{I\in D_m} \int\limits_{\mathbb{R}} \mathbf{V}_{I^-}(y) \left[\int\limits_y^\infty (\tau-y)^{\beta-1}\overline{F}_I(\tau)d\tau\right]^q dy$$

[by Theorem 1.1.6]

$$\le c \sum_{I\in D_m} \left[\int\limits_{\mathbb{R}} \left(\int\limits_I w_1^{1-p'}(t)dt\right)^{1-p} (\overline{F}_I(y))^p w_2(y)dy\right]^{q/p}$$

$$= c \sum_{I\in D_m} \left(\int\limits_I w_1^{1-p'}(t)dt\right)^{-q/p'} \left[\int\limits_{\mathbb{R}} (\overline{F}_I(y))^p w_2(y)dy\right]^{q/p}$$

[by Minkowski's integral inequality]

$$\le c \sum_{I\in D_m} \left(\int\limits_I w_1^{1-p'}(t)dt\right)^{-q/p'} \left[\int\limits_I \left(\int\limits_{\mathbb{R}} f^p(t,y)w_2(y)dy\right)^{1/p} dt\right]^q$$

[by Theorem 3.1.1]

$$\le c\|f\|^q_{L^p_w(\mathbb{R}^2)}.$$

Necessity. To prove necessity first we show that the boundedness of $\overline{M}^+_\alpha W_\beta$ from $L^p_w(\mathbb{R}^2)$ to $L^q_v(\mathbb{R}^2)$ implies that

$$\int\limits_a^{a+h}\int\limits_{b-s}^{b+s} w^{1-p'}(x,y)dxdy < \infty; \tag{3.3.4}$$

$$I(a,b,h,s) := \int\limits_a^{a+h}\int\limits_{b+s}^{\infty} \frac{w^{1-p'}(x,y)}{(y-b)^{(1-\beta)p'}}dxdy < \infty \tag{3.3.5}$$

for all $a, b \in \mathbb{R}$ and $h, s > 0$. Let us show that (3.3.5) holds. Suppose the contrary: $I(a, b, h, s) = \infty$ for some a, b, h, s. Then by the duality arguments there exists a non-negative $g : \mathbb{R} \to \mathbb{R}$ with a support in $(a, a+h) \times (b-s, \infty)$ such that

$$\int_a^{a+h}\int_{b+s}^{\infty} g^p(x, y)w(x, y)dxdy < \infty$$

and

$$\int_a^{a+h}\int_{b+s}^{\infty} g(x, y)(y-b)^{\alpha-1}dxdy = \infty.$$

On the other hand, if $z_1 \in (a-h, a)$, $z_2 \in (b-s, b+s)$, then

$$(M_\alpha^+ W_\beta)g(z_1, z_2) \geq (2h)^{\alpha-1}\int_{z_1}^{z_1+2h}\int_{b+s}^{\infty}\frac{g(t,\tau)}{(\tau-z_2)^{1-\beta}}dtd\tau$$

$$\geq 2^{\alpha+\beta-2}h^{\alpha-1}\int_a^{a+h}\int_{b+s}^{\infty} g(t,\tau)(\tau-b)^{\beta-1}dtd\tau = \infty.$$

Here we used the following observation: if $z_1 \in (a-h, a)$, $z_2 \in (b-s, b+s)$, $\tau > b+s$, then

$$(a, a+h) \subset (z_1, z_1+2h); \qquad (\tau - z_2) \leq 2(\tau - b).$$

Consequently $(a-h, a) \times (b-s, b+s) \subset \{(x, y) : (M_\alpha^+ W_\beta)g(x, y) > \lambda\}$ for every positive λ. Now from the boundedness of $M_\alpha^+ W_\beta$ from $L^p_w(\mathbb{R}^2)$ to $L^q_v(\mathbb{R}^2)$ we have that

$$\int_{a-h}^{a}\int_{b-s}^{b+s} v(x, y)dxdy \leq \int\limits_{\{(x,y):\ (M_\alpha^+ W_\beta)g(x,y)>\lambda\}} v(x, y)dxdy$$

$$\leq \frac{c}{\lambda^q}\left(\int_a^{a+h}\int_{b+s}^{\infty} g^p(x, y)w(x, y)dxdy\right)^{q/p} < \infty.$$

Since λ is arbitrary positive number, we conclude that

$$\int_{a-h}^{a}\int_{b-s}^{b+s} v(x, y)dxdy = 0,$$

which is not possible unless $v = 0$ on $(a-h, a) \times (b-s, b+s)$. This completes the proof of (3.3.5). The proof of (3.3.4) is similar to that of (3.3.5) and therefore, it is omitted.

Now let $f \geq 0$, $a, b \in \mathbb{R}$, $h, s > 0$ and let $x \in (a-h, a)$, $y \in (b-s, b+s)$. We have

$$(M_\alpha^+ W_\beta) f(x,y) \geq (2h)^{\alpha-1} \int_x^{x+2h} \int_y^{\infty} \frac{f(t,\tau)}{(\tau - y)^{1-\beta}} dt d\tau$$

$$\geq c_{\alpha,\beta} h^{\alpha-1} \int_a^{a+h} \int_{b+s}^{\infty} \frac{f(t,\tau)}{(\tau - b)^{1-\beta}} dt d\tau := J(a,b,h,s,\alpha,\beta).$$

Hence, by using the two–weight weak (p,q) type inequality for $M_\alpha^+ W_\beta$ we obtain that

$$\int_{a-h}^{a} \int_{b-s}^{b+s} v(x,y) dx dy \leq \int_{\{(x,y):\ (M_\alpha^+ W_\beta) f(x,y) > J(a,b,h,s,\alpha,\beta)\}} v(x,y) dx dy$$

$$\leq c(J(a,b,h,s,\alpha,\beta))^{-q} \left(\iint_{\mathbb{R}^2} f^p(x,y) w(x,y) dx dy \right)^{q/p}$$

$$= c \left(h^{\alpha-1} \int_a^{a+h} \int_{b+s}^{\infty} \frac{f(t,\tau)}{(\tau - b)^{1-\beta}} dt d\tau \right)^{-q} \|f\|^q_{L^p_w(\mathbb{R}^2)},$$

where the positive constant c does not depend on h, s, a, b, f. Let us now take

$$f(x,y) = w^{1-p'}(x,y) \chi_{(a,a+h)}(x) \chi_{(b+s,\infty)}(y) (y-b)^{(\beta-1)(p'-1)}.$$

Then

$$\|f\|^q_{L^p_w(\mathbb{R}^2)} = \left(\int_a^{a+h} \int_{b+s}^{\infty} w^{1-p'}(x,y) (y-b)^{(\beta-1)p'} dx dy \right)^{q/p}.$$

From this observation we immediately get that $E^{(2)}_{+,+} < \infty$.

Analogously we can obtain $E^{(1)}_{+,+} < \infty$ and the proof is complete. □

The next three statements follow analogously.

Theorem 3.3.2. *Let $1 < p < q < \infty$ and let $0 < \alpha, \beta < 1$. Then $M_\alpha^+ R_\beta$ is bounded from $L^p_w(\mathbb{R}^2)$ to $L^q_v(\mathbb{R}^2)$ if and only if*

(i)

$$E^{(1)}_{+,-} := \sup_{\substack{a,b\in\mathbb{R}\\ h,s>0}} h^{\alpha-1}\left(\int_{a-h}^{a}\int_{b-s}^{b+s} v(x,y)dxdy\right)^{1/q}\left(\int_{a}^{a+h}\int_{-\infty}^{b-s}\frac{w^{1-p'}(x,y)}{(b-y)^{(1-\beta)p'}}dxdy\right)^{1/p'} < \infty;$$

(ii)

$$E^{(2)}_{+,-} := \sup_{\substack{a,b\in\mathbb{R}\\ h,s>0}} h^{\alpha-1}\left(\int_{a-h}^{a}\int_{b+s}^{\infty}\frac{v(x,y)}{(y-b)^{(1-\beta)q}}dxdy\right)^{1/q}\left(\int_{a}^{a+h}\int_{b-s}^{b+s} w^{1-p'}(x,y)dxdy\right)^{1/p'} < \infty$$

provided that $w(x,y) = w_1(x)w_2(y)$ *with* $w_1 \in RD^{(d)}(\mathbb{R})$.

Theorem 3.3.3. *Let* $1 < p < q < \infty$ *and let* $0 < \alpha,\ \beta < 1$. *Suppose that* $w(x,y) = w_1(x)w_2(y)$, *where* $w_1 \in RD^{(d)}(\mathbb{R})$. *Then* $M^-_\alpha R_\beta$ *is bounded from* $L^p_w(\mathbb{R}^2)$ *to* $L^q_v(\mathbb{R}^2)$ *if and only if*

(i)

$$E^{(1)}_{-,-} := \sup_{\substack{a,b\in\mathbb{R}\\ h,s>0}} h^{\alpha-1}\left(\int_{a}^{a+h}\int_{b-s}^{b+s} v(x,y)dxdy\right)^{1/q}\left(\int_{a-h}^{a}\int_{-\infty}^{b-s}\frac{w^{1-p'}(x,y)dxdy}{(b-y)^{(1-\beta)p'}}\right)^{1/p'} < \infty;$$

(ii)

$$E^{(2)}_{-,-} := \sup_{\substack{a,b\in\mathbb{R}\\ h,s>0}} h^{\alpha-1}\left(\int_{a}^{a+h}\int_{b+s}^{\infty}\frac{v(x,y)dxdy}{(y-b)^{(1-\beta)q}}\right)^{1/q}\left(\int_{a-h}^{a}\int_{b-s}^{b+s} w^{1-p'}(x,y)dxdy\right)^{1/p'} < \infty.$$

Theorem 3.3.4. *Let* $1 < p < q < \infty$ *and let* $0 < \alpha,\ \beta < 1$. *Suppose that* $w(x,y) = w_1(x)w_2(y)$, *where* $w_1 \in RD^{(d)}(\mathbb{R})$. *Then* $M^-_\alpha W_\beta$ *is bounded from* $L^p_w(\mathbb{R}^2)$ *to* $L^q_v(\mathbb{R}^2)$ *if and only if*

(i)

$$E^{(1)}_{-,+} := \sup_{\substack{a,b\in\mathbb{R}\\ h,s>0}} h^{\alpha-1}\left(\int_{a}^{a+h}\int_{-\infty}^{b-s}\frac{v(x,y)dxdy}{(b-y)^{(1-\beta)q}}\right)^{1/q}\left(\int_{a-h}^{a}\int_{b-s}^{b+s} w^{1-p'}(x,y)dxdy\right)^{1/p'} < \infty;$$

(ii)

$$E^{(2)}_{-,+} := \sup_{\substack{a,b\in\mathbb{R}\\ h,s>0}} h^{\alpha-1}\left(\int_{a}^{a+h}\int_{b-s}^{b+s} v(x,y)dxdy\right)^{1/q}\left(\int_{a-h}^{a}\int_{b+s}^{\infty}\frac{w^{1-p'}(x,y)dxdy}{(y-b)^{(1-\beta)q}}\right)^{1/p'} < \infty.$$

3.4 One–sided Potentials with Product Kernels

This section deals with one–sided potentials with multiple kernels:

$$(W_{\alpha,\beta}f)(x,y) := \int_x^\infty \int_y^\infty \frac{f(t,\tau)dtd\tau}{(t-x)^{1-\alpha}(\tau-y)^{1-\beta}},$$

$$(R_{\alpha,\beta}f)(x,y) := \int_{-\infty}^x \int_{-\infty}^y \frac{f(t,\tau)dtd\tau}{(x-t)^{1-\alpha}(y-\tau)^{1-\beta}},$$

$$(W_\alpha R_\beta f)(x,y) := \int_x^\infty \int_{-\infty}^y \frac{f(t,\tau)dtd\tau}{(t-x)^{1-\alpha}(y-\tau)^{1-\beta}},$$

$$(R_\alpha W_\beta f)(x,y) := \int_{-\infty}^x \int_y^\infty \frac{f(t,\tau)dtd\tau}{(x-t)^{1-\alpha}(\tau-t)^{1-\beta}},$$

where $(x,y) \in \mathbb{R}^2$, $0 < \alpha, \beta < 1$.

Together with these operators we are interested in the double Riemann–Liouville transform defined on $\mathbb{R}_+$:

$$(\mathcal{R}_{\alpha,\beta}f)(x,y) := \int_0^x \int_0^y \frac{f(t,\tau)dtd\tau}{(x-t)^{1-\alpha}(y-\tau)^{1-\beta}}, \quad x,y>0,$$

where $0 < \alpha, \beta < 1$.

Criteria governing the trace inequality for $\mathcal{R}_{\alpha,\beta}$ were established in Chapter 2.

By applying Theorems 3.3.1–3.3.4 and Theorem 3.1.5 we can easily obtain the proofs of the next four statements, so we omit the details.

Theorem 3.4.1. *Let $1 < p < q < \infty$ and let $0 < \alpha,\ \beta < 1$. Suppose that $v \in A^+_{\infty,x}(\mathbb{R})$. Suppose also that $w(x,y) = w_1(x)w_2(y)$ with $w_1 \in RD^{(d)}(\mathbb{R})$. Then $W_{\alpha,\beta}$ is bounded from $L^p_w(\mathbb{R}^2)$ to $L^q_v(\mathbb{R}^2)$ if and only if* $\max\left\{E^{(1)}_{+,+}, E^{(2)}_{+,+}\right\} < \infty$.

Theorem 3.4.2. *Let $1 < p < q < \infty$ and let $0 < \alpha,\ \beta < 1$. Suppose that $v \in A^-_{\infty,x}(\mathbb{R})$ and $w(x,y) = w_1(x)w_2(y)$ with $w_1 \in RD^{(d)}(\mathbb{R})$. Then $R_{\alpha,\beta}$ is bounded from $L^p_w(\mathbb{R}^2)$ to $L^q_v(\mathbb{R}^2)$ if and only if* $\max\left\{E^{(1)}_{-,-}, E^{(2)}_{-,-}\right\} < \infty$.

Theorem 3.4.3. *Let $1 < p < q < \infty$ and let $0 < \alpha,\ \beta < 1$. Suppose that $v \in A^+_{\infty,x}(\mathbb{R})$ and $w(x,y) = w_1(x)w_2(y)$ with $w_1 \in RD^{(d)}(\mathbb{R})$. Then $W_\alpha R_\beta$ is bounded from $L^p_w(\mathbb{R}^2)$ to $L^q_v(\mathbb{R}^2)$ if and only if* $\max\left\{E^{(1)}_{+,-}, E^{(2)}_{+,-}\right\} < \infty$.

Theorem 3.4.4. *Let $1 < p < q < \infty$ and let $0 < \alpha,\ \beta < 1$. Assume that $v \in A^{-}_{\infty,x}(\mathbb{R})$. Let $w(x,y) = w_1(x)w_2(y)$ with $w_1 \in RD^{(d)}(\mathbb{R})$. Then $R_\alpha W_\beta$ is bounded from $L^p_w(\mathbb{R}^2)$ to $L^q_v(\mathbb{R}^2)$ if and only if* $\max\left\{E^{(1)}_{-,+}, E^{(2)}_{-,+}\right\} < \infty$.

Let us now consider the case when the right–hand side weight is equal to 1.

Lemma 3.4.1. *Let $1 < p < q < \infty$ and let $0 < \alpha,\ \beta < 1/p$. Suppose that $w \equiv 1$. Then the following conditions are equivalent:*

(i) $E_{+,+} := \max\left\{E^{(1)}_{+,+}, E^{(2)}_{+,+}\right\} < \infty$;

(ii) $E_{-,-} := \max\left\{E^{(1)}_{-,-}, E^{(2)}_{-,-}\right\} < \infty$;

(iii) $E_{+,-} := \max\left\{E^{(1)}_{+,-}, E^{(2)}_{+,-}\right\} < \infty$;

(iv) $E_{-,+} := \max\left\{E^{(1)}_{-,+}, E^{(2)}_{-,+}\right\} < \infty$;

(v) $F := \sup\left(\int\limits_I\int\limits_J v(x,y)dxdy\right)|I|^{(\alpha-1/p)q}|J|^{(\beta-1/p)q} < \infty$, *where the supremum is taken over all one-dimensional intervals I and J.*

Proof. Let $w \equiv 1$. In this case we have that

$$E^{(1)}_{+,+} = c_p \sup_{\substack{a,b\in\mathbb{R}\\ h,s>0}} h^{\alpha-1/p}\left(\int\limits_a^{h}\int\limits_{-\infty}^{b-s}\frac{v(x,y)}{(b-y)^{(1-\beta)q}}dxdy\right)^{1/q}\cdot s^{1/p'},$$

$$E^{(2)}_{+,+} := \sup_{\substack{a,b\in\mathbb{R}\\ h,s>0}} h^{\alpha-1/p}\left(\int\limits_{a-h}^{a}\int\limits_{b-s}^{b+s} v(x,y)dxdy\right)^{1/q}\left(\int\limits_{b+s}^{\infty}\frac{dy}{(y-b)^{(1-\beta)p'}}dxdy\right)^{1/p'}$$

$$= c_{\beta,p}\sup_{\substack{a,b\in\mathbb{R}\\ h,s>0}} h^{\alpha-1/p}s^{\beta-1/p}\left(\int\limits_{a-h}^{a}\int\limits_{b-s}^{b+s} v(x,y)dxdy\right) := c_{\beta,p}F.$$

Here we have used the condition $\beta < 1/p$.

To prove the relation $E^{(1)}_{+,+} \approx F$ observe that for $a, b \in \mathbb{R}$ and $h, s > 0$,

$$h^{(\alpha-1/p)q}s^{q/p'}\int\limits_{a-h}^{a}\int\limits_{-\infty}^{b-s}\frac{v(x,y)}{(b-y)^{(1-\beta)q}}dxdy$$

$$= h^{(\alpha-1/p)q}s^{q/p'}\int\limits_{a-h}^{a}\left(\sum_{k=0}^{\infty}\int\limits_{2^k s<b-y<2^{k+1}s}\frac{v(x,y)dy}{(b-y)^{(1-\beta)q}}\right)dx$$

$$\le h^{(\alpha-1/p)q} s^{q/p'} \sum_{k=0}^{\infty} (2^k s)^{(\beta-1)q} \int_{a-h}^{a} \int_{b-2^{k+1}s}^{b-2^k s} v(x,y)dxdy$$

$$\le F^q \sum_{k=0}^{\infty} 2^{-kq/p'} \le c_{p,q} F^q.$$

Hence

$$E^{(1)}_{+,+} \le c_{p,q,\beta} F.$$

The inequality $F \le cE^{(1)}_{+,+}$ with some positive constant c is obvious; so the proof is complete. □

The next theorem is a direct consequence of the statements proved above.

Theorem 3.4.5. *Let $1 < p < q < \infty$ and let $0 < \alpha,\ \beta < 1/p$. The following statements are equivalent:*

(i) $M^{+,+}_{\alpha,\beta}$ *is bounded from $L^p(\mathbb{R}^2)$ to $L^q_v(\mathbb{R}^2)$;*

(ii) $M^{-,-}_{\alpha,\beta}$ *is bounded from $L^p(\mathbb{R}^2)$ to $L^q_v(\mathbb{R}^2)$;*

(iii) $M^{+,-}_{\alpha,\beta}$ *is bounded from $L^p(\mathbb{R}^2)$ to $L^q_v(\mathbb{R}^2)$;*

(iv) $M^{-,+}_{\alpha,\beta}$ *is bounded from $L^p(\mathbb{R}^2)$ to $L^q_v(\mathbb{R}^2)$;*

(v) $M^+_\alpha W_\beta$ *is bounded from $L^p(\mathbb{R}^2)$ to $L^q_v(\mathbb{R}^2)$;*

(vi) $M^+_\alpha R_\beta$ *is bounded from $L^p(\mathbb{R}^2)$ to $L^q_v(\mathbb{R}^2)$;*

(vii) $M^-_\alpha W_\beta$ *is bounded from $L^p(\mathbb{R}^2)$ to $L^q_v(\mathbb{R}^2)$;*

(viii) $M^-_\alpha R_\beta$ *is bounded from $L^p(\mathbb{R}^2)$ to $L^q_v(\mathbb{R}^2)$;*

(ix) $W_{\alpha,\beta}$ *is bounded from $L^p(\mathbb{R}^2)$ to $L^q_v(\mathbb{R}^2)$ provided that either $v \in A^+_{\infty,x}(\mathbb{R})$ or $v \in A^+_{\infty,y}(\mathbb{R})$;*

(x) $R_{\alpha,\beta}$ *is bounded from $L^p(\mathbb{R}^2)$ to $L^q_v(\mathbb{R}^2)$ provided that either $v \in A^-_{\infty,x}(\mathbb{R})$ or $v \in A^-_{\infty,y}(\mathbb{R})$;*

(xi) $W_\alpha R_\beta$ *is bounded from $L^p(\mathbb{R}^2)$ to $L^q_v(\mathbb{R}^2)$ provided that either $v \in A^+_{\infty,x}(\mathbb{R})$ or $v \in A^-_{\infty,y}(\mathbb{R})$;*

(xii) $R_\alpha W_\beta$ *is bounded from $L^p(\mathbb{R}^2)$ to $L^q_v(\mathbb{R}^2)$ provided that either $v \in A^-_{\infty,x}(\mathbb{R})$ or $v \in A^+_{\infty,y}(\mathbb{R})$;*

(xiii) $F < \infty$*, where F is defined in Lemma* 3.4.1.

To formulate our next results recall that (see Definition 1.6.1) a weight function $\rho : \mathbb{R}^2_+ \to \mathbb{R}_+$ belongs to the class $DC(y)$ if there is a constant $b > 1$ such that for all $t > 0$ and almost every $x \in \mathbb{R}_+$,

$$\rho_x([0, 2t]) \le b\rho_x(J_t),$$

where J_t denotes the following two types of intervals $[0, t]$ and $[t, 2t]$.

Let us introduce the following two-dimensional operator:

$$\mathcal{R}_\alpha H(f)(x,y) := \int_0^x \int_0^y \frac{f(t,\tau)dtd\tau}{(x-t)^{1-\alpha}}, \quad x,y>0, \quad 0<\alpha<1.$$

In fact, $\mathcal{R}_\alpha H = \mathcal{R}_{\alpha,1}$.

Proposition 3.4.1. *Let $0<\alpha<1$, $1<p\le q<\infty$ and let $\alpha\in[1/p-1/q,1/p)\cup(1/p,1)$. Suppose that $w^{1-p'}\in DC(y)$. Assume that one of the following conditions holds:*

(a) *$\alpha\in[1/p-1/q,1/p)$ and u and w are monotonic weights on $\mathbb{R}_+$ in the first variable for the second fixed one;*

(b) *$\alpha\in(1/p,1)$ and w is increasing on $\mathbb{R}_+$ in the first variable for the second fixed one;*

(c) *$\alpha\in(1/p,1)$, w is decreasing on $\mathbb{R}_+$ in the first variable for the second fixed one and $w^{1-p'}\in DC(x)$.*

Then $\mathcal{R}_\alpha H$ is bounded from $L^p_w(\mathbb{R}^2_+)$ to $L^q_u(\mathbb{R}^2_+)$ is and only if

$$\widetilde{G} := \sup_{a,b>0}\left[\int_a^\infty\int_b^\infty u(x,y)x^{(\alpha-1)q}dxdy\right]^{1/q}\left[\int_0^a\int_0^b w^{1-p'}(x,y)dxdy\right]^{\frac{1}{p'}}<\infty, \quad (3.4.1)$$

Moreover, $\|\mathcal{R}_\alpha H\|\approx\widetilde{G}$.

Proof. *Necessity* follows immediately substituting the test function

$$f(x,y)=\chi_{(0,a)}(x)\chi_{(0,b)}(y)w^{1-p'}(x,y)$$

in the two–weight inequality for $\mathcal{R}_\alpha H$ and observing that

$$\int_0^a\int_0^b w^{1-p'}(x,y)dxdy<\infty$$

for all $a,b>0$.

Sufficiency. By Lemma 1.6.1 there exists a constant η such that for all $t>0$ and for a. e. $x>0$ the inequality

$$\int_0^{2t} w^{1-p'}(x,y)dy \ge \eta\int_0^t w^{1-p'}(x,y)dy$$

holds. In the sequel we will use the notation:

$$\widetilde{u}_j(x) := \int_{2^j}^{2^{j+1}} u(x,y)dy; \quad F_j(t) := \int_0^{2^{j+1}} f(t,\tau)d\tau.$$

Let $f \geq 0$. We have that

$$I := \int_0^\infty \int_0^\infty u(x,y)\big((\mathcal{R}_\alpha H f)(x,y)\big)^q dxdy$$

$$\leq \int_0^\infty \int_0^\infty u(x,y) \left(\int_0^x (x-t)^{\alpha-1} \left(\int_0^y f(t,\tau)d\tau \right) dt \right)^q dxdy$$

$$\leq \sum_{j\in\mathbb{Z}} \int_0^\infty \widetilde{u}_j(x) \left(\int_0^x (x-t)^{\alpha-1} F_j(t)dt \right)^q dx.$$

Further, according to Proposition 3.1.1 and the estimate

$$\widetilde{G} \geq \left[\int_a^\infty \left(\int_{2^j}^{2^{j+1}} u(x,y)dy \right) x^{(\alpha-1)q}dx \right]^{\frac{1}{q}} \left[\int_0^a \left(\int_0^{2^j} w^{1-p'}(x,y)dy \right)^{(1-p')(1-p)} dx \right]^{\frac{1}{p'}}$$

we find that

$$I \leq c\widetilde{G}^q \sum_{j\in\mathbb{Z}} \left[\int_0^\infty \left(\int_0^{2^j} w^{1-p'}(x,y)dy \right)^{1-p} F_j^p(x)dx \right]^{\frac{q}{p}}$$

$$\leq c\widetilde{G}^q \left[\int_0^\infty \sum_{j\in\mathbb{Z}} \left(\int_0^{2^j} w^{1-p'}(x,y)dy \right)^{1-p} F_j^p(x)dx \right]^{\frac{q}{p}}.$$

On the other hand, by using the condition $w^{1-p'} \in RD(y)$ (see Lemma 1.6.1) we obtain that

$$\sup_{\substack{x>0\\ k\in\mathbb{Z}}} \sum_{j=k}^\infty \left(\int_0^{2^j} w^{1-p'}(x,y)dy \right)^{1-p} \left(\sum_{j=-\infty}^k \int_{2^j}^{2^{j+1}} w^{1-p'}(x,y)dy \right)^{p-1} < \infty.$$

Therefore Theorem 3.1.4 and Hölder's inequality yield that

$$I \leq c\widetilde{G}^q \left[\int_0^\infty \sum_{j\in\mathbb{Z}} \left(\int_0^{2^j} w^{1-p'}(x,y)dy \right)^{1-p} \left(\sum_{k=-\infty}^j \int_{2^k}^{2^{k+1}} f(x,t)dt \right)^p dx \right]^{\frac{q}{p}}$$

$$\le c\widetilde{G}^q \left[\int_0^\infty \sum_{k\in\mathbb{Z}} \left(\int_{2^k}^{2^{k+1}} w^{1-p'}(x,\tau)d\tau\right)^{1-p} \left(\int_{2^k}^{2^{k+1}} f(x,t)dt\right)^p dx\right]^{\frac{q}{p}}$$

$$\le c\widetilde{G}^q \left[\int_0^\infty \sum_{j\in\mathbb{Z}} \left(\int_{2^j}^{2^{j+1}} w^{1-p'}(x,y)dy\right)^{1-p}\right.$$

$$\left.\times \left(\int_{2^j}^{2^{j+1}} w(x,t)f^p(x,t)dt\right)\left(\int_{2^j}^{2^{j+1}} w^{1-p'}(x,t)dt\right)^{p-1} dx\right]^{q/p} \le c\widetilde{G}^q \|f\|^q_{L^p(\mathbb{R}^2_+)}.$$

The proof is complete. □

Proposition 3.4.2. *Let $1 < p \le q < \infty$ and let $\alpha \in \left[\frac{1}{p} - \frac{1}{q}, \frac{1}{p}\right) \cup \left(\frac{1}{p}, 1\right)$. Suppose that $w(x,y) = w_1(x)w_2(y)$ and that one of the following conditions holds:*

(a) *v and w are monotonic in the first variable for the second fixed one, and $\alpha \in \left[\frac{1}{p} - \frac{1}{q}, \frac{1}{p}\right)$;*

(b) *w_1 is increasing and $\alpha \in \left(\frac{1}{p}, 1\right)$;*

(c) *w_1 is decreasing, $w_1 \in DC(\mathbb{R}_+)$ and $\alpha \in \left(\frac{1}{p}, 1\right)$.*

Then $\mathcal{R}_\alpha II$ is bounded from $L^p_w(\mathbb{R}^2_+)$ to $L^q_u(\mathbb{R}^2_+)$ if and only if (3.4.1) holds.

Proof. *Sufficiency.* (*a*) First assume that $S := \int\limits_0^\infty w_1^{1-p'} = \infty$. Let us take the sequence of positive numbers $\{x_k\}_{k=-\infty}^{+\infty}$ so that

$$2^k = \int_0^{x_k} w_2^{1-p'}(t)dt, \quad k \in \mathbb{Z}. \tag{3.4.2}$$

It is obvious that $\{x_k\}$ is increasing, $\mathbb{R}_+ = \bigcup\limits_{k\in\mathbb{Z}} [x_k, x_{k+1})$ and

$$2^k = \int_{x_k}^{x_{k+1}} w_2^{1-p'}(t)dt.$$

Let $f \ge 0$. We have that

$$I := \|\mathcal{R}_\alpha f\|^q_{L^q_v(R^2_+)} = \sum_{k\in\mathbb{Z}} \int_0^\infty \int_{x_k}^{x_{k+1}} u(x,y) \left(\int_0^x \int_0^y \frac{f(t,\tau)}{(x-t)^{1-\alpha}} dt d\tau\right)^q dxdy$$

$$\leq \sum_{k\in\mathbb{Z}} \left(\int_{x_k}^{x_{k+1}} u(x,y)dy \right) \left(\int_0^x (x-t)^{\alpha-1} \left(\int_0^{x_{k+1}} f(t,\tau)d\tau \right) dt \right)^q dx$$

$$= \sum_{k\in\mathbb{Z}} \int_0^\infty \overline{u}_k(x) \left(\int_0^x \frac{F_k(t)dt}{(x-t)^{1-\alpha}} \right)^q dx,$$

where

$$\overline{u}_k(x) := \int_{x_k}^{x_{k+1}} u(x,y)dy; \quad F_k(t) := \int_0^{x_{k+1}} f(t,\tau)d\tau.$$

By Proposition 3.1.1 and condition (3.4.1) we conclude that

$$I \leq c\widetilde{G}^q \sum_{j\in\mathbb{Z}} \left[\int_0^\infty w_1(x) \left(\int_0^{x_j} w_2^{1-p'}(y)dy \right)^{1-p} (F_j(x))^p \, dx \right]^{q/p}$$

$$\leq c\widetilde{G}^q \left[\int_0^\infty w_1(x) \sum_{j\in\mathbb{Z}} \left(\int_0^{x_j} w_2^{1-p'}(y)dy \right)^{1-p} \left(\sum_{k=-\infty}^{j} \int_{x_k}^{x_{k+1}} f(x,\tau)d\tau \right)^p dx \right]^{\frac{q}{p}}.$$

On the other hand, (3.4.2) yields that

$$\sum_{k=n}^{\infty} \left(\int_0^{x_k} w_2^{1-p'}(y)dy \right)^{1-p} \left(\sum_{k=-\infty}^{n} \int_{x_k}^{x_{k+1}} w_2^{1-p'}(y)dy \right)^{p-1} \leq c.$$

Hence, by Theorem 3.1.4 and Hölder's inequality we conclude that

$$I \leq c\widetilde{G}^q \Bigg[\int_0^\infty w_1(x) \sum_{j\in\mathbb{Z}} \left(\int_{x_j}^{x_{j+1}} w_2^{1-p'}(y)dy \right)^{1-p} \left(\int_{x_j}^{x_{j+1}} f(x,\tau)d\tau \right)^p dx \Bigg]^{q/p}$$

$$\leq c\widetilde{G}^q \left[\int_0^\infty w_1(x) \sum_{j\in\mathbb{Z}} \left(\int_{x_j}^{x_{j+1}} w_2(\tau) f^p(x,\tau)d\tau \right) dx \right]^{\frac{q}{p}} = c\widetilde{G}^q \|f\|^q_{L^p_w(\mathbb{R}_+)}.$$

If $S < \infty$, then without loss of generality we can assume that $S = 1$. In this case we choose the sequence $\{x_k\}_{k=-\infty}^0$ for which (3.4.2) holds for all $k \leq 0$. Arguing as in the case $S = \infty$ and using Theorem 1.1.5 instead of Theorem 1.1.4 we obtain sufficiency also in this case.

Necessity can be derived in the same way as in the proof of Proposition 3.4.1, and the proof is complete. □

Theorem 3.4.6. *Let* $1 < p \le q < \infty$ *and let* $\alpha, \beta \in \left[\frac{1}{p} - \frac{1}{q}, \frac{1}{p}\right) \cup \left(\frac{1}{p}, 1\right)$ *with* $\alpha, \beta \neq 0$. *Suppose that* v *and* w *are weights on* $\mathbb{R}^2_+$. *Let* $w^{1-p'} \in DC(x) \cap DC(y)$ *and let one of the following conditions hold:*

(a) *v and w are monotonic in each variable for another fixed one, and* $\alpha, \beta \in \left[\frac{1}{p} - \frac{1}{q}, \frac{1}{p}\right)$;

(b) *w is monotonic in each variable for another fixed one, and* $\alpha, \beta \in \left(\frac{1}{p}, 1\right)$.

(c) *v is monotonic in the first variable for the second fixed one, w is monotonic in each variable for another fixed one,* $\alpha \in \left[\frac{1}{p} - \frac{1}{q}, \frac{1}{p}\right)$ *and* $\beta \in \left(\frac{1}{p}, 1\right)$.

(d) *v is monotonic in the second variable for the first fixed one, w is monotonic in each variable for another fixed one,* $\beta \in \left[\frac{1}{p} - \frac{1}{q}, \frac{1}{p}\right)$ *and* $\alpha \in \left(\frac{1}{p}, 1\right)$.

Then $\mathcal{R}_{\alpha,\beta}$ *is bounded from* $L^p_w(\mathbb{R}^2_+)$ *to* $L^q_v(\mathbb{R}^2_+)$ *if and only if*

$$H := \sup_{a,b>0} \left[\int_a^\infty \int_b^\infty \frac{v(x,y)dxdy}{x^{(1-\alpha)q} y^{(1-\beta)q}} \right]^{1/q} \left[\int_0^a \int_0^b w^{1-p'}(x,y)dxdy \right]^{1/p'} < \infty. \qquad (3.4.3)$$

Proof. (a) Suppose that v and w are increasing in each variable for another fixed one (other cases can be proved in a similar manner). Let $f \ge 0$. We have that

$$\|\mathcal{R}_{\alpha,\beta} f\|^q_{L^q_v(\mathbb{R}^2_+)} \le c \left[\int_0^\infty \int_0^\infty v(x,y) \left(\int_0^x \int_0^{y/2} \frac{f(t,\tau)dtd\tau}{(x-t)^{1-\alpha}(y-\tau)^{1-\beta}} \right)^q dxdy \right.$$

$$\left. + \int_0^\infty \int_0^\infty v(x,y) \left(\int_0^x \int_{y/2}^y \frac{f(t,\tau)dtd\tau}{(x-t)^{1-\alpha}(y-\tau)^{1-\beta}} \right)^q dxdy \right] := c(I_1 + I_2)$$

It is easy to see that Proposition 3.4.1 yields that

$$I_1 \le c \left[\int_0^\infty \int_0^\infty \frac{v(x,y)}{y^{(1-\beta)q}} \left(\int_0^x \int_0^y \frac{f(t,\tau)dtd\tau}{(x-t)^{1-\alpha}} \right)^q dxdy \right] \le cH^q \|f\|^q_{L^p_w(\mathbb{R}^2_+)}.$$

For I_2 we obtain that

$$I_2 \le c \left[\int_0^\infty \int_0^\infty v(x,y) \left(\int_0^{x/2} \int_{y/2}^y \cdots \right)^q dxdy + \int_0^\infty \int_0^\infty v(x,y) \left(\int_{x/2}^x \int_{y/2}^y \cdots \right)^q dxdy \right]$$

$$=: c[I_{21} + I_{22}].$$

For I_{21} we find that

$$I_{21} \le c \int_0^\infty \int_0^\infty \frac{\widetilde{v}(y,x)}{x^{(1-\alpha)q}} \left(\int_0^y \int_0^x \frac{\widetilde{f}(\tau,t)}{(y-\tau)^{1-\beta}} d\tau dt \right)^q dydx,$$

where $\widetilde{v}(y,x) = v(x,y)$, $\widetilde{f}(\tau,t) = f(t,\tau)$.

Therefore, by Proposition 3.4.1 we conclude that

$$I_{21} \le cH^q \|f\|^q_{L^p_w(\mathbb{R}^2_+)}.$$

It remains to estimate I_{22}. First we introduce the numbers $p_1^* = \frac{p}{1-\alpha p}$, $p_2^* = \frac{p}{1-\beta p}$. Then $\left(\frac{p_1^*}{q}\right)' = \frac{p}{p-q+\alpha pq}$, $\left(\frac{p_2^*}{q}\right)' = \frac{p}{p-q+\beta pq}$. Further, it is easy to see that

$$H \ge cv^{1/q}(4a,4b)w^{-1/p}(a,b)a^{\alpha+1/q-1/p}b^{\beta+1/q-1/p}, \tag{3.4.4}$$

where the positive constant c depends only on α, β, p and q.

Using the notation:

$$f_{k,j}(x,y) := f(x,y)\chi_{[2^{k-1},2^{k+1})}(x)\chi_{[2^{j-1},2^{j+1})}(y),$$

$$F_{k,j,x}(\tau) := \int_0^x \frac{f_{k,j}(t,\tau)}{(x-t)^{1-\alpha}} dt,$$

$$\widetilde{F}_{k,j}(t) := \int_0^\infty f^p_{k,j}(t,y)dy,$$

$$\mathbb{V}_j(x) := \left(\int_{2^j}^{2^{j+1}} v^{(p_2^*/q)'}(x,y)dy \right)^{\frac{1}{\left(\frac{p_2^*}{q}\right)'}},$$

we have that

$$I_{22} = \iint_{\mathbb{R}^2} v(x,y) \left(\int_{\frac{x}{2}}^x \int_{\frac{y}{2}}^y \frac{f(t,\tau)dtd\tau}{(x-t)^{1-\alpha}(y-\tau)^{1-\beta}} \right)^q dxdy$$

$$\le \sum_{k,j} \int_{2^k}^{2^{k+1}} \int_{2^j}^{2^{j+1}} v(x,y) \left(\int_0^x \int_0^y \frac{f_{k,j}(t,\tau)dtd\tau}{(x-t)^{1-\alpha}(y-\tau)^{1-\beta}} \right)^q dxdy$$

$$= \sum_{k,j} \int_{2^k}^{2^{k+1}} \left(\int_{2^j}^{2^{j+1}} v(x,y) \left(\int_0^y (y-\tau)^{\beta-1} \left(\int_0^x \frac{f_{k,j}(t,\tau)dt}{(x-t)^{1-\alpha}} \right) d\tau \right)^q dy \right) dx$$

$$\left[\text{by Hölder's inequality with the exponent } \frac{p_2^*}{q}\right]$$

$$\leq \sum_{k,j} \int_{2^k}^{2^{k+1}} \left(\int_0^\infty \left(\int_0^y (y-\tau)^{\beta-1} F_{k,j,x}(\tau) d\tau \right)^{p_2^*} dy \right)^{\frac{q}{p_2^*}} \left(\int_{2^j}^{2^{j+1}} v_2^{(p^*/q)'}(x,y) dy \right)^{1/(p_2^*/q)'} dx$$

$$\left[\text{applying Theorem 3.1.2}\right]$$

$$\leq c \sum_{k,j} \int_{2^k}^{2^{k+1}} \left(\int_0^\infty (F_{k,j,x}(y))^p \, dy \right)^{\frac{q}{p}} \mathbb{V}_j(x) dx$$

$$= c \sum_{k,j} \int_{2^k}^{2^{k+1}} \left(\int_0^\infty \left(\int_0^x \frac{f_{k,j}(t,y)dt}{(x-t)^{1-\alpha}} \right)^p dy \right)^{\frac{q}{p}} \mathbb{V}_j(x) dx$$

$$\left[\text{by Minkowski's integral inequality}\right]$$

$$\leq c \sum_{k,j} \int_{2^k}^{2^{k+1}} \left(\int_0^x (x-t)^{\alpha-1} \left(\int_0^\infty f_{k,j}^p(t,y) dy \right)^{\frac{1}{p}} dt \right)^q \mathbb{V}_j(x) dx$$

$$\left[\text{using Hölder's inequality with the exponent } \frac{p_1^*}{q}\right]$$

$$\leq \sum_{k,j} \left(\int_{2^k}^{2^{k+1}} \left(\int_0^x (x-t)^{\alpha-1} \widetilde{F}_{k,j}(t) dt \right)^{p_1^*} dx \right)^{\frac{q}{p_1^*}} \left(\int_{2^k}^{2^{k+1}} (\mathbb{V}_j(x))^{\left(\frac{p_1^*}{q}\right)'} dx \right)^{\frac{1}{\left(\frac{p_1^*}{q}\right)'}}$$

$$\left[\text{due to Theorem 3.1.2}\right]$$

$$\leq c \sum_{k,j} \left(\int_0^\infty \int_0^\infty f_{k,j}^p(x,y) dx dy \right)^{\frac{q}{p}} \left(\int_{2^k}^{2^{k+1}} (\mathbb{V}_j(x))^{\left(\frac{p_1^*}{q}\right)'} dx \right)^{\frac{1}{\left(\frac{p_1^*}{q}\right)'}}.$$

Now observe that the monotonicity of $v(x,y)$ in each variable and (3.4.4) yield that

$$S_{k,j} = \left(\int_{2^k}^{2^{k+1}} \left(\int_{2^j}^{2^{j+1}} v^{(p_2^*/q)'}(x,y) dy \right)^{(p_1^*/q)'/(p_2^*/q)'} dx \right)^{1/(p_1^*/q)'}$$

$$\leq v(2^{k+1}, 2^{j+1}) \, 2^{k/(p_1^*/q)'} 2^{j/(p_2^*/q)'}$$

$$\leq v(2^{k+1},2^{j+1})\,2^{k(1-q/p+\alpha q)}2^{j(1-q/p+\beta q)}\leq cw^{q/p}(2^{k-1},2^{j-1}).$$

Hence, since w is increasing in each variable we obtain that

$$I_{22}\leq c\sum_{k,j}w^{q/p}\left(2^{k-1},2^{j-1}\right)\left(\int\limits_{2^{k-1}}^{2^{k+1}}\int\limits_{2^{j-1}}^{2^{j+1}}f^p(x,y)dxdy\right)^{q/p}\leq c\|f\|^q_{L^p_w(\mathbb{R}^2_+)}.$$

(b) The only difference between the cases (a) and (b) is the estimate of I_{22}. Let us assume that w is increasing in both variables. Due to Hölder's inequality and the monotonicity of w we have that

$$I_{22}=\int\limits_0^\infty\int\limits_0^\infty\left(\int\limits_{\frac{x}{2}}^{x}\int\limits_{\frac{y}{2}}^{y}\frac{f(t,\tau)dtd\tau}{(x-t)^{1-\alpha}(y-\tau)^{1-\beta}}\right)^q dxdy$$

$$\leq\int\limits_0^\infty\int\limits_0^\infty v(x,y)\left(\int\limits_{\frac{x}{2}}^{x}\int\limits_{\frac{y}{2}}^{y}f^p(t,\tau)w(t,\tau)dtd\tau\right)^{\frac{q}{p}}$$

$$\times\left(\int\limits_{\frac{x}{2}}^{x}\int\limits_{\frac{y}{2}}^{y}\frac{w^{1-p'}(t,\tau)dtd\tau}{(x-t)^{(1-\alpha)p'}(y-\tau)^{(1-\beta)p'}}\right)^{\frac{q}{p'}}dxdy$$

$$\leq c\sum_{k,j}\int\limits_{2^k}^{2^{k+1}}\int\limits_{2^j}^{2^{j+1}}v(x,y)\left(\int\limits_{\frac{x}{2}}^{x}\int\limits_{\frac{y}{2}}^{y}f^p(t,\tau)w(t,\tau)dtd\tau\right)^{\frac{q}{p}}$$

$$\times w^{-\frac{q}{p}}\left(\frac{x}{2},\frac{y}{2}\right)\,x^{q\frac{(\alpha-1)p'+1}{p'}}y^{q\frac{(\beta-1)p'+1}{p'}}dxdy$$

$$\leq\sum_{k,j}\left(\int\limits_{2^{k-1}}^{2^{k+1}}\int\limits_{2^{j-1}}^{2^{j+1}}f^p(t,\tau)w(t,\tau)dtd\tau\right)^{q/p}\left(\int\limits_{2^k}^{2^{k+1}}\int\limits_{2^j}^{2^{j+1}}v(x,y)x^{q(\alpha-1)}y^{q(\beta-1)}dxdy\right)$$

$$\times w^{-\frac{q}{p}}(2^{k-1},2^{j-1})\,2^{k\frac{q}{p'}}2^{j\frac{q}{p'}}\leq cH^q\sum_{k,j}\left(\int\limits_{2^{k-1}}^{2^{k+1}}\int\limits_{2^{j-1}}^{2^{j+1}}f^p(t,\tau)w(t,\tau)dtd\tau\right)^{q/p}$$

$$\leq cH^q\|f\|^q_{L^p_w(\mathbb{R}^2_+)}.$$

(c) Suppose that v is increasing in the first variable for the second fixed one and w is increasing in each variable. We notice that I_1 and I_2 can be estimated in the same manner as is the case (a).

Let us introduce the notation:

$$E(y,t) := \left(\int\limits_{y/2}^{y} f^p(t,\tau)d\tau \right)^{1/p} ; \quad E_k(y,t) = E(y,t) \cdot \chi_{[2^{k-1},2^{k+1})}(t).$$

For I_{22}, we have that

$$I_{22} = \int\limits_0^\infty \int\limits_0^\infty v(x,y) \left(\int\limits_{\frac{x}{2}}^{x} \int\limits_{y/2}^{y} \frac{f(t,\tau)dtd\tau}{(x-t)^{1-\alpha}(y-\tau)^{1-\beta}} \right)^q dxdy$$

$$\left[\text{by Hölder's inequality and the fact } \frac{1}{p} < \beta < 1\right]$$

$$\leq c \int\limits_0^\infty \int\limits_0^\infty v(x,y) y^{(\beta-1/p)q} \left(\int\limits_{\frac{x}{2}}^{x} (x-t)^{\alpha-1} \left(\int\limits_{y/2}^{y} f^p(t,\tau)d\tau \right)^{1/p} dt \right)^q dxdy$$

$$= c \int\limits_0^\infty y^{(\beta-1/p)q} \left(\int\limits_0^\infty v(x,y) \left(\int\limits_{\frac{x}{2}}^{x} (x-t)^{\alpha-1} E(y,t)dt \right)^q dx \right) dy$$

$$\leq c \int\limits_0^\infty y^{(\beta-1/p)q} \left(\int\limits_{2^k}^{2^{k+1}} v(x,y) \left(\int\limits_0^{x} (x-t)^{\alpha-1} E_k(y,t)dt \right)^q dx \right) dy$$

$$\left[\text{applying the monotonicity of } v, \text{ one–dimensional Hölder's inequality}\right.$$

$$\left.\text{with the exponent } p_1^*/q,\ p_1^* = \frac{p}{1-\alpha p}, \text{ and Theorem 3.1.2}\right]$$

$$c \sum_{k\in\mathbb{Z}} \int\limits_0^\infty y^{(\beta-1/p)q} v(2^{k+1},y) 2^{k(1-q/p+\alpha q)} \left[\int\limits_{2^{k-1}}^{2^{k+1}} F^p(y,x)dx \right]^{q/p} dy$$

$$= c \sum_{k,j\in\mathbb{Z}} \int\limits_{2^j}^{2^{j+1}} y^{(\beta-1/p)q} v(2^{k+1},y) 2^{k(1-q/p+\alpha q)} \left[\int\limits_{2^{k-1}}^{2^{k+1}} \int\limits_{y/2}^{y} f^p(x,\tau)dxd\tau \right]^{\frac{q}{p}} dy.$$

On the other hand,

$$H^q \geq \left(\int\limits_a^\infty\int\limits_b^\infty \frac{v(x,y)dxdy}{x^{(1-\alpha)q}y^{(1-\beta)q}}\right)\left(\int\limits_0^a\int\limits_0^b w^{1-p'}(x,y)dxdy\right)^{q/p'}$$

$$\geq c\left(\int\limits_{4a}^{8a}\int\limits_{2b}^{4b} v(x,y)dxdy\right) a^{(\alpha-1)q}b^{(\beta-1)q}w^{-q/p}(a,b)(ab)^{q/p'}.$$

Hence,

$$\int\limits_{2a}^{4a}\int\limits_{b}^{2b} v(x,y)dxdy \leq cw^{q/p}\left(\frac{a}{2},\frac{b}{2}\right)a^{(\alpha-1/p)q}b^{(\beta-1/p)q}$$

for all $a, b > 0$. Consequently,

$$I_{22} \leq c\sum_{k,j\in\mathbb{Z}}\left(\int\limits_{2^j}^{2^{j+1}} y^{(\beta-1/p)q}v(2^{k+1},y)2^{k(1-q/p+\alpha q)}dy\right)\left[\int\limits_{2^{k-1}}^{2^{k+1}}\int\limits_{2^{j-1}}^{2^{j+1}} f^p(x,\tau)dxd\tau\right]^{q/p}$$

$$\leq c\sum_{k,j\in\mathbb{Z}}\left(\int\limits_{2^{k+1}}^{2^{k+2}}\int\limits_{2^j}^{2^{j+1}} v(x,y)dxdy\right)2^{k(\alpha-1/p)q}2^{j(\beta-1/p)q}\left[\int\limits_{2^{k+1}}^{2^{k+2}}\int\limits_{2^{j-1}}^{2^{j+1}} f^p(x,\tau)dxd\tau\right]^{q/p}$$

$$\leq c\sum_{k,j\in\mathbb{Z}}\left(w^{q/p}(2^{k-1},2^{j-1})\int\limits_{2^{k+1}}^{2^{k+2}}\int\limits_{2^{j-1}}^{2^{j+1}} f^p(x,\tau)dxd\tau\right)^{q/p}$$

[using the monotonicity of w in each variable]

$$\leq c\iint\limits_{\mathbb{R}^2} f^p(x,\tau)w(x,\tau)dxd\tau$$

and (c) is also proved.

(d) This case is similar to that of (c); so we omit the details. We only remark that *necessity* can be obtained by putting the test function

$$f(x,y) = \chi_{(0,a)\times(0,b)}(x,y)w^{1-p'}(x,y)$$

in the two–weight inequality for the operator $\mathcal{R}_{\alpha,\beta}$. □

The proofs of the next statements are similar to that of the previous one.

Theorem 3.4.7. *Let $1 < p \le q < \infty$ and let $\alpha, \beta \in \left[\frac{1}{p} - \frac{1}{q}, \frac{1}{p}\right) \cup \left(\frac{1}{p}, 1\right)$. Suppose that a two-dimensional weight $w(x,y)$ has the form $w(x,y) = w_1(x)w_2(y)$. Let one of the following conditions hold:*

(a) *$\alpha, \beta \in \left[\frac{1}{p} - \frac{1}{q}, \frac{1}{p}\right)$, v is monotonic in each variable for another fixed one, and w_1 and w_2 are monotonic weights on $\mathbb{R}_+$;*

(b) *$\alpha, \beta \in \left(\frac{1}{p}, 1\right)$, w_1 and w_2 are monotonic, and if w_i is decreasing for some $i = \{1,2\}$, then we suppose that $w_i \in DC(\mathbb{R})$;*

(c) *$\alpha \in \left[\frac{1}{p} - \frac{1}{q}, \frac{1}{p}\right)$, $\beta \in \left(\frac{1}{p}, 1\right)$, v is monotonic in the first variable, w_1 and w_2 are monotonic, and if w_2 is decreasing, then we assume that $w_2 \in DC(\mathbb{R})$;*

(d) *$\beta \in \left[\frac{1}{p} - \frac{1}{q}, \frac{1}{p}\right)$, $\alpha \in \left(\frac{1}{p}, 1\right)$, v is monotonic in the second variable, w_1 and w_2 are monotonic, and if w_1 is decreasing, then we assume, in addition, that $w_1 \in DC(\mathbb{R})$.*

Then $\mathcal{R}_{\alpha,\beta}$ is bounded from $L^p_w(\mathbb{R}^2_+)$ to $L^q_v(\mathbb{R}^2_+)$ if and only if (3.4.3) holds.

3.5 One–weight Inequalities

In this section we establish one-weight criteria for the following operators: $M^{+,+}_{\alpha,\alpha}$, $M^{-,-}_{\alpha,\alpha}$, $M^{+,-}_{\alpha,\alpha}$, $M^{-,+}_{\alpha,\alpha}$, $W_{\alpha,\alpha}$, $R_{\alpha,\alpha}$, $W_\alpha R_\alpha$, $R_\alpha W_\alpha$.

Our main result of this section is the following statement.

Theorem 3.5.1. *Let $1 < p < \infty$ and let $0 < \alpha < 1/p$. We set $p^* = \frac{p}{1-\alpha p}$. Then the following statements are equivalent:*

(i) *$M^{+,+}_{\alpha,\alpha}$ is bounded from $L^p_{u^p}(\mathbb{R}^2)$ to $L^{p^*}_{u^{p^*}}(\mathbb{R}^2)$;*

(ii) *$W_{\alpha,\alpha}$ is bounded from $L^p_{u^p}(\mathbb{R}^2)$ to $L^{p^*}_{u^{p^*}}(\mathbb{R}^2)$;*

(iii) *the next two conditions hold simultaneously*

$$(a)\quad s^{\alpha-1}\left(\int\limits_{b-s}^{b} u^{p^*}(x,y)dy\right)^{1/p^*}\left(\int\limits_{b}^{b+s} u^{-p'}(x,y)dy\right)^{1/p'} \le c \tag{3.5.1}$$

for all $b \in \mathbb{R}$, $s > 0$ and a.e. $x \in \mathbb{R}$,

$$(b)\quad h^{\alpha-1}\left(\int\limits_{a-h}^{a} u^{p^*}(x,y)dx\right)^{1/p^*}\left(\int\limits_{a}^{a+h} u^{-p'}(x,y)dx\right)^{1/p'} \le c \tag{3.5.2}$$

for all $a \in \mathbb{R}$, $h > 0$ and a.e. $y \in \mathbb{R}$;

(iv)

$$\sup_{\substack{a,b\in\mathbb{R}\\ h,s>0}} \frac{1}{(hs)^{1-\alpha}} \left(\int_{a-h}^{a} \int_{b-s}^{b} u^{p^*}(x,y)dxdy \right)^{\frac{1}{p^*}} \left(\int_{a}^{a+h} \int_{b}^{b+s} u^{-p'}(x,y)dxdy \right)^{\frac{1}{p'}} < \infty. \quad (3.5.3)$$

Proof. Let $f \geq 0$. By the inequality

$$(M^{+,+}_{\alpha,\alpha} f)(x,y) \leq (W_{\alpha,\alpha} f)(x,y), \qquad f \geq 0,$$

it follows the implication (ii) $\Longrightarrow$ (i).

Let us now show that (iii) $\Longrightarrow$ (ii). Indeed, using Theorem 3.1.9 twice and Minkowski's integral inequality we find that

$$\left(\iint_{\mathbb{R}^2} (W_{\alpha,\alpha} f)^{p^*}(x,y) u^{p^*}(x,y) dxdy \right)^{\frac{1}{p^*}}$$

$$= \left(\int_{\mathbb{R}} \left(\int_{\mathbb{R}} u^{p^*}(x,y) \left(\int_{x}^{\infty} (t-x)^{\alpha-1} \left(\int_{y}^{\infty} (\tau-y)^{\alpha-1} f(t,\tau) d\tau \right) dt \right)^{p^*} dx \right) dy \right)^{\frac{1}{p^*}}$$

$$\leq c \left(\int_{\mathbb{R}} \left(\int_{\mathbb{R}} u^{p}(x,y) \left(\int_{y}^{\infty} (\tau-y)^{\alpha-1} f(x,\tau) d\tau \right)^{p} dx \right)^{\frac{p^*}{p}} dy \right)^{\frac{1}{p^*}}$$

$$\leq c \left(\int_{\mathbb{R}} \left(\int_{\mathbb{R}} u^{p^*}(x,y) \left(\int_{y}^{\infty} (\tau-y)^{\alpha-1} f(x,\tau) d\tau \right)^{p^*} dy \right)^{\frac{p}{p^*}} dx \right)^{\frac{1}{p}}$$

$$\leq c \left(\iint_{\mathbb{R}^2} u^{p}(x,y) f^{p}(x,y) dxdy \right)^{1/p}.$$

Next, we show that (i) $\Longrightarrow$ (iv).

First note that the boundedness of $M^{+,+}_{\alpha,\alpha}$ from $L^p_{u^p}(\mathbb{R}^2)$ to $L^{p^*}_{u^{p^*}}(\mathbb{R}^2)$ implies that u^{p^*}, $u^{-p'} \in L_{\rm loc}(\mathbb{R}^2)$. Further,

$$\left(\iint_{\mathbb{R}^2} (M^{+,+}_{\alpha,\alpha} f)^{p^*}(x,y) u^{p^*}(x,y) dxdy \right)^{1/p^*}$$

$$\geq \left(\int_{a-h}^{a} \int_{b-s}^{b} \left(\frac{1}{(2h2s)^{1-\alpha}} \int_{x}^{x+2h} \int_{y}^{y+2s} f(t,\tau)dtd\tau \right)^{p^*} u^{p^*}(x,y)dxdy \right)^{1/p^*}$$

$$\geq c(hs)^{\alpha-1} \left(\int_{a-h}^{a} \int_{b-s}^{b} u^{p^*}(x,y)dxdy \right)^{1/p^*} \left(\int_{a}^{a+h} \int_{b}^{b+s} f(x,y)dxdy \right).$$

Now substituting $f(x,y) = u^{1-p'}(x,y)\chi_{(a,a+h)\times(b,b+s)}(x,y)$ in the one-weight inequality we find that (3.5.3) holds. It remains to show that (iv) $\Longrightarrow$ (iii). This follows by the Lebesgue differentiation theorem with respect to each variable for another fixed one and the condition $\alpha = 1/p - 1/p^*$. The proof is complete. □

Analogously, it follows the next three statements:

Theorem 3.5.2. *Let $1 < p < \infty$, $0 < \alpha < 1/p$. Suppose that $p^* = \frac{p}{1-\alpha p}$. Then the following conditions are equivalent:*

(i) *$M^{-,-}_{\alpha,\alpha}$ is bounded from $L^p_{u^p}(\mathbb{R}^2)$ to $L^{p^*}_{u^{p^*}}(\mathbb{R}^2)$;*

(ii) *$R_{\alpha,\alpha}$ is bounded from $L^p_{u^p}(\mathbb{R}^2)$ to $L^{p^*}_{u^{p^*}}(\mathbb{R}^2)$;*

(iii) *the next two conditions hold simultaneously:*

(a)

$$s^{\alpha-1} \left(\int_{b}^{b+s} u^{p^*}(x,y)dy \right)^{1/p^*} \left(\int_{b-s}^{b} u^{-p'}(x,y)dy \right)^{1/p'} < c$$

for all $b \in \mathbb{R}$, $s > 0$ and a.e. $x \in \mathbb{R}$,

(b)

$$h^{\alpha-1} \left(\int_{a}^{a+h} u^{p^*}(x,y)dx \right)^{1/p^*} \left(\int_{a-h}^{a} u^{-p'}(x,y)dx \right)^{1/p'} \leq c$$

for all $a \in \mathbb{R}$, $h > 0$ and a.e $y \in \mathbb{R}$;

(iv)

$$\sup_{\substack{a,b\in\mathbb{R} \\ h,s>0}} \frac{1}{(hs)^{1-\alpha}} \left(\int_{a}^{a+h} \int_{b}^{b+s} u^{p^*}(x,y)dxdy \right)^{\frac{1}{p^*}} \left(\int_{a-h}^{a} \int_{b-s}^{b} u^{-p'}(x,y)dxdy \right)^{\frac{1}{p'}} < \infty.$$

Theorem 3.5.3. *Let $1 < p < \infty$, $0 < \alpha < 1/p$. $p^* = \frac{p}{1-\alpha p}$. Then the following conditions are equivalent:*

(i) *$M^{+,-}_{\alpha,\alpha}$ is bounded from $L^p_{u^p}(\mathbb{R}^2)$ to $L^{p^*}_{u^{p^*}}(\mathbb{R}^2)$;*

(ii) *$W_\alpha R_\alpha$ is bounded from $L^p_{u^p}(\mathbb{R}^2)$ to $L^{p^*}_{u^{p^*}}(\mathbb{R}^2)$;*

(iii) *the next two conditions hold simultaneously:*

(a)

$$s^{\alpha-1}\left(\int\limits_{b}^{b+s} u^{p^*}(x,y)dy\right)^{1/p^*}\left(\int\limits_{b-s}^{b} u^{-p'}(x,y)dy\right)^{1/p'} \leq c$$

for all $b \in \mathbb{R}$, $s > 0$ *and a.e.* $x \in \mathbb{R}$;

(b)

$$h^{\alpha-1}\left(\int\limits_{a-h}^{a} u^{p^*}(x,y)dx\right)^{1/p^*}\left(\int\limits_{a}^{a+h} u^{-p'}(x,y)dx\right)^{1/p'} \leq c$$

for all $a \in \mathbb{R}$, $h > 0$ *and a.e.* $y \in \mathbb{R}$;

(iv)

$$\sup_{\substack{a,b\in\mathbb{R}\\ h,s>0}} \frac{1}{(hs)^{1-\alpha}}\left(\int\limits_{a-h}^{a}\int\limits_{b}^{b+s} u^{p^*}(x,y)dxdy\right)^{\frac{1}{p^*}}\left(\int\limits_{a}^{a+h}\int\limits_{b-s}^{b} u^{-p'}(x,y)dxdy\right)^{\frac{1}{p'}} < \infty.$$

Theorem 3.5.4. *Let* $1 < p < \infty$, $0 < \alpha < 1/p$. $p^* = \frac{p}{1-\alpha p}$. *Then the following statements are equivalent:*

(i) $M_{\alpha,\alpha}^{-,+}$ *is bounded from* $L^p_{u^p}(\mathbb{R}^2)$ *to* $L^{p^*}_{u^{p^*}}(\mathbb{R}^2)$;

(ii) $R_\alpha W_\alpha$ *is bounded from* $L^p_{u^p}(\mathbb{R}^2)$ *to* $L^{p^*}_{u^{p^*}}(\mathbb{R}^2)$;

(iii) *the next two conditions hold simultaneously:*

(a)

$$s^{\alpha-1}\left(\int\limits_{b-s}^{b} u^{p^*}(x,y)dy\right)^{1/p^*}\left(\int\limits_{b}^{b+s} u^{-p'}(x,y)dy\right)^{1/p'} \leq c$$

for all $b \in \mathbb{R}$, $s > 0$ *and a.e.* $x \in \mathbb{R}$;

(b)

$$h^{\alpha-1}\left(\int\limits_{a}^{a+h} u^{p^*}(x,y)dx\right)^{1/p^*}\left(\int\limits_{a-h}^{a} u^{-p'}(x,y)dx\right)^{1/p'} \leq c$$

for all $a \in \mathbb{R}$, $h > 0$ *and a.e.* $y \in \mathbb{R}$;

(iv)

$$\sup_{\substack{a,b\in\mathbb{R}\\ h,s>0}} (hs)^{1-\alpha}\left(\int\limits_{a}^{a+h}\int\limits_{b-s}^{b} u^{p^*}(x,y)dxdy\right)^{\frac{1}{p^*}}\left(\int\limits_{a-h}^{a}\int\limits_{b}^{b+s} u^{-p'}(x,y)dxdy\right)^{\frac{1}{p'}} < \infty.$$

In particular, these results imply the following corollary:

Corollary 3.5.1. *Let* $1 < p < \infty$ *and* $0 < \alpha < 1/p$. $p^* = \frac{p}{1-\alpha p}$. *Then the following statements hold:*

(i) *if* u *is increasing in each variable for another fixed one, then* $W_{\alpha,\alpha}$ *is bounded from* $L^p_{u^p}(\mathbb{R}^2)$ *to* $L^{p^*}_{u^{p^*}}(\mathbb{R}^2)$;

(ii) *if* u *is decreasing in each variable for another fixed one, then* $R_{\alpha,\alpha}$ *is bounded from* $L^p_{u^p}(\mathbb{R}^2)$ *to* $L^{p^*}_{u^{p^*}}(\mathbb{R}^2)$;

(iii) *if* u *is increasing in the first variable for another fixed one and is decreasing in the second variable for the first fixed one, then* $W_\alpha R_\alpha$ *is bounded from* $L^p_{u^p}(\mathbb{R}^2)$ *to* $L^{p^*}_{u^{p^*}}(\mathbb{R}^2)$;

(iv) *if* u *is decreasing in the first variable for the second fixed one and is increasing in the second variable for the first fixed one, then* $R_\alpha W_\alpha$ *is bounded from* $L^p_{u^p}(\mathbb{R}^2)$ *to* $L^{p^*}_{u^{p^*}}(\mathbb{R}^2)$.

3.6 Weighted Strichartz Estimates for Semilinear Wave Equations

In this section we derive sufficient conditions on a pair of weights (V, W) on $\mathbb{R}^2_+$ governing the two–weight Strichartz estimate

$$\|V(t-|x|,t+|x|)\omega\|_{L^q(t\geq|x|)} \leq C\|W(t-|x|,t+|x|)F\|_{L^{q'}(t\geq|x|)}, \tag{3.6.1}$$

for the solution of the inhomogeneous wave equation

$$\begin{cases} \Box\omega(t,x) = F(t,x), & (t,x) \in \mathbb{R}^{1+n}_+ \\ 0 = \omega(0,\cdot) = \partial_t\omega(0,\cdot) \end{cases}. \tag{3.6.2}$$

Here $\Box = \frac{\partial^2}{\partial t^2} - \Delta_x$ denotes the D'Alemberian and n is odd.

Our main result is the following statement

Theorem 3.6.1. *Let* n *be odd and let* $\frac{2n}{n-1} < q \leq \frac{2(n+1)}{n-1}$. *Suppose that* F *is spherically symmetric function with supp* $F \subset \{(t,x) \in \mathbb{R}^{1+n}_+ : |x| < t\}$. *Assume that the two-dimensional weights* V *and* W *are increasing in each variable for another fixed one. In addition, suppose that* $W^{-q} \in DC(s) \cap DC(\tau)$ *or* $W(s,\tau) = W_1(s)W_2(\tau)$. *If* ω *solves* (3.6.2), *then the condition*

$$\sup_{a,b>0}\left(\int_a^\infty\int_b^\infty \frac{V^q(s,\tau)}{(s\tau)^{q(n-1)(1/2-1/q)}}dsd\tau\right)\left(\int_0^a\int_0^b W^{-q}(s,\tau)dsd\tau\right) < \infty \tag{3.6.3}$$

implies inequality (3.6.1) *with the constant* C *depending only on* V, W, q *and* n.

Proof. First notice that ω has the following form (see, e.g., [106])

$$\omega(t,r)=r^{-\frac{n-1}{2}}\int\limits_0^t\int\limits_{|t-r-s|}^{t+r-s}P_m(\mu)F(s,\rho)\rho^{\frac{n-1}{2}}d\rho ds,$$

where $P_m(\mu)$ are the Legendre polynomials of degree $m=\frac{n-3}{2}$ and $\mu=(r^2+\rho^2-(t-s)^2)/2r\rho$ satisfies $-1\le\mu\le 1$ in the domain of integration. By multiplying ω by $k(t,r)v(t-|x|,t+|x|)$ and integrating with respect to $dxdt=c_n drdt$, we see that it is enough to show that

$$\int\limits_0^\infty\int\limits_0^\infty\int\limits_0^t\int\limits_{|t-r-s|}^{t+r-s}\frac{|k(t,r)|r^{\frac{n-1}{q'}}|F(s,\rho)|\rho^{\frac{n-1}{q'}}v(t-r,t+r)d\rho dsdrdt}{(r\rho)^\gamma}\tag{3.6.4}$$

is bounded by $\|k\|_{L^{q'}}\|FW(t-|x|,t+|x|)\|_{L^{q'}}$, where $\gamma=(n+1)(1/2-1/q)$. Let us introduce the notation: $u=t+r$, $v=t-r$, $\xi=s-\rho$, $\eta=s+\rho$, $G(\xi,\eta)=F(s,\rho)\rho^{\frac{n-1}{q'}}W(s-\rho,s+\rho)$, $H(u,v)=|k(t,r)|r^{\frac{n-1}{q'}}$. Then it is clear that $u-v=2r$, $\xi-\eta=2\rho$. We want now to prove that

$$\iiiint\limits_{0<\eta\le v\le\xi\le u}\frac{G(\xi,\eta)H(u,v)V(u,v)}{(u-v)^\gamma(\xi-\eta)^\gamma W(\xi,\eta)}d\xi d\eta dudv\le c\|G\|_{L^{q'}}\|H\|_{L^{q'}}.$$

For the latter inequality it is enough to show that

$$\left(\int\limits_0^\infty\int\limits_0^\infty\left(\int\limits_0^v\int\limits_0^u G(\xi,\eta)(u-\xi)^{-\gamma}(v-\eta)^{-\gamma}d\xi d\eta\right)^q V(u,v)dudv\right)^{1/q}$$

$$\le c\left[\int\limits_0^\infty\int\limits_0^\infty W^{q'}(\xi,\eta)G^{q'}(\xi,\eta)d\xi d\eta\right]^{1/q'}.$$

Applying Theorems 3.4.6 and 3.4.7 we conclude that this inequality holds and the proof is complete. □

In a similar way follows the next statement.

Theorem 3.6.2. *Let n be odd and let $2<q<\frac{2n}{n-1}$. Assume that the function F is spherically symmetric and supp $F\subset\{(t,x)\in\mathbb{R}_+^{1+n}:|x|<t\}$. Assume also that the two-dimensional weight W is increasing in each variable for another fixed one. In addition, suppose that $W^{-q}\in DC(s)\cap DC(\tau)$ or $W(s,\tau)=W_1(s)W_2(\tau)$. If ω solves (3.6.2), then condition (3.6.3) implies inequality (3.6.1) with the constant C depending only on V, W, q and n.*

Now we give some corollaries of the statements formulated above:

Corollary 3.6.1. *Let n be odd and let $2 < q \le \frac{2(n+1)}{(n-1)}$. Suppose that* supp $F \subset \{(t,x) \in R_+^{1+n} : |x| < t\}$. *If ω solves (3.6.2), then*

$$\|(t^2 - |x|^2)^{-\alpha}\omega\|_{L^q(t \ge |x|)} \le C_\gamma \|(t^2 - |x|^2)^{\beta} F\|_{L^{q'}(t \ge |x|)},$$

where $\beta < 1/q$, $\alpha + \beta + \gamma = 2/q$, $\gamma = (n-1)(1/2 - 1/q)$.

This result is due to V. Georgiev, H. Lindblad and C. Sogge [55].

Corollary 3.6.2. *Let n be odd and let $q = \frac{2(n+1)}{n-1}$. Suppose that F is spherically symmetric and supported in the light cone $\{(t,x) \in \mathbb{R}_+^{1+n} : |x| < t\}$. Then the inequality*

$$\Big\|(t^2 - |x|^2)^{\gamma - 1/q} \log^{\beta} \frac{4T^2}{t^2 - |x|^2}\omega\Big\|_{L^q(t+|x| \le T)}$$

$$\le C\Big\|(t^2 - |x|^2)^{1/q} \log^{\lambda} \frac{4T^2}{t^2 - |x|^2} F\Big\|_{L^{q'}(t+|x| \le T)}$$

holds, where $\beta = \lambda - 4/q$, $\lambda > 3/q$ and $\gamma = (n-1)(1/2 - 1/q)$.

From this corollary we obtain the following statements:

Proposition 3.6.1. *Let n be odd and let $T \ge 2$, $q = \frac{2(n+1)}{n-1}$. Assume that F is spherically symmetric and* supp $F \subset \{(t,x) \in \mathbb{R}_+^{1+n} : t^2 - |x|^2 \ge 1\}$. *Then the inequality*

$$\|(t^2 - |x|^2)^{1/q}\omega\|_{L^q(\{|x| < t < T/2\})} \le c(\log T)^{4/q} \|(t^2 - |x|^2)^{1/q} F\|_{L^{q'}(\mathbb{R}_+^{1+n})}$$

holds, where the constant c does not depend on T.

Proposition 3.6.2. *Let n be odd and let $T > 1$. Suppose that $q = \frac{2(n+1)}{n-1}$. Assume that F is spherically symmetric and* supp $F \subset \{(t,x) \in \mathbb{R}_+^{1+n} : t - |x| > 1\}$. *Then the inequality*

$$\|(t - |x|)^{1/q}\omega\|_{L^q(\{t - |x| < T\})} \le c(\log T)^{2/q} \|(t - |x|)^{1/q} F\|_{L^{q'}(\{t - |x| < T\})}$$

holds and the constant c does not depend on T.

3.7 Notes and Comments on Chapter 3

The solution of the one-weight problem for one–sided fractional maximal functions and integrals was given in [7] (see also the survey paper [111]). Necessary and sufficient conditions guaranteeing one–weight inequalities for the Riesz potentials with product kernels were established in [85], [86].

The two–weight problem for the Riemann–Liouville and Weyl transforms with parameters less than one was solved in [53, Chapter 3], [40], [39, Chapter 2] for the non–diagonal case. Criteria governing the trace inequality for one–sided potentials were obtained in [117], [89], [146], [147] (see also [94], [95]).

Two–weight Sawyer–type conditions for one–sided fractional maximal functions and integrals were derived in [113], [109].

Two–weight Strichartz estimates with power–type weights have been established in [54], [55], [81]. In these papers existence of global weak solution for the semilinear wave equation

$$\begin{cases} \Box\omega = |u|^p, & (t,x) \in \mathbb{R}_+^{1+n}, \\ u(0,x) = \varepsilon f(x), & \partial_t u(0,x) = \varepsilon g(x), \end{cases}$$

where ε is small and p is greater that the critical exponent p_c in the sense of Strauss (see [166], [167], [78]), have been established.

The results of this chapter except Section 3.6 is published for the first time. Section 3.6 is taken from [56].

Chapter 4

Strong Fractional Maximal Functions and Multiple Riesz Potentials

In this chapter criteria governing the two–weight inequality

$$\|K_{\alpha,\beta}f\|_{L^q_v(\mathbb{R}^2)} \le c\|f\|_{L^p_w(\mathbb{R}^2)},$$

are established, where $K_{\alpha,\beta}$ is one of the following two-dimensional operators:

$$(M_{\alpha,\beta}f)(x,y) := \sup_{I\times J\ni(x,y)} \frac{1}{|I|^{1-\alpha}|J|^{1-\beta}} \int_I \int_J |f(t,\tau)|dtd\tau;$$

$$(I_{\alpha,\beta}f)(x,y) := \iint_{\mathbb{R}^2} |x-t|^{\alpha-1}|y-\tau|^{\beta-1} f(t,\tau)dtd\tau;$$

$$(I_\alpha J_\beta f)(x,y) := \int_{\mathbb{R}} \int_{|\tau|<2|y|} |x-t|^{\alpha-1}|y-\tau|^{\beta-1} f(t,\tau)dtd\tau;$$

$$(M_\alpha I_\beta f)(x,y) := \sup_{I\ni x} \frac{1}{|I|^{1-\alpha}} \int_I \left| \int_{\mathbb{R}} |y-\tau|^{\beta-1} f(t,\tau)d\tau \right| dt,$$

and I and J are bounded intervals in $\mathbb{R}$. From these results we have necessary and sufficient conditions guaranteeing the trace inequality for these operators. For the operator $I_{\alpha,\beta}$ we additionally assume that the weight v satisfies the Muckenhoupt one-dimensional $A_\infty(\mathbb{R})$ condition with respect to only one variable uniformly to another one. Moreover, criteria for the two–weight inequality for these operators are obtained, provided that the weight on the right-hand side is a product of two one-dimensional weights. Inequalities of Fefferman–Stein type for strong fractional maximal functions and potentials with product kernels are also established.

In this chapter the set D of dyadic intervals in $\mathbb{R}$ and the class $DR^{(d)}(\mathbb{R})$ of weights satisfying the dyadic reverse doubling condition will be needed, definitions of which are given in Chapter 3.

4.1 Single Kernel Operators

Let I_α be the Riesz potential operator given by

$$I_\alpha f(x) := \int_{\mathbb{R}^n} \frac{f(y)}{|x-y|^{n-\alpha}} dy, \quad x \in \mathbb{R}^n. \tag{4.1.1}$$

For the following statement we refer to [46], [52] (see also [53, Chapter 3]).

Theorem 4.1.1. *Let $1 < p < q < \infty$ and let $0 < \alpha < n$. Then I_α is bounded from $L^p_w(\mathbb{R}^n)$ to $L^q_v(\mathbb{R}^n)$ if and only if*

$$A_1 := \sup_{\substack{x\in\mathbb{R}^n \\ r>0}} \Bigg(\int_{B(x,2r)} v \Bigg)^{1/q} \Bigg(\int_{|x-y|>r} |x-y|^{(\alpha-n)p'} w^{1-p'}(y) dy \Bigg)^{1/p'} < \infty$$

and

$$A_2 := \sup_{\substack{x\in\mathbb{R}^n \\ r>0}} \Bigg(\int_{B(x,2r)} w^{1-p'} \Bigg)^{1/p'} \Bigg(\int_{|x-y|>r} |x-y|^{(\alpha-n)q} v(y) dy \Bigg)^{1/q} < \infty,$$

where $p' = p/(p-1)$ and $B(x,r)$ is a ball with center x and radius r. Moreover, there exist positive constants c_1 and c_2 depending only on p, q and α such that

$$c_1 \max\{A_1, A_2\} \le \|I_\alpha\| \le c_2 \max\{A_1, A_2\}.$$

We also mention the next result obtained in [4].

Theorem 4.1.2. *Let $1 < p < q < \infty$, $0 < \alpha < n/p$. Then the operator I_α is bounded from $L^p(\mathbb{R}^n)$ to $L^q_v(\mathbb{R}^n)$ if and only if*

$$A' := \sup_{\substack{x\in\mathbb{R}^n \\ r>0}} \Bigg(\int_{B(x,r)} v(x) dx \Bigg) r^{q(\alpha-n/p)} < \infty.$$

For $p = q$ the following two–weight criterion of pointwise type was established in [116] (see also [172] for more general case):

Theorem 4.1.3. *Let $1 < p < \infty$, $0 < \alpha < 1/p$. Then the operator I_α defined by (4.1.1) is bounded from $L^p(\mathbb{R}^n)$ to $L^p_v(\mathbb{R}^n)$ if and only if $I_\alpha v \in L^{p'}_{loc}(\mathbb{R}^n)$ and*

$$A_3 := \operatorname*{ess\,sup}_{x\in\mathbb{R}^n} \left(\frac{I_\alpha [I_\alpha v]^{p'}(x)}{I_\alpha v(x)} \right)^{1/p'} < \infty.$$

Moreover,

$$c_1 A_3 \le \|I_\alpha\| \le c_2 A_3,$$

where the constants c_1 and c_2 depend only on p and α.

We shall need the following two–weight criterion for the Hardy–type operator

$$Hf(x) := \int\limits_{\{t\in\mathbb{R}^n:|t|<|x|\}} f(t)dt, \quad x\in\mathbb{R}^n.$$

(see, e.g, [100], [39]):

Theorem 4.1.4. *Let $1<p\le q<\infty$. Then H is bounded from $L^p_u(\mathbb{R}^n)$ to $L^q_\sigma(\mathbb{R}^n)$ if and only if*

$$A_4 := \sup_{t>0}\bigg(\int\limits_{|x|>t}\sigma(x)dx\bigg)^{1/q}\bigg(\int\limits_{|x|<t}u^{1-p'}(x)dx\bigg)^{1/p'}<\infty.$$

Moreover, there exist positive constants c_1 and c_2 depending only on p and q such that $c_1 A_4 \le \|H\| \le c_2 A_4$.

The next statement is a direct consequence of the classical Hardy inequality [68].

Proposition 4.1.1. *Let $1<p<\infty$. Then inequality*

$$\int\limits_{\mathbb{R}^n}(Hf(x))^p|x|^{-p}dx \le c\int\limits_{\mathbb{R}^n}(f(x))^p dx$$

holds for all non–negative measurable f on $\mathbb{R}^n$.

Taking into account the duality arguments we have the next statement for the operator

$$H'f(x) = \int\limits_{\{t\in\mathbb{R}^n:|t|>|x|\}} f(t)dt, \quad x\in\mathbb{R}^n:$$

Theorem 4.1.5. *Let $1<p\le q<\infty$. Then H' is bounded from $L^p_u(\mathbb{R}^n)$ to $L^q_\sigma(\mathbb{R}^n)$ if and only if*

$$A_5 := \sup_{t>0}\bigg(\int\limits_{|x|<t}\sigma(x)dx\bigg)^{1/q}\bigg(\int\limits_{|x|>t}u^{1-p'}(x)dx\bigg)^{1/p'}<\infty.$$

The next statement concerning the truncated Riesz potential

$$J_\alpha f(x) := \int\limits_{\{y\in\mathbb{R}^n:|y|<2|x|\}}\frac{f(y)}{|x-y|^{n-\alpha}}dy, \quad x\in\mathbb{R}^n,$$

was derived in [152] for $p = q$ (for the simple proof in the case $1 < p \le q < \infty$ see [39], Section 5.1):

Theorem 4.1.6. *Let $1 < p \le q < \infty$. Suppose that $\alpha > n/p$. Then the operator J_α is bounded from $L^p(\mathbb{R}^n)$ to $L^q_v(\mathbb{R}^n)$ if and only if*

$$A_6 := \sup_{t>0} \Big(\int\limits_{|x|>t} \frac{v(x)}{|x|^{(n-\alpha)p}} dx \Big)^{1/q} t^{n/p'} < \infty.$$

Moreover, there exist positive constants c_1 and c_2 depending only on p and q such that

$$c_1 A_6 \le \|J_\alpha\| \le c_2 A_6.$$

We shall also need following two well–known Hardy–type inequalities in the discrete form (see, e.g., [20], [48], [100]).

Theorem 4.1.7. *Let $1 < p \le q < \infty$ and let $\{a_n\}$, $\{b_n\}$ be positive sequences. Then the two–weight inequality*

$$\Big(\sum_{n=-\infty}^{\infty} \Big| \sum_{k=n}^{\infty} g_k \Big|^q a_n^q \Big)^{1/q} \le c \Big(\sum_{n=-\infty}^{\infty} |g_n|^p b_n^p \Big)^{1/p}$$

with the positive constant c independent of $\{g_k\}$ $\big(g_k \in l^p_{b_n^p}(Z)\big)$, holds if and only if

$$\sup_{n\in\mathbb{Z}} \Big(\sum_{k=-\infty}^{n} a_k^q \Big)^{1/q} \Big(\sum_{k=n}^{\infty} b_k^{-p'} \Big)^{1/p'} < \infty.$$

Theorem 4.1.8. *Let $1 < p \le q < \infty$ and let m be an integer. Suppose that $\{a_n\}_{n=m}^{\infty}$, $\{b_n\}_{n=m}^{\infty}$ are positive sequences. Then the two–weight inequality*

$$\Big(\sum_{n=m}^{\infty} \Big| \sum_{k=n}^{\infty} g_k \Big|^q a_n^q \Big)^{1/q} \le c \Big(\sum_{n=m}^{\infty} |g_n|^p b_n^p \Big)^{1/p}$$

holds if and only if

$$\sup_{\substack{m\in\mathbb{Z}\\ n\ge m}} \Big(\sum_{k=m}^{n} a_k^q \Big)^{1/q} \Big(\sum_{k=n}^{\infty} b_k^{-p'} \Big)^{1/p'} < \infty.$$

Definition 4.1.1. We say that the weight ρ on $\mathbb{R}$ satisfies the $A_\infty(\mathbb{R})$ condition ($\rho \in A_\infty(\mathbb{R})$) if there exist constants $c, \delta > 0$ such that for all intervals I and measurable sets $E \subset I$ the inequality

$$\frac{\rho(E)}{\rho(I)} \le c \Big(\frac{|E|}{|I|} \Big)^\delta$$

holds. Further, we say that a two-dimensional weight u belongs to the class $A_\infty(\mathbb{R})$ with respect to the first variable uniformly to the second one ($u \in A_\infty^{(x)}(\mathbb{R})$) if the inequality

$$\frac{u_y(E)}{u_y(I)} \leq c\left(\frac{|E|}{|I|}\right)^\delta$$

holds for all $y \in \mathbb{R}$, all intervals $I \subset \mathbb{R}$ and measurable sets $E \subset I$. The class $A_\infty^{(y)}(\mathbb{R})$ is defined analogously.

It is known (see [72], [34], [49, Chapter IV]) that $\rho \in A_\infty(\mathbb{R})$ if and only if ρ belongs to the Muckenhoupt $A_p(\mathbb{R})$ class for some $p \geq 1$.

The next result was derived by B. Muckenhoupt and R. L. Wheeden [126]

Proposition 4.1.2. *Let $1 < q < \infty$. Suppose that $0 < \alpha < 1$ and $\rho \in A_\infty(\mathbb{R})$. Then there exists a positive constant c depending only on α, p and w such that the inequality*

$$\|I_\alpha f\|_{L^q_\rho(\mathbb{R})} \leq c\|M_\alpha f\|_{L^q_\rho(\mathbb{R})}$$

holds.

4.2 Two–weight Problem for Strong Fractional Maximal Functions

Here we study the two–weight problem for the strong fractional maximal operator $M_{\alpha,\beta}$.

We begin with the following lemma which will be useful for us:

Lemma 4.2.1. *Let $0 < \alpha < 1$, $0 < \beta < 1$. Suppose that $k \in \mathbb{Z}$ and $f \geq 0$. Then the following inequalities hold:*

$$(M_\alpha^{(2^k)} I_\beta)f(x,y) \leq \frac{c_\alpha}{|R(0,2^{k+2})|} \int\limits_{R(0,2^{k+2})} K_t(x,y)dt; \tag{4.2.1}$$

$$M_{\alpha,\beta}^{(2^k)} f(x,y) \leq \frac{c_{\alpha,\beta}}{|R(0,2^{k+2})|^2} \int\limits_{R(0,2^{k+2})} \int\limits_{R(0,2^{k+2})} S_{t,\tau}(x,y)dtd\tau, \tag{4.2.2}$$

where

$$(M_\alpha^{(2^k)} I_\beta)f(x,y) := \sup_{\substack{I \ni x \\ |I| \leq 2^k}} \frac{1}{|I|^{1-\alpha}} \int\limits_I \int\limits_{\mathbb{R}} |y-\tau|^{\beta-1} f(t,\tau)dtd\tau;$$

$$M_{\alpha,\beta}^{(2^k)} f(x,y) := \sup_{\substack{I\times J \ni (x,y) \\ |I|,|J| \leq 2^k}} \frac{1}{|I|^{1-\alpha}|J|^{1-\beta}} \int\limits_I \int\limits_J f(t,\tau)dtd\tau;$$

$$R(0,r) = \{t : -r \leq t \leq r\}, \quad r > 0;$$

$$K_t(x,y) := \sup_{\substack{I-t\ni x\\ I\in D}} \frac{1}{|I|^{1-\alpha}} \int\limits_{I-t} [I_\beta f(s,\cdot)](y)ds;$$

$$I-t := \{x-t : x\in I\};$$

$$S_{t,\tau}(x,y) := \sup_{\substack{I-t\ni x\\ I\in D}} \sup_{\substack{J-\tau\ni y\\ J\in D}} \frac{1}{|I|^{1-\alpha}} \frac{1}{|J|^{1-\beta}} \int\limits_{I-t}\int\limits_{J-\tau} f(s,\epsilon)dsd\epsilon.$$

Proof. We shall need the following observation (see, e.g., [49, p. 431]): Let j and I be an integer and an interval respectively such that $2^{j-1} < |I| \le 2^j$. Suppose that $k\in\mathbb{Z}$, $j\le k$. Let E be the set of those $t\in R(0,2^{k+2})$ for which there is some $I_1\in D-t$ with $|I_1| = 2^{j+1}$ and such that $I\subset I_1$. Then

$$|E| \ge 2^{k+2}, \tag{4.2.3}$$

where $D-t := \{I-t : I\in D\}$.

To prove (4.2.1) we take f and $(x,y)\in\mathbb{R}^2$. Then there exists an interval I ($|I|\le 2^k$) such that $x\in I$ and

$$\frac{2}{|I|^{1-\alpha}} \int\limits_I (I_\beta f(s,\cdot))(y)ds \ge (M_\alpha^{(2^k)} I_\beta) f(x,y).$$

Let j be an integer such that $2^{j-1} < |I| \le 2^j$. In this case $j\le k$. Let E be the set defined above. Further, for every $t\in E$ there exists an interval $Q\in D-t$ such that $|Q| = 2^{j+1}$, $I\subset Q$ and

$$\frac{1}{2}(M_\alpha^{(2^k)} I_\beta) f(x,y) \le \frac{1}{|I|^{1-\alpha}} \int\limits_I (I_\beta f(s,\cdot))(y)ds$$

$$\le \frac{c}{|Q|^{1-\alpha}} \int\limits_Q (I_\beta f(s,\cdot))(y)ds \le cK_t(x,y).$$

By (4.2.3) we have also that $|E| \ge |R(0,2^{k+2})|/2$. Hence

$$(M_\alpha^{(2^k)} I_\beta) f(x,y) \le \frac{c}{|E|} \int\limits_E K_t(x,y)dt \le \frac{c}{|R(0,2^{k+2})|} \int\limits_{R(0,2^{k+2})} K_t(x,y)dt.$$

To prove (4.2.2) we argue as above. For given f and $(x,y)\in\mathbb{R}^2$ we take intervals $R_1\ni x$ and $R_2\ni y$, $|R_1|,|R_2|\le 2^k$, such that

$$\frac{2}{|R_1|^{1-\alpha}|R_2|^{1-\beta}} \iint\limits_{R_1\times R_2} f(s,\epsilon)dsd\epsilon > M_{\alpha,\beta}^{(2^k)} f(x,y).$$

Now, let us take integers j and i so that $2^{j-1} < |R_1| \le 2^j$ and $2^{i-1} < |R_2| \le 2^i$.

It is obvious that $j, i \leq k$. Moreover, let us define the sets E_1 and E_2 in the following way:

$$E_1 := \{t \in R(0, 2^{k+2}) : \exists I \in D - t,\ |I| = 2^{j+1},\ R_1 \subset I\};$$

$$E_2 := \{\tau \in R(0, 2^{k+2}) : \exists J \in D - \tau,\ |J| = 2^{i+1},\ R_2 \subset J\}.$$

Then for all $t \in E_1$ and $\tau \in E_2$ we have that

$$\frac{1}{2} M^{(2^k)}_{\alpha,\beta} f(x,y) \leq \frac{1}{|R_1|^{1-\alpha}|R_2|^{1-\beta}} \iint\limits_{R_1 \times R_2} f(s,\epsilon)(y)\,ds\,d\epsilon \leq c S_{t,\tau}(x,y).$$

According to (4.2.3) we have that $|E_1|, |E_2| \geq |R(0, 2^{k+2})|/2$. Hence,

$$M^{(2^k)}_{\alpha,\beta} f(x,y) \leq \frac{c}{|E_1 \times E_2|} \iint\limits_{E_1 \times E_2} S_{t,\tau}(x,y)\,dt\,d\tau$$

$$\leq \frac{c}{|R(0,2^{k+2})|^2} \int\limits_{R(0,2^{k+2})} \int\limits_{R(0,2^{k+2})} S_{t,\tau}(x,y)\,dt\,d\tau.$$

The lemma is proved. □

Theorem 4.2.1. *Let $1 < p < q < \infty$ and let $0 < \alpha, \beta < 1$. Suppose that $w(x,y) = w_1(x)w_2(y)$ with $w_1^{1-p'}, w_2^{1-p'} \in RD^{(d)}(\mathbb{R})$.*

Then $M_{\alpha,\beta}$ is bounded from $L^p_w(\mathbb{R}^2)$ to $L^q_v(\mathbb{R}^2)$ if and only if

$$B := \sup_{I,J} |I|^{\alpha-1}|J|^{\beta-1} \Big(\iint\limits_{I \times J} v(x,y)\,dx\,dy \Big)^{1/q} \Big(\iint\limits_{I \times J} w^{1-p'}(x,y)\,dx\,dy \Big)^{1/p'} < \infty,$$

where the supremum are taken over all arbitrary bounded intervals I and J in $\mathbb{R}$.

Proof. First we prove the theorem for the operator

$$N_{\alpha,\beta} f(x,y) := \sup_{\substack{I \times J \ni (x,y) \\ I,J \in D}} |I|^{\alpha-1}|J|^{\beta-1} \iint\limits_{I \times J} |f(t,\tau)|\,dt\,d\tau.$$

We can without loss of generality assume that f is a non-negative bounded function with compact support. In this case $N_{\alpha,\beta} f(x,y) < \infty$ for all $(x,y) \in \mathbb{R}^2$.

For every $(x,y) \in \mathbb{R}^2$ let us take the pair of dyadic intervals $\{I(x), J(y)\}$, $x \in I(x)$, $y \in J(y)$ such that

$$\frac{2}{|I(x)|^{1-\alpha}|J(y)|^{1-\beta}} \iint\limits_{I(x) \times J(y)} f(t,\tau)\,dt\,d\tau \geq N_{\alpha,\beta} f(x,y). \tag{4.2.4}$$

Let us introduce the set

$$F_{I,J} = \{(x,y) \in \mathbb{R}^2 : x \in I, y \in J \text{ and (4.2.4) holds for } I \text{ and } J\}$$

for each $(I,J), I,J \in D$. In this case for a fixed point (x,y) the minimal dyadic rectangle $I \times J$ satisfying (4.2.4) (i.e., for any proper dyadic rectangle $I_1 \times J_1$, $I_1 \subset I$, $J_1 \subset J$ (4.2.4) fails) exists but the latter is not unique. For these reasons it is enough that (4.2.4) holds for every $(x,y) \in F_{I,J}$; $\mathbb{R}^2 = \cup_{I,J\in D} F_{I,J}$ and $F_{I,J} \subset I \times J$.

In the sequel we will use the notation:

$$\overline{V}_{I,J} := |I|^{(\alpha-1)q}|J|^{(\beta-1)q} \iint\limits_{I\times J} v(x,y)dxdy; \;\; \overline{F}_{I,J} := \iint\limits_{I\times J} f(t,\tau)dtd\tau.$$

We have that

$$\iint\limits_{\mathbb{R}^2} [N_{\alpha,\beta} f(x,y)]^q v(x,y)dxdy \le 2^q \sum_{I,J\in D} \overline{V}_{I,J} \overline{F}^p_{I,J}$$

$$\left[\text{by the condition } B < \infty\right]$$

$$\le 2^q B^q \sum_{I,J\in D} \Big(\int_I w_1^{1-p'}(t)dt\Big)^{-q/p'} \Big(\int_J w_2^{1-p'}(t)dt\Big)^{-q/p'} \Big(\int_I\int_J f(t,\tau)dtd\tau\Big)^q$$

$$= c \sum_{J\in D} \Big(\int_J w_2^{1-p'}(t)dt\Big)^{-q/p'} \sum_{I\in D} \Big(\int_I w_1^{1-p'}(t)dt\Big)^{-q/p'} \Big[\int_I \Big(\int_J f(t,\tau)d\tau\Big)dt\Big]^q$$

$$\left[\text{due to Theorem 3.1.1 with respect to the intervals } I\right]$$

$$\le c \sum_{J\in D} \Big(\int_J w_2^{1-p'}(t)dt\Big)^{q/p} \Big(\int_{\mathbb{R}} w_1(t) \Big(\int_J f(t,\tau)d\tau\Big)^p dt\Big)^{q/p}$$

$$\left[\text{by generalized Minkowski's inequality and Theorem 3.1.1}\right]$$

$$\le c \sum_{J\in D} \Big(\int_J w_2^{1-p'}(t)dt\Big)^{q/p} \Big(\int_J \Big(\int_{\mathbb{R}} f^p(t,\tau)w_1(t)dt\Big)^{1/p} d\tau\Big)^q \le c\|f\|^q_{L^p_w(\mathbb{R}^2)}.$$

Using (4.2.2) and the fact that the condition $B < \infty$ is translation invariant, we can derive sufficiency of the theorem.

To show *necessity* we observe that

$$\|M_{\alpha,\beta} f\|_{L^q_v(\mathbb{R}^2)} \ge |I|^{\alpha-1}|J|^{\beta-1} \Big(\iint\limits_{I\times J} v(x,y)dxdy\Big)^{1/q} \Big(\iint\limits_{I\times J} f(t,\tau)dtd\tau\Big)$$

for any pair of intervals (I,J). It remains now to use the two–weight inequality for the test function

$$f(x,y) = \chi_I(x)\chi_J(y)w^{1-p'}(x,y). \qquad \square$$

We have also the following Fefferman–Stein– type inequality for the operator $M_{\alpha,\beta}$:

Theorem 4.2.2. *Let $1 < p < q < \infty$. Suppose that $1/p - 1/q < \alpha, \beta < 1/p$. Then there exists a positive constant c such that*

$$\Big(\iint_{\mathbb{R}^2} (M_{\alpha,\beta} f)^q(x,y) v(x,y)dxdy\Big)^{1/q} \le c\Big(\iint_{\mathbb{R}^2} |f(x,y)|^p (\widetilde{M}_{\alpha,\beta} v)^{p/q}(x,y)dxdy\Big)^{1/p},$$

where

$$(\widetilde{M}_{\alpha,\beta} v)(x,y) := \sup_{\substack{I\ni x\\ J\ni y}} |I|^{(\alpha-1/p)q}|J|^{(\beta-1/p)q} \iint_{I\times J} v(t,\tau)dtd\tau$$

and the positive constant does not depend on f and v.

Proof. First we prove the statement for the dyadic strong fractional maximal function $N_{\alpha,\beta} f(x,y)$. Suppose that $f \ge 0$, f is bounded and has a compact support. Arguing as in the proof of Theorem 4.2.1 for given $(x,y) \in \mathbb{R}^2$ we take dyadic intervals $I(x)$ and $J(y)$ $(x \in I(x),\ y \in J(y))$ so that (4.2.4) holds. Using the notation from the proof of Theorem 4.2.1 we find that

$$S := \iint_{\mathbb{R}^2} (N_{\alpha,\beta} f)^q(x,y)v(x,y)dxdy \le 2^q \sum_{I,J\in D} \overline{V}_{I,J}\overline{F}^q_{I,J}$$

$$= 2^q \sum_{I,J\in D} \Big(\iint_{I\times J} f(x,y)\Big(\iint_{I\times J} v(t,\tau)dtd\tau\Big)^{1/q} dxdy\Big)^q |I|^{(\alpha-1)q}|J|^{(\beta-1)q}$$

$$= 2^q \sum_{I,J\in D} \frac{1}{(|I||J|)^{q/p'}} \Big(\iint_{I\times J} f(x,y)\Big(|I|^{q(\alpha-1/p)}|J|^{q(\beta-1/p)} \iint_{I\times J} v(t,\tau)dtd\tau\Big)^{1/q} dxdy\Big)^q$$

$$\le 2^q \sum_{I,J\in D} |I|^{-q/p'}|J|^{-q/p'}\Big(\iint_{I\times J} f(x,y)(\widetilde{N}_{\alpha,\beta} v)^{1/q}(x,y)dxdy\Big)^q,$$

where

$$\widetilde{N}_{\alpha,\beta} v(x,y) := \sup_{\substack{I\ni x J\ni y\\ I,J\in D}} |I|^{q(\alpha-1/p)}|J|^{q(\beta-1/p)} \iint_{I\times J} v(t,\tau)dtd\tau.$$

By applying Theorem 3.1.1 twice with $w(x) = 1$ and Minkowski's inequality we obtain that

$$S \le c\sum_{I\in D} |I|^{-q/p'}\Big(\int_{\mathbb{R}}\Big(\int_I f(x,y)(\widetilde{N}_{\alpha,\beta} v)^{1/q}(x,y)dx\Big)^p dy\Big)^{q/p}$$

$$\leq c\sum_{I\in D}|I|^{-q/p'}\bigg(\int_I\bigg(\int_{\mathbb{R}} f^p(x,y)(\widetilde{N}_{\alpha,\beta}v)^{p/q}(x,y)dy\bigg)^{1/p}dx\bigg)^q$$

$$\leq c\bigg(\iint_{\mathbb{R}^2} f^p(x,y)(\widetilde{N}_{\alpha,\beta}v)^{p/q}(x,y)dxdy\bigg)^{q/p}.$$

Let us now pass to the operator $M_{\alpha,\beta}$. Keeping the notation of Lemma 4.2.1 we find that

$$D_{t,\tau}:=\iint_{\mathbb{R}^2}(S_{t,\tau}(x,y))^q v(x,y)dxdy$$

$$=\iint_{\mathbb{R}^2}\bigg(\sup_{\substack{I-t\ni x,J-\tau\ni y\\ I,J\in D}}|I|^{\alpha-1}|J|^{\beta-1}\int_{I-t}\int_{J-\tau}f(s,\varepsilon)dsd\varepsilon\bigg)^q v(x,y)dxdy$$

$$=\iint_{\mathbb{R}^2}\bigg(\sup_{\substack{I\ni x,J\ni y\\ I,J\in D}}|I|^{\alpha-1}|J|^{\beta-1}\iint_{I\times J}f(s-t,\varepsilon-\tau)dsd\varepsilon\bigg)^q v(x-t,y-\tau)dxdy$$

$$=\iint_{\mathbb{R}^2}(Nf(\cdot-t,\cdot-\tau))^q(x,y)v(x-t,y-\tau)dxdy$$

$$\leq c\bigg(\iint_{\mathbb{R}^2}(f(x-t,y-\tau))^p(\widetilde{N}_{\alpha,\beta}v(\cdot-t,\cdot-\tau))^{p/q}(x,y)dxdy\bigg)^{q/p}$$

$$=c\bigg(\iint_{\mathbb{R}^2}(f(x-t,y-\tau))^p\bigg(\sup_{I\ni x;J\ni y;I,J\in D}|I|^{q(\alpha-1/p)}|J|^{q(\beta-1/p)}$$

$$\times\iint_{I\times J}v(s-t,\varepsilon-\tau)dsd\varepsilon\bigg)^{p/q}dxdy\bigg)^{q/p}$$

$$=c\bigg(\iint_{\mathbb{R}^2}(f(x-t,y-\tau))^p\bigg(\sup_{\substack{I\ni x;J\ni y\\ I,J\in D}}|I-t|^{q(\alpha-1/p)}|J-\tau|^{q(\beta-1/p)}$$

$$\times\iint_{(I-t)\times(J-\tau)}v(s,\varepsilon)dsd\varepsilon\bigg)^{p/q}dxdy\bigg)^{q/p}$$

$$\leq c\bigg(\iint_{\mathbb{R}^2}f^p(x,y)(\widetilde{N}_{\alpha,\beta}v(\cdot-t,\cdot-\tau))^{p/q}(x,y)dxdy\bigg)^{q/p}$$

$$\leq c\bigg(\iint_{\mathbb{R}^2}(f(x,y))^p(\widetilde{M}_{\alpha,\beta}v)^{p/q}(x,y)dxdy\bigg)^{q/p}$$

and the proof follows. □

4.3 Mixed Multiple Operators

Here we investigate the two–weight problem for the operators $M_\alpha I_\beta$ and $I_\alpha J_\beta$.

Let us begin with the operator $M_\alpha I_\beta$.

Theorem 4.3.1. *Let $1 < p < q < \infty$ and let $0 < \alpha, \beta < 1$. Suppose that $w(x,y) = w_1(x)w_2(y)$ with $w_1^{1-p'} \in RD^{(d)}(\mathbb{R})$. Then $M_\alpha I_\beta$ is bounded from $L^p_w(\mathbb{R}^2)$ to $L^q_v(\mathbb{R}^2)$ if and only if*

$$B_1 := \sup_{\substack{a\in\mathbb{R}\\ r>0\\ I\subset\mathbb{R}}} |I|^{\alpha-1}\Big(\int_I\int_{|y-a|<r} w^{1-p'}(x,y)dxdy\Big)^{1/p'}\Big(\int_I\int_{|y-a|>r}\frac{v(x,y)}{|y-a|^{(1-\beta)q}}dxdy\Big)^{1/q}<\infty;$$

$$B_2 := \sup_{\substack{a\in\mathbb{R}\\ r>0\\ I\subset\mathbb{R}}} |I|^{\alpha-1}\Big(\int_I\int_{|y-a|>r} \frac{w^{1-p'}(x,y)}{|y-a|^{(1-\beta)p'}}dxdy\Big)^{1/p'}\Big(\int_I\int_{|y-a|<r} v(x,y)dxdy\Big)^{1/q}<\infty,$$

where I is arbitrary bounded interval in $\mathbb{R}$.

Proof. *Sufficiency.* First let us show sufficiency for the dyadic version of the operator $M_\alpha I_\beta$:

$$N_\alpha I_\beta f(x,y) = \sup_{\substack{I\ni x\\ I\in D}} |I|^{\alpha-1}\int_I |I_\beta(f(t,\cdot))(y)|dt.$$

Let $f \geq 0$. Without loss of generality we can assume that f is bounded and has compact support. Therefore $N_\alpha I_\beta$ is finite for all $(x,y) \in \mathbb{R}^2$. We claim that if we prove the theorem for such functions, then by an elementary discussion we can pass to an arbitrary $f \in L^p_w(\mathbb{R}^2)$. Indeed, let $f \in L^p_w(\mathbb{R}^2)$ and let us take the sequence $f_n = f\chi_{B(0,n)}\chi_{\{f<n\}}$. Then $f_n \to f$ in $L^p_w(\mathbb{R}^2)$ and also pointwise. Moreover, $f_n(x,y) \leq f(x,y)$. On the other hand, $(M_\alpha I_\beta)f_n$ is a Cauchy sequence, because

$$\|(M_\alpha I_\beta)f_n - (M_\alpha I_\beta)f_m\|_{L^q_v(\mathbb{R}^2)} \leq \|(M_\alpha I_\beta)(f_n - f_m)\|_{L^q_v(\mathbb{R}^2)} \leq c\|f_n - f_m\|_{L^p_w(\mathbb{R}^2)}.$$

Consequently, by the completeness of the space $L^q_v(\mathbb{R}^2)$ there exists $g \in L^q_v(\mathbb{R}^2)$ such that

$$\|(M_\alpha I_\beta)f_n - g\|_{L^q_v(\mathbb{R}^2)} \to 0.$$

Hence, there exists a subsequence $(M_\alpha I_\beta)f_{n_k}$ which converges in norm and almost everywhere to g. But f_{n_k} converges to f in $L^p_w(\mathbb{R}^2)$ and also a.e.. This leads to the inequality

$$\|g\|_{L^q_v(\mathbb{R}^2)} \leq c\|f\|_{L^p_w(\mathbb{R}^2)}, \tag{4.3.1}$$

where c does not depend on f.

Now we will argue for the subsequence f_{n_k}. For all $(x,y)\in\mathbb{R}^2$ with $x\in I$ we have (f_{n_k} is non-decreasing)

$$|I|^{\alpha-1}\iint\limits_{I\times\mathbb{R}}\frac{f(t,\tau)}{|y-\tau|^{1-\beta}}dtd\tau=\lim_{k\to\infty}|I|^{\alpha-1}\iint\limits_{I\times\mathbb{R}}\frac{f_{n_k}(t,\tau)}{|y-\tau|^{1-\beta}}dtd\tau$$

$$\le\lim_{k\to\infty}\sup_{I,I\ni x}|I|^{\alpha-1}\iint\limits_{I\times\mathbb{R}}\frac{f_{n_k}(t,\tau)}{|y-\tau|^{1-\beta}}dtd\tau=\lim_{k\to\infty}(M_\alpha I_\beta)f_{n_k}(x,y)$$

and the last limit exists, because it converges a.e. to g. Hence

$$(M_\alpha I_\beta)f(x,y)\le\lim_{k\to\infty}(M_\alpha I_\beta)f_{n_k}(x,y)=g(x,y)$$

for almost all (x,y). Finally, by (4.3.1) we see that

$$\|(M_\alpha I_\beta)f\|_{L^q_v(\mathbb{R}^2)}\le c\|f\|_{L^p_w(\mathbb{R}^2)}.$$

Further, for $(x,y)\in\mathbb{R}^2$ there exists dyadic interval $I_y(x)\subset\mathbb{R}$ containing x such that

$$\frac{2}{|I_y(x)|^{1-\alpha}}\int\limits_{I_y(x)}I_\beta(f(t,\cdot))(y)dt\ge N_\alpha I_\beta f(x,y). \tag{4.3.2}$$

Let

$$F_I:=\{(x,y)\in\mathbb{R}^2: x\in I \; and \; I \text{ is "minimal" for which (4.3.2) holds}\}.$$

Let D_m be a subset of D for which $F_I\neq\emptyset$. By the main property of dyadic intervals it is clear that $F_I\cap F_J=\emptyset$ when $I\neq J$. Notice also that from the inequality

$$(N_\alpha I_\beta)f(x,y)\le\frac{2}{|I_y(x)|^{1-\alpha}}\int\limits_{I_y(x)}I_\beta f(t,\cdot)(y)dt\le c|I_y(x)|^\alpha,$$

with the constant c depending on the L^∞ norm and the support of f, it follows the existence of minimal dyadic interval containing x for which (4.3.2) holds.

From the definition of maximal operator and the latter inequality it is obvious that $\mathbb{R}^2=\cup_{I\in D_m}F_I$. It is also clear that $F_I\subset I\times\mathbb{R}$ and that F_I is a measurable set for every I. Let us use the notation:

$$V_I(y):=|I|^{(\alpha-1)q}\int\limits_I v(x,y)dx;\quad \overline{F}_I(\tau):=\int\limits_I f(t,\tau)dt.$$

It yields that

$$\iint\limits_{\mathbb{R}^2}[(N_\alpha I_\beta)f(x,y)]^q v(x,y)dxdy$$

$$\leq 2^q \sum_{I\in D_m} \iint_{F_I} \left[\frac{1}{|I|^{1-\alpha}} \int_I I_\beta(f(t,\cdot))(y)dt\right]^q v(x,y)dxdy$$

$$\leq 2^q \sum_{I\in D_m} \int_{\mathbb{R}} V_I(y)\left[\int_I \left(\int_{\mathbb{R}} |y-\tau|^{\beta-1} f(t,\tau)d\tau\right)dt\right]^q dy$$

[change the order of integration in the inner integral]

$$= 2^q \sum_{I\in D_m} \int_{\mathbb{R}} V_I(y)\left[\int_{\mathbb{R}} |y-\tau|^{\beta-1}\overline{F}_I(\tau)d\tau\right]^q dy$$

[by Theorem 4.1.1 and the fact that in the case $n=1$ the first integrals of A_1 and A_2 in Theorem 4.1.1 can be taken over the intervals $(x-r, x+r)$ (see [52])]

$$\leq c(\max\{B_1,B_2\})^q \sum_{I\in D_m} \left[\int_{\mathbb{R}} \left(\int_I w_1^{1-p'}(t)dt\right)^{1-p} (\overline{F}_I(y))^p w_2(y)dy\right]^{q/p}$$

$$= c \sum_{I\in D_m} \left(\int_I w_1^{1-p'}(t)dt\right)^{-q/p'} \left[\int_{\mathbb{R}} (\overline{F}_I(y))^p w_2(y)dy\right]^{q/p}$$

[due to Minkowski's integral inequality and Theorem 3.1.1]

$$\leq c \sum_{I\in D_m} \left(\int_I w_1^{1-p'}(t)dt\right)^{-q/p'} \left[\int_I \left(\int_{\mathbb{R}} f^p(t,y)w_2(y)dy\right)^{1/p} dt\right]^q \leq c\|f\|^q_{L^p_w(\mathbb{R}^2)}.$$

Hence $N_\alpha I_\beta$ is bounded from $L^p_w(\mathbb{R}^2)$ to $L^q_v(\mathbb{R}^2)$.

Let us now pass to the operator $M_\alpha I_\beta$. Assume that $k \in \mathbb{Z}$. Due to Lemma 4.2.1 we have that

$$D_t := \iint_{\mathbb{R}^2} (K_t(x,y))^q v(x,y)dxdy$$

$$= \int_{\mathbb{R}^2} \left(\sup_{\substack{x\in I\\ I\in D}} \frac{1}{|I|^{1-\alpha}} \iint_{I\times\mathbb{R}} |y-\epsilon|^{\beta-1} f(s-t,\epsilon)d\epsilon ds\right)^q v(x-t,y)dxdy$$

[by the condition $\max\{B_1,B_2\}<\infty$ and the fact that this condition is translation invariant]

$$\leq c(\max\{B_1,B_2\})^q \left(\iint_{\mathbb{R}^2} (f(x-t,y))^p w(x-t,y)dxdy\right)^{q/p}$$

$$= c\left(\iint_{\mathbb{R}^2} (f(x,y))^p w(x,y)dxdy\right)^{q/p}.$$

Consequently,

$$(D_t)^{1/q} \le c\Big(\iint\limits_{\mathbb{R}^2} (f(x,y))^p w(x,y)dxdy\Big)^{1/p},$$

where the constant c does not depend on t.

By applying Lemma 4.2.1, Minkowski's inequality and the latter inequality we find that

$$\Big(\iint\limits_{\mathbb{R}^2} [(M_\alpha^{(2^k)} I_\beta) f(x,y)]^q v(x,y)dxdy\Big)^{1/q}$$

$$\le c\Big(\iint\limits_{\mathbb{R}^2} \Big(\frac{1}{|R(0,2^{k+2})|} \int\limits_{R(0,2^{k+2})} K_t(x,y)dt\Big)^q v(x,y)dxdy\Big)^{1/q}$$

$$\le c|R(0,2^{k+2})|^{-1} \int\limits_{R(0,2^{k+2})} \Big(\iint\limits_{\mathbb{R}^2} K_t^q(x,y)v(x,y)dxdy\Big)^{1/q} dt$$

$$= c|R(0,2^{k+2})|^{-1} \int\limits_{R(0,2^{k+2})} (D_t)^{1/q}dt \le c\Big(\iint\limits_{\mathbb{R}^2} (f(x,y))^p w(x,y)dxdy\Big)^{1/p}.$$

Letting now k goes to the infinity we finally obtain sufficiency.

To prove necessity we take the function

$$f(x,y) = \chi_I(x)\chi_{\{y:|y-a|>r\}}(y) w^{1-p'}(x,y)|a-y|^{(p'-1)(\beta-1)},$$

where $a \in \mathbb{R}$ and I is an interval in $\mathbb{R}$. Then we have that

$$\iint\limits_{\mathbb{R}^2} v(x,y)[(M_\alpha I_\beta) f(x,y)]^q dxdy$$

$$\ge \int\limits_I \int\limits_{|y-a|<r} v(x,y)|I|^{(\alpha-1)q}\Big(\int\limits_I \int\limits_{|\tau-a|>r} \frac{f(t,\tau)}{|y-\tau|^{1-\beta}} dtd\tau\Big)^q dxdy$$

$$\big[\text{ if } |y-a|<r \text{ and } |\tau-a|>r, \text{ then } |\tau-y| \le 2|\tau-a| \,\big]$$

$$\ge c\Big(\int\limits_I \int\limits_{|y-a|<r} v(x,y)dxdy\Big)|I|^{(\alpha-1)q}\Big(\int\limits_I \int\limits_{|\tau-a|>r} \frac{w^{1-p'}(t,\tau)}{|\tau-a|^{(1-\beta)p'}} dtd\tau\Big)^q.$$

On the other hand,

$$\|f\|^q_{L^p_w(\mathbb{R}^2)} = \Big(\int\limits_I \int\limits_{|y-a|>r} w^{1-p'}(x,y)|y-a|^{(\beta-1)p'} dxdy\Big)^{q/p}.$$

From the two–weight inequality we derive the condition $B_2 < \infty$. To prove that $B_1 < \infty$ we take the function

$$f(x,y) = \chi_I(x)\chi_{\{y:|y-a|<r\}}(y)w^{1-p'}(x,y)$$

and argue as above. The proof is complete. □

Theorem 4.3.2. *Let $1 < p < q < \infty$. Suppose that $0 < \alpha < 1$ and $\beta > 1/p$. Then the two–weight inequality*

$$\bigg(\iint\limits_{\mathbb{R}^2} |(I_\alpha J_\beta f)(x,y)|^q v(x,y)dxdy\bigg)^{1/q} \le c\bigg(\iint\limits_{\mathbb{R}^2} |f(x,y)|^p u(x)dxdy\bigg)^{1/p}$$

holds if and only if

(i)

$$B_3 \equiv \sup_{\substack{a\in\mathbb{R}\\ r>0\\ k\in\mathbb{Z}}} \bigg(\int\limits_{|x-a|>r} \int\limits_{2^k<|y|<2^{k+1}} \frac{v(x,y)}{|x-a|^{(1-\alpha)q}} dxdy\bigg)^{1/q} \times \bigg(\int\limits_{|x-a|<r} u^{1-p'}(x)dx\bigg)^{1/p'} 2^{k(\beta-1/p)} < \infty;$$

(ii)

$$B_4 \equiv \sup_{\substack{a\in\mathbb{R}\\ r>0\\ k\in\mathbb{Z}}} \bigg(\int\limits_{|x-a|<r} \int\limits_{2^k<|y|<2^{k+1}} v(x,y)dxdy\bigg)^{1/q} \times \bigg(\int\limits_{|x-a|>r} \frac{u^{1-p'}(x)}{|x-a|^{(\alpha-1)p'}}dx\bigg)^{1/p'} 2^{k(\beta-1/p)} < \infty.$$

To prove Theorem 4.3.2 we need the next statement concerning the operator

$$(I_\alpha H)f(x,y) := \int\limits_{\mathbb{R}} \int\limits_{|\tau|<|y|} \frac{f(t,\tau)}{|x-t|^{1-\alpha}} dtd\tau, \quad x,y \in \mathbb{R}.$$

Proposition 4.3.1. *Let $1 < p < q < \infty$, $0 < \alpha < 1$. Suppose that $w(x,y) = w_1(x)w_2(y)$. Then the operator $I_\alpha H$ is bounded from $L^p_w(\mathbb{R}^2)$ to $L^q_v(\mathbb{R}^2)$ if and only if*

(i)

$$D_1 := \sup_{\substack{a\in\mathbb{R}\\ r>0\\ s>0}} \bigg(\int\limits_{|x-a|<r} \int\limits_{|y|<s} w^{1-p'}(x,y)dxdy\bigg)^{1/p'} \bigg(\int\limits_{|x-a|>r} \int\limits_{|y|>s} \frac{v(x,y)}{|x-a|^{(1-\alpha)q}} dxdy\bigg)^{1/q} < \infty;$$

(ii)

$$D_2 := \sup_{\substack{a\in\mathbb{R}\\ r>0\\ s>0}} \bigg(\int\limits_{|x-a|>r} \int\limits_{|y|<s} \frac{w^{1-p'}(x,y)}{|x-a|^{(1-\alpha)p'}} dxdy \bigg)^{1/p'} \bigg(\int\limits_{|x-a|<r} \int\limits_{|y|>s} v(x,y)dxdy \bigg)^{1/q} < \infty.$$

Proof. *Sufficiency.* First suppose that $S := \int_{\mathbb{R}} w_2^{1-p'}(y)dy = \infty$.

Let $\{x_k\}_{k=-\infty}^{+\infty}$ be a sequence of positive numbers for which the equality

$$2^k = \int\limits_{|y|<x_k} w_2^{1-p'}(y)dy \tag{4.3.3}$$

holds for all $k \in \mathbb{Z}$. It is clear that $\{x_k\}$ is increasing and

$$\mathbb{R} \setminus \{0\} = \cup_{k\in\mathbb{Z}} E_k,$$

where $E_k := \{y \in \mathbb{R} : x_k \le |y| < x_{k+1}\}$. Moreover, it is easy to verify that $2^k = \int_{E_k} w_2^{1-p'}$. Let $f \ge 0$. We have that

$$I := \iint\limits_{\mathbb{R}^2} v(x,y)\big(I_\alpha H f(x,y)\big)^q dxdy$$

$$= \sum_{k\in\mathbb{Z}} \iint\limits_{\mathbb{R}\times E_k} v(x,y) \bigg(\int\limits_{\mathbb{R}} \int\limits_{|\tau|<|y|} \frac{f(t,\tau)}{|x-t|^{1-\alpha}} dt d\tau \bigg)^q dxdy$$

$$\le \sum_{k\in\mathbb{Z}} \int\limits_{\mathbb{R}} \bigg(\int\limits_{E_k} v(x,y)dy \bigg) \bigg(\int\limits_{\mathbb{R}} \bigg(\int\limits_{|\tau|<x_{k+1}} \frac{f(t,\tau)}{|x-t|^{1-\alpha}} d\tau \bigg) dt \bigg)^q dx$$

$$= \sum_{k\in\mathbb{Z}} \int\limits_{\mathbb{R}} \overline{v}_k(x) \bigg(\int\limits_{\mathbb{R}} |x-t|^{\alpha-1} F_k(t)dt \bigg)^q dx := J,$$

where

$$\overline{v}_k(x) := \int\limits_{E_k} v(x,y)dy; \;\; F_k(t) := \int\limits_{|\tau|<x_{k+1}} f(t,\tau)d\tau.$$

Further, by Theorem 4.1.1 it follows that

$$J \le c\big(\max\{D_1, D_2\}\big)^q \sum_{j\in\mathbb{Z}} \bigg[\int\limits_{\mathbb{R}} w_1(x) \bigg(\int\limits_{|y|<x_j} w_2^{1-p'}(y)dy \bigg)^{1-p} (F_j(x))^p dx \bigg]^{q/p}$$

$$\le c\bigg[\int\limits_{\mathbb{R}} w_1(x) \sum_{j\in\mathbb{Z}} \bigg(\int\limits_{|y|<x_j} w_2^{1-p'}(y)dy \bigg)^{1-p} \bigg(\sum_{k=-\infty}^{j} \int\limits_{E_k} f(x,\tau)d\tau \bigg)^p dx \bigg]^{q/p}.$$

On the other hand, (4.3.3) yields that

$$\sum_{k=n}^{+\infty}\Big(\int\limits_{|\tau|<x_k} w_2^{1-p'}(\tau)d\tau\Big)^{1-p}\Big(\sum_{k=-\infty}^{n}\int\limits_{x_k<|\tau|<x_{k+1}} w_2^{1-p'}(\tau)d\tau\Big)^{p-1}$$

$$=\Big[\sum_{k=n}^{+\infty}\Big(\int\limits_{|\tau|<x_k} w_2^{1-p'}(\tau)d\tau\Big)^{1-p}\Big]\Big(\int\limits_{|\tau|<x_{n+1}} w_2^{1-p'}(y)dy\Big)^{p-1}$$

$$=\Big(\sum_{k=n}^{+\infty}2^{k(1-p)}\Big)2^{(n+1)(p-1)}=c<\infty$$

for all $n\in\mathbb{Z}$. Hence, by Theorem 1.1.4 and Hölder's inequality with respect to the integral $\int_{x_j<|\tau|<x_{j+1}} f(x,\tau)d\tau$ we conclude that

$$I\le c\Big[\int\limits_{\mathbb{R}} w_1(x)\sum_{j\in\mathbb{Z}}\Big(\int\limits_{x_j<|y|<x_{j+1}} w_2^{1-p'}(y)dy\Big)^{1-p}\Big(\int\limits_{x_j<|\tau|<x_{j+1}} f(x,\tau)d\tau\Big)^{p}dx\Big]^{q/p}$$

$$\le c\Big[\int\limits_{\mathbb{R}} w_1(x)\sum_{j\in\mathbb{Z}}\Big(\int\limits_{x_j<|\tau|<x_{j+1}} w_2(\tau)f^p(x,\tau)d\tau\Big)dx\Big]^{q/p}=c\|f\|^q_{L^p_w(\mathbb{R}^2)}.$$

If $S<\infty$, then without loss of generality we can assume that$S=1$. In this case we choose the sequence $\{x_k\}_{k=-\infty}^{0}$ for which (4.3.3) holds for all $k\le 0$. Arguing as in the case $S=\infty$ and using Lemma 1.1.5 instead of Lemma 1.1.4, we obtain the desired result.

Necessity. If we substitute the function

$$f(x,y)=\chi_{\{x:|x-a|>r\}}(x)\chi_{\{y:|y|>t\}}(y)w^{1-p'}(x,y)|x-a|^{(p'-1)(\alpha-1)},$$

$$f(x,y)=\chi_{\{x:|x-a|<r\}}(x)\chi_{\{y:|y|>t\}}(y)w^{1-p'}(x,y)$$

($a\in\mathbb{R}$; $r,t>0$) in the two–weight inequality for $I_\alpha H$, then we are done.

The proof is complete. □

Proof of Theorem 4.3.2. Let $f\ge 0$. Represent $I_\alpha J_\beta$ as follows:

$$I_\alpha J_\beta f(x,y)=\int\limits_{\mathbb{R}}\int\limits_{|\tau|<|y|/2}\frac{f(t,\tau)}{|x-t|^{1-\alpha}|\tau-y|^{1-\beta}}dtd\tau$$

$$+\int\limits_{\mathbb{R}}\int\limits_{|y|/2<|\tau|<2|y|}\frac{f(t,\tau)}{|x-t|^{1-\alpha}|\tau-y|^{1-\beta}}dtd\tau:=S^{(1)}_{\alpha,\beta}f(x,y)+S^{(2)}_{\alpha,\beta}f(x,y).$$

We have that

$$\|I_\alpha J_\beta f\|^q_{L^q_v(\mathbb{R}^2)} \le c\big[\|S^{(1)}_{\alpha,\beta} f\|^q_{L^q_v(\mathbb{R}^2)} + \|S^{(2)}_{\alpha,\beta} f\|^q_{L^q_v(\mathbb{R}^2)}\big] := c[S^{(1)} + S^{(2)}].$$

If $|\tau| < |y|/2$, then $|y-\tau|^{\beta-1} \le c_\beta |y|^{\beta-1}$. Therefore, due to Proposition 4.3.1 we find that

$$S^{(1)} \le c \iint\limits_{\mathbb{R}^2} v(x,y)|y|^{q(\beta-1)} \bigg(\int\limits_{\mathbb{R}} \int\limits_{|\tau|<|y|/2} \frac{f(t,\tau)}{|x-t|^{1-\alpha}} dt d\tau \bigg)^q dxdy$$

$$\le c \bigg(\iint\limits_{\mathbb{R}^2} (f(x,y))^p u(x) dxdy \bigg)^{q/p}.$$

Here we have used the inequality

$$\max\{D_1', D_2'\} \le c \max\{B_3, B_4\},$$

where

$$D_1' := \sup_{\substack{a\in\mathbb{R}\\ r>0\\ s>0}} s^{1/p'} \bigg(\int\limits_{|x-a|>r} \int\limits_{|y|>s} \frac{v(x,y)}{|x-a|^{(1-\alpha)q}|y|^{(1-\beta)q}} dxdy \bigg)^{1/q} \bigg(\int\limits_{|x-a|<r} u^{1-p'}(x) dx \bigg)^{1/p'};$$

$$D_2' := \sup_{\substack{a\in\mathbb{R}\\ r>0\\ s>0}} s^{1/p'} \bigg(\int\limits_{|x-a|<r} \int\limits_{|y|>s} \frac{v(x,y)}{|y|^{(1-\beta)q}} dxdy \bigg)^{1/q} \bigg(\int\limits_{|x-a|>r} \frac{u^{1-p'}(x)}{|x-a|^{(1-\alpha)p'}} dx \bigg)^{1/p'}.$$

To check the latter inequality we take $r > 0$ and $s > 0$. Then $s \in [2^n, 2^{n+1})$ for some $n \in \mathbb{Z}$. Hence

$$s^{q/p'} \bigg(\int\limits_{|x-a|<r} u^{1-p'}(x) dx \bigg)^{q/p'} \int\limits_{|x-a|>r} \int\limits_{|y|>s} \frac{v(x,y)}{|x-a|^{(1-\alpha)q}|y|^{(1-\beta)q}} dxdy$$

$$\le c 2^{nq/p'} \bigg(\int\limits_{|x-a|<r} u^{1-p'}(x) dx \bigg)^{q/p'} \int\limits_{|x-a|>r} \int\limits_{|y|>2^n} \frac{v(x,y)}{|x-a|^{(1-\alpha)q}|y|^{(1-\beta)q}} dxdy$$

$$= c 2^{nq/p'} \bigg(\int\limits_{|x-a|<r} u^{1-p'}(x) dx \bigg)^{q/p'}$$

$$\times \sum_{k=n}^{\infty} 2^{k(\beta-1)q} \int\limits_{|x-a|>r} \int\limits_{2^k<|y|<2^{k+1}} \frac{v(x,y)}{|x-a|^{(1-\alpha)q}} dxdy \le c B_3^q 2^{nq/p'} \sum_{k=n}^{\infty} 2^{-kq/p'} = c B_3^q.$$

Hence $D_1' \le cB_3$. Analogously it can be proved that $D_2' \le cB_4$.

Further, Hölder's inequality, Theorem 4.1.1 and simple calculations yield

$$S^{(2)} \le \int\limits_{\mathbb{R}^2} v(x,y)\Big(\int\limits_{\mathbb{R}} |x-t|^{\alpha-1}\Big(\int\limits_{|y|/2<|\tau|<2|y|} f^p(t,\tau)d\tau\Big)^{1/p}$$

$$\times\Big(\int\limits_{|y|/2<|\tau|<2|y|} |y-\tau|^{(\beta-1)p'}d\tau\Big)^{1/p'} dt\Big)^q dxdy$$

$$= c\sum_{j\in\mathbb{Z}} \int\limits_{\mathbb{R}} \int\limits_{2^j<|y|<2^{j+1}} v(x,y)|y|^{q(\beta-1/p)}$$

$$\times\Big(\int\limits_{\mathbb{R}} |x-t|^{\alpha-1}\Big(\int\limits_{|y|/2<|\tau|<2|y|} f^p(t,\tau)d\tau\Big)^{1/p} dt\Big)^q dx$$

$$\le c\sum_{j\in\mathbb{Z}} \int\limits_{\mathbb{R}} \Big(\int\limits_{2^j<|y|<2^{j+1}} v(x,y)|y|^{q(\beta-1/p)}dy\Big)$$

$$\times\Big(\int\limits_{\mathbb{R}} |x-t|^{\alpha-1}\Big(\int\limits_{2^{j-1}<|\tau|<2^{j+2}} f^p(t,\tau)d\tau\Big)^{1/p} dt\Big)^q dx$$

$$\le c(\max\{B_3,B_4\})^q \sum_{j\in\mathbb{Z}} \Big(\int\limits_{\mathbb{R}} \int\limits_{2^{j-1}<|x|<2^{j+2}} u(x) f^p(x,y)dxdy\Big)^{q/p}$$

$$\le c\Big(\iint\limits_{\mathbb{R}^2} f^p(x,y)u(x)dxdy\Big)^{q/p}.$$

Necessity follows in the same manner as in the proof of Theorem 4.3.1 has been obtained taking appropriate test functions; therefore we omit the details. The proof is complete. □

Taking into account the proof of Theorem 4.2.2 we can formulate the next statement for the classical fractional maximal function

$$M_\gamma f(x) := \sup_{I\ni x} |I|^{\gamma-1} \int\limits_I |f(y)|dy, \;\; x\in\mathbb{R},$$

where the supremum is taken over all bounded intervals I containing x. The proof is omitted.

Lemma 4.3.3. *Let $1 < p < q < \infty$. Suppose that ρ is an one-dimensional weight. Assume that $0 < \gamma < 1/p$. Then the inequality*

$$\Big(\int_{\mathbb{R}} (M_\gamma f(x))^q \rho(x)dx\Big)^{1/q} \leq c\Big(\int_{\mathbb{R}} |f(x)|^p (\widetilde{M}_\gamma \rho)^{p/q}(x)dx\Big)^{1/p}$$

holds with a positive constant c independent of f and ρ, where

$$(\widetilde{M}_\gamma \rho)^{p/q}(x) = \sup_{I \ni x} |I|^{q(\gamma - 1/p)} \int_I \rho(x)dx.$$

Proposition 4.1.2 and Lemma 4.3.3 yield the following corollary:

Corollary 4.3.1. *Let $1 < p < q < \infty$ and let $\rho \in A_\infty(\mathbb{R})$. Suppose that $0 < \gamma < 1/p$. Then the inequality*

$$\Big(\int_{\mathbb{R}} |I_\gamma f(x)|^q \rho(x)dx\Big)^{1/q} \leq c\Big(\int_{\mathbb{R}} |f(x)|^p (\widetilde{M}_\gamma \rho)^{p/q}(x)dx\Big)^{1/p}$$

holds with a positive constant c independent of f.

Theorem 4.3.3. *Let $1 < p < q < \infty$ and let $1/p - 1/q < \alpha, \beta < 1/p$. Suppose that $v \in A_\infty^{(y)}$. Then the following inequality holds:*

$$\Big(\iint_{\mathbb{R}^2} (M_\alpha I_\beta f)^q(x,y)v(x,y)dxdy\Big)^{1/q} \leq c\Big(\iint_{\mathbb{R}^2} |f(x,y)|^p (\widetilde{M}_{\alpha,\beta} v)^{p/q}(x,y)dxdy\Big)^{1/p}$$

with a positive constant c independent of f, where $\widetilde{M}_{\alpha,\beta}$ is defined in Theorem 4.2.2.

Proof. As before we assume without loss of generality that f is non-negative bounded and has a compact support. Using the notation in the proof of Theorem 4.3.1 find that

$$S := \iint_{\mathbb{R}^2} (N_\alpha I_\beta f)^q(x,y)v(x,y)dxdy \leq 2^q \sum_{I \in D_m} \int_{\mathbb{R}} \overline{V}_I(y)\Big[\int_{\mathbb{R}} |y-\tau|^{\beta-1}\overline{F}_I(\tau)d\tau\Big]^q dy$$

$$= 2^q \sum_{I \in D_m} |I|^{(\alpha-1)q} \int_{\mathbb{R}} \Big(\int_I v(x,y)dx\Big)\Big(\int_{\mathbb{R}} \overline{F}_I(\tau)|y-\tau|^{\beta-1}d\tau\Big)^q dy$$

$$\Big[\text{by Corollary 4.3.1 and the fact } \int_I v(x,\cdot)dx \in A_\infty(\mathbb{R})$$

$$\text{with the constant independent of } I\Big]$$

$$\leq c \sum_{I \in D_m} |I|^{(\alpha-1)q} \Bigg[\int_{\mathbb{R}} \Big(\int_I f(t,y)dt \Big)^p \Big[\widetilde{M}_\beta \Big(\int_I v(x,\cdot)dx \Big) \Big]^{p/q} (y)dy \Bigg]^{q/p}$$

[by generalized Hölder's inequality]

$$\leq c \sum_{I \in D_m} |I|^{(\alpha-1)q} \Bigg(\int_I \Bigg[\int_{\mathbb{R}} f^p(t,y) \Big[\widetilde{M}_\beta \Big(\int_I v(x,\cdot)dx \Big) \Big]^{p/q} (y)dy \Bigg]^{1/p} dt \Bigg)^q$$

$$= c \sum_{I \in D_m} \frac{1}{|I|^{(1-\alpha)q}} \Bigg(\int_I \Bigg[\int_{\mathbb{R}} f^p(t,y) \Bigg[\sup_{\substack{J \ni y \\ J \in D_m}} \frac{1}{|J|^{(1/p-\beta)q}} \int_I \int_J v(x,\tau)dxd\tau \Bigg]^{p/q} dy \Bigg]^{1/p} dt \Bigg)^q$$

$$= c \sum_{I \in D_m} \frac{1}{|I|^{q/p'}} \Bigg(\int_I \Bigg[\int_{\mathbb{R}} f^p(t,y) \Bigg[\sup_{\substack{J \ni y \\ J \in D_m}} \frac{1}{|I|^{(1/p-\alpha)q}|J|^{(1/p-\beta)q}}$$

$$\times \iint_{I \times J} v(x,\tau)dxd\tau \Bigg]^{\frac{p}{q}} dy \Bigg]^{\frac{1}{p}} dt \Bigg)^q$$

[by Theorem 3.1.1 with $w \equiv 1$]

$$\leq c \Bigg(\iint_{\mathbb{R}^2} (f(x,y))^p (\widetilde{M}_{\alpha,\beta} v)^{p/q}(x,y)dxdy \Bigg)^{q/p}.$$

Moreover, arguments similar to those of the proof of the previous theorem complete the proof. □

4.4 Solution of the Trace Problem

This section deals with the trace inequalities of the multiple operators studied above.

Theorem 4.4.1. *Let $1 < p < q < \infty$ and let $0 < \alpha, \beta < 1/p$. Then the following statements are equivalent:*

(i) $M_\alpha I_\beta$ is bounded from $L^p(\mathbb{R}^2)$ to $L^q_v(\mathbb{R}^2)$;

(ii) $M_{\alpha,\beta}$ is bounded from $L^p(\mathbb{R}^2)$ to $L^q_v(\mathbb{R}^2)$;

(iii)

$$B_5 := \sup_{I,J} \Bigg(\iint_{I \times J} v(x,y)dxdy \Bigg) |I|^{q(\alpha-1/p)} |J|^{q(\beta-1/p)} < \infty, \tag{4.4.1}$$

where I and J are arbitrary bounded intervals in $\mathbb{R}$.

Proof. To prove the implication (iii) ⇒ (i) observe that condition (4.1.1) implies that $\max\{B_1, B_2\} < \infty$ for $w \equiv 1$ (see Theorem 4.3.1 for the definition of B_1 and B_2). Indeed, simple calculation shows that

$$B_2 = c \sup_{\substack{a\in\mathbb{R};r>0\\ I\subset\mathbb{R}}} |I|^{\alpha-1/p} r^{\beta-1/p} \Big(\int_I \int_{|y-a|<r} v(x,y)dxdy \Big)^{1/q}$$

while

$$B_1 = c \sup_{\substack{a\in\mathbb{R};r>0\\ I\subset\mathbb{R}}} |I|^{\alpha-1/p} r^{1/p'} \Big(\int_I \int_{|y-a|>r} \frac{v(x,y)}{|y-a|^{(1-\beta)q}} dxdy \Big)^{1/q}.$$

Further,

$$|I|^{(\alpha-1/p)q} r^{q/p'} \int_I \int_{|y-a|>r} \frac{v(x,y)}{|y-a|^{(1-\beta)q}} dxdy$$

$$= |I|^{(\alpha-1/p)q} r^{q/p'} \sum_{k=0}^{\infty} \int_I \int_{2^k r<|y-a|<2^{k+1}r} \frac{v(x,y)}{|y-a|^{(1-\beta)q}} dxdy$$

$$\le |I|^{(\alpha-1/p)q} r^{q/p'} \sum_{k=0}^{\infty} (2^k r)^{(\beta-1)q} \int_I \int_{2^k r<|y-a|<2^{k+1}r} v(x,y)dxdy$$

$$\le cB_5 r^{q/p'} \sum_{k=0}^{\infty} (2^k r)^{-q/p'} = cB_5 \sum_{k=0}^{\infty} 2^{-kq/p'} = cB_5.$$

Hence, $B_2 \le cB_5^{1/q}$, and consequently, $\max\{B_1, B_2\} \le cB_5^{1/q}$ when $w \equiv 1$. Now it remains to apply Theorem 4.3.1. It is easy to see that the implication (ii) ⇒ (iii) follows from the boundedness of the operator $M_{\alpha,\beta}$ on the class of functions

$$f_{I,J}(x,y) = \chi_I(x)\chi_J(y).$$

Due to the inequality $M_{\alpha,\beta} f(x,y) \le M_\alpha I_\beta f(x,y)$ $(f \ge 0)$ we find that (i) ⇒ (ii). Summing us we have proved that (iii) ⇒ (i) ⇒ (ii) ⇒ (iii) so we are done. □

Theorem 4.4.2. *Let $1 < p < q < \infty$. Suppose that $0 < \alpha < 1$ and $\beta > 1/p$. Then the operator $I_\alpha J_\beta$ is bounded from $L^p(\mathbb{R}^2)$ to $L^q_v(\mathbb{R}^2)$ if and only if*

$$\sup_{\substack{a\in\mathbb{R}\\ r>0\\ k\in\mathbb{Z}}} \Big(\int_{a-r}^{a+r} \int_{2^k<|y|<2^{k+1}} v(x,y)dxdy \Big)^{1/q} r^{\alpha-1/p} 2^{k(\beta-1/p)} < \infty.$$

Proof of this statement follows immediately from Theorem 4.3.2 taking $u \equiv 1$; the details are omitted. □

In the diagonal $p = q$ case we have the following theorem:

Theorem 4.4.3. *Let $1 < p < \infty$. Suppose that $0 < \alpha < 1/p < \beta$. Then the operator $I_\alpha J_\beta$ is bounded from $L^p(\mathbb{R}^2)$ to $L^p_v(\mathbb{R}^2)$ if and only if $I_\alpha \mathcal{V}_j \in L^{p'}_{loc}(\mathbb{R})$ for all $j \in \mathbb{Z}$ and there exists a positive constant c such that for almost all $x \in \mathbb{R}$ and all $j \in \mathbb{Z}$ the inequality*

$$I_\alpha[I_\alpha \mathcal{V}_j]^{p'}(x) \le c I_\alpha[\mathcal{V}_j](x)$$

holds, where I_α is the one-dimensional potential (see (4.1.1) *for $n = 1$) and*

$$\mathcal{V}_j(x) \equiv \int\limits_{2^j<|y|<2^{j+1}} v(x,y)|y|^{\beta p-1}dy.$$

Proof. The theorem can be proved easier than in the case of Theorem 4.3.2 using Proposition 4.1.1. Indeed, let us first show *sufficiency*. Keeping the notation from of proof of Theorem 4.3.2 and applying Theorem 4.1.3 and Proposition 4.1.1 we have the following chain of inequalities:

$$\|S^{(1)}_{\alpha,\beta} f\|^p_{L^p_v(\mathbb{R}^2)} \le c \iint\limits_{\mathbb{R}^2} v(x,y)|y|^{(\beta-1)p}\Big(\int\limits_{\mathbb{R}}\int\limits_{|\tau|<|y|/2} \frac{f(t,\tau)}{|x-t|^{1-\alpha}}dtd\tau\Big)^p dxdy$$

$$= c\sum_{j\in\mathbb{Z}}\int\limits_{\mathbb{R}}\int\limits_{2^j<|y|<2^{j+1}} v(x,y)|y|^{(\beta-1)p}\Big(\int\limits_{\mathbb{R}}\int\limits_{|\tau|<|y|/2} \frac{f(t,\tau)}{|x-t|^{1-\alpha}}dtd\tau\Big)^p dxdy$$

$$\le c\sum_{j\in\mathbb{Z}} 2^{j(1-p)}\int\limits_{\mathbb{R}}\mathcal{V}_j(x)\Big(\int\limits_{\mathbb{R}}|x-t|^{\alpha-1}\Big(\int\limits_{|\tau|<2^j} f(t,\tau)d\tau\Big)dt\Big)^p dx$$

$$\le \sum_{j\in\mathbb{Z}} 2^{j(1-p)}\int\limits_{\mathbb{R}}\Big(\int\limits_{|\tau|<2^j} f(t,\tau)d\tau\Big)^p dt = c\int\limits_{\mathbb{R}}\sum_{j} 2^{j(1-p)}\Big(\int\limits_{|\tau|<2^j} f(t,\tau)d\tau\Big)^p dt$$

$$\le c\int\limits_{\mathbb{R}}\Big(\sum_{j}\int\limits_{2^j<|x|<2^{j+1}}|x|^{-p}\Big(\int\limits_{|\tau|<|x|} f(t,\tau)d\tau\Big)^p dx\Big)dt$$

$$= c\int\limits_{\mathbb{R}}\Big(\int\limits_{\mathbb{R}}|x|^{-p}\Big(\int\limits_{|\tau|<|x|} f(t,\tau)d\tau\Big)^p dx\Big)dt \le c\iint\limits_{\mathbb{R}^2} f^p(t,\tau)dtd\tau,$$

where $f \ge 0$. Using again Theorem 4.1.3 we find that

$$\|S^{(2)}_{\alpha,\beta} f\|^p_{L^p_v(\mathbb{R}^2)} \le c\iint\limits_{\mathbb{R}^2} v(x,y)\Big(\int\limits_{\mathbb{R}}|x-t|^{\alpha-1}\Big(\int\limits_{|y|/2<|\tau|<|y|} f^p(t,\tau)d\tau\Big)^{1/p}$$

$$\times\Big(\int\limits_{|y|/2<|\tau|<|y|} |y-\tau|^{(\beta-1)p'}d\tau\Big)^{1/p'}dt\Big)^{p}dxdy$$

$$\le c\sum_j\int\limits_{\mathbb{R}}\mathcal{V}_j(x)\Big(\int\limits_{\mathbb{R}}|x-t|^{\alpha-1}\Big(\int\limits_{2^{j-1}<|\tau|<2^{j+1}} f^p(t,\tau)d\tau\Big)^{1/p}dt\Big)^{p}dx$$

$$\le c\sum_j\int\limits_{\mathbb{R}}\int\limits_{2^{j-1}<|\tau|<2^{j+1}} f^p(t,\tau)dtd\tau \le c\|f\|^p_{L^p(\mathbb{R}^2)}.$$

Necessity. To prove necessity we observe that the boundedness of the operator $I_\alpha J_\beta$ from $L^p(\mathbb{R}^2)$ to $L^p_v(\mathbb{R}^2)$ implies the inequality

$$\int\limits_{\mathbb{R}}\mathcal{V}_j(x)\Big(\int\limits_{\mathbb{R}}|x-t|^{\alpha-1}g(t)dt\Big)^{p}dx \le c\|g\|^p_{L^p(\mathbb{R})},$$

for all non-negative functions g defined on $\mathbb{R}$ and all $j\in\mathbb{Z}$. This is possible if we put the functions of the form $f_j(x,y)=g(x)\chi_{\{y:2^{j-2}<|y|<2^{j-1}\}}(y)$ in the inequality

$$\|I_\alpha J_\beta f\|_{L^p_v(\mathbb{R}^2)}\le c\|f\|_{L^p(\mathbb{R}^2)}.$$

Further, by Theorem 4.1.3 we conclude that $I_\alpha\mathcal{V}_j\in L^{p'}_{loc}(\mathbb{R})$ for all $j\in\mathbb{Z}$ and

$$\sup_{\substack{x\in\mathbb{R},\\ j\in\mathbb{Z}}}\left(\frac{I_\alpha[I_\alpha\mathcal{V}_j]^{p'}(x)}{I_\alpha\mathcal{V}_j(x)}\right)^{1/p'}<\infty.$$

The proof is complete. □

4.5 Riesz Potentials with Product Kernels

We begin this section with the following statement:

Theorem 4.5.1. *Let $1<p<q<\infty$ and let $0<\alpha,\beta<1$. Suppose that $w(x,y)=w_1(x)w_2(y)$ with $w_1^{1-p'}\in RD^{(d)}(\mathbb{R})$ and $v\in A^{(x)}_\infty(\mathbb{R})$ uniformly in the second variable. Then $I_{\alpha,\beta}$ is bounded from $L^p_w(\mathbb{R}^2)$ to $L^q_v(\mathbb{R}^2)$ if and only if $\max\{B_1,B_2\}<\infty$ (see Theorem 4.3.1 for the definition of B_1 and B_2).*

Proof. The proof follows easily by applying Theorem 4.3.1 and Proposition 4.1.2 to the first variable uniformly with respect to the second one. □

Theorem 4.5.1 yields the next statement:

Theorem 4.5.2. *Let $1<p<q<\infty$, $0<\alpha,\beta<1/p$. Suppose that a two-dimensional weight $v(x,y)$ belongs either to $A^{(x)}_\infty(\mathbb{R})$ or to $A^{(y)}_\infty(\mathbb{R})$. Then the following statements are equivalent:*

(i) $I_{\alpha,\beta}$ *is bounded from* $L^p(\mathbb{R}^2)$ *to* $L^q_v(\mathbb{R}^2)$;
(ii) (4.4.1) *holds.*

Proof. First we prove the implication (ii) $\Rightarrow$ (i). We have already seen (see the proof of Theorem 4.4.1) that $\max\{B_1, B_2\} \le cB_5^{1/q}$. Further, let $v \in A_\infty^{(x)}$. Then the result follows directly from Theorem 4.5.1. Let now $v \in A_\infty^{(y)}$. Then, taking into account the previous case we have that

$$\iint\limits_{\mathbb{R}^2} v(x,y)(I_{\alpha,\beta}f)^q(x,y)dxdy = \iint\limits_{\mathbb{R}^2} v^*(y,x)(I_{\beta,\alpha}f_1)^q(y,x)dydx \le c\|f\|^q_{L^p(\mathbb{R}^2)},$$

where $v^*(y,x) = v(x,y)$, $f_1(\tau,t) = f(t,\tau)$. It remains to notice that the condition $v \in A_\infty^{(y)}$ is equivalent to the fact $v^* \in A_\infty^{(x)}$. The implication (ii) $\Rightarrow$ (i) is proved.

The fact that (i) $\Rightarrow$ (ii) follows from Theorem 4.4.1 by using the obvious inequality $M_{\alpha,\beta}f(x,y) \le I_{\alpha,\beta}f(x,y)$, $f \ge 0$. The proof is complete. □

Theorem 4.5.3. *Let* $1 < p < q < \infty$ *and let* $1/p - 1/q < \alpha, \beta < 1/p$. *Suppose that* $v \in A_\infty^{(x)}(\mathbb{R}) \cap A_\infty^{(y)}(\mathbb{R})$. *Then the following inequality*

$$\Big(\iint\limits_{\mathbb{R}^2} |(I_{\alpha,\beta}f)(x,y)|^q v(x,y)dxdy\Big)^{1/q} \le c\Big(\iint\limits_{\mathbb{R}^2} |f(x,y)|^p(\widetilde{M}_{\alpha,\beta}v)^{p/q}(x,y)dxdy\Big)^{1/p}$$

holds, where $\widetilde{M}_{\alpha,\beta}$ *is defined in Theorem 4.2.2 and the positive constant* c *does not depend on* f.

Proof. We have that

$$\iint\limits_{\mathbb{R}^2} |I_{\alpha,\beta}f(x,y)|^q v(x,y)dxdy$$

$$\le \int\limits_{\mathbb{R}}\Big(\int\limits_{\mathbb{R}} v(x,y)\Big(\int\limits_{\mathbb{R}} |x-t|^{\alpha-1}\Big(\int\limits_{\mathbb{R}} \frac{|f(t,\tau)|}{|y-\tau|^{1-\beta}}d\tau\Big)dt\Big)^q dy\Big)dx$$

$$\Big[\text{ by Proposition 4.1.2 since } v \in A_\infty^{(x)}(\mathbb{R})\Big]$$

$$\le c\iint\limits_{\mathbb{R}^2} (M_\alpha I_\beta f(x,y))^q v(x,y)dxdy$$

$$\Big[\text{by Theorem 4.3.3 since } v \in A_\infty^{(y)}(\mathbb{R})\Big]$$

$$\le c\Big(\iint\limits_{\mathbb{R}^2} |f(x,y)|^p(\widetilde{M}_{\alpha,\beta}v)^{p/q}(x,y)dxdy\Big)^{q/p}.$$

The proof is complete. □

To formulate our further results we need some definitions:

Definition 4.5.1. We say that a two-dimensional weight function $\rho(x,y)$ defined on $\mathbb{R}^2$ satisfies the reverse doubling condition at 0 with respect to y uniformly to x, if there exist constants $\eta_1 > 1$ and $\eta_2 > 1$ such that for all $r > 0$ and almost every $x \in \mathbb{R}$ the inequality

$$\int\limits_{B(0,\eta_1 r)} \rho(x,y)dy \geq \eta_2 \int\limits_{B(0,r)} \rho(x,y)dy$$

holds. In this case we write $\rho \in RD_0(y,\eta_1,\eta_2)$ (or $\rho \in RD_0(y)$).

Definition 4.5.2. A weight $\rho(x,y)$ belongs to the class $RD_\infty(y,\mu_1,\mu_2)$ (or $RD_\infty(y)$), where $\mu_1,\mu_2 > 1$, if the inequality

$$\int\limits_{\mathbb{R}^2 \setminus B(0,r/\mu_1)} \rho(x,y)dy \geq \mu_2 \int\limits_{\mathbb{R}^2 \setminus B(0,r)} \rho(x,y)dy$$

holds for all $r > 0$ and almost every $x \in \mathbb{R}$.

Analogously, we can define the classes $RD_0(x)$, $RD_\infty(x)$.

It is known that (see, e.g., [168], [176] for the one-dimensional case) that if ρ satisfies the doubling condition on $\mathbb{R}$ with respect to the second variable uniformly to x (resp. with respect to the first variable uniformly to y), that is,

$$\rho_x\big(I(a,2r)\big) \leq b\rho_x\big(I(a,r)\big)$$

$$\Big(\text{ resp. } \rho_y\big(I(a,2r)\big) \leq b\rho_y\big(I(a,r)\big)\Big),$$

then $\rho \in RD_0(y)$ (resp. $\rho \in RD_0(x)$).

Our next aim in this section to prove the following statements (two theorems and a corollary):

Theorem 4.5.4. *Let $1 < p < q < \infty$, $0 < \alpha,\beta < 1$. Suppose that the $w^{1-p'} \in RD_0(y,\eta_1,\eta_2)$ and $\frac{w^{1-p'}(x,y)}{|y|^{(1-\beta)p'}} \in RD_\infty(y,\mu_1,\mu_2)$. Assume also that w satisfies the condition: there exists a positive constant c such that for a.e. $y \in \mathbb{R}$ and a.e. $t \in \mathbb{R}$ the inequality*

$$\int\limits_{\frac{|y|}{\eta_1}<|\tau|<\mu_1^2|y|} \frac{w^{1-p'}(t,\tau)}{|y-\tau|^{(1-\beta)p'}}d\tau \leq \frac{c}{|y|^{(1-\beta)p'}} \int\limits_{\frac{|y|}{\eta_1^2}<|\tau|<\frac{|y|}{\eta_1}} w^{1-p'}(t,\tau)d\tau \tag{4.5.1}$$

holds. Then $I_{\alpha,\beta}$ is bounded from $L^p_w(\mathbb{R}^2)$ to $L^q_v(\mathbb{R}^2)$ if and only if the following four conditions are satisfied simultaneously:

(i)

$$\sup_{\substack{a\in\mathbb{R}\\ r>0\\ s>0}}\bigg(\int\limits_{|x-a|<r}\int\limits_{|y|>s}\frac{w^{1-p'}(x,y)}{|y|^{(1-\beta)p'}}dxdy\bigg)^{1/p'}\bigg(\int\limits_{|x-a|>r}\int\limits_{|y|<s}\frac{v(x,y)}{|x-a|^{(1-\alpha)q}}dxdy\bigg)^{1/q}<\infty;$$

(ii)

$$\sup_{\substack{a\in\mathbb{R}\\ r>0\\ s>0}}\bigg(\int\limits_{|x-a|>r}\int\limits_{|y|>s}\frac{w^{1-p'}(x,y)}{|x-a|^{(1-\alpha)p'}|y|^{(1-\beta)p'}}dxdy\bigg)^{1/p'}\bigg(\int\limits_{|x-a|<r}\int\limits_{|y|<s}v(x,y)dxdy\bigg)^{1/q}<\infty;$$

(iii)

$$\sup_{\substack{a\in\mathbb{R}\\ r>0\\ s>0}}\bigg(\int\limits_{|x-a|<r}\int\limits_{|y|<s}w^{1-p'}(x,y)dxdy\bigg)^{1/p'}\bigg(\int\limits_{|x-a|>r}\int\limits_{|y|>s}\frac{v(x,y)}{|x-a|^{(1-\alpha)q}|y|^{(1-\beta)q}}dxdy\bigg)^{1/q}<\infty;$$

(iv)

$$\sup_{\substack{a\in\mathbb{R}\\ r>0\\ s>0}}\bigg(\int\limits_{|x-a|>r}\int\limits_{|y|<s}\frac{w^{1-p'}(x,y)}{|x-a|^{(1-\alpha)p'}}dxdy\bigg)^{1/p'}\bigg(\int\limits_{|x-a|<r}\int\limits_{|y|>s}\frac{v(x,y)}{|y|^{(1-\beta)q}}dxdy\bigg)^{1/q}<\infty.$$

Theorem 4.5.5. *Let $1<p<q<\infty$, $0<\alpha,\beta<1$. Suppose that a two-dimensional weight w is a product of two one-dimensional weights: $w(x,y)=w_1(x)w_2(y)$. Assume that, in addition, the condition*

$$\int\limits_{\frac{|y|}{2}<|\tau|<2|y|}\frac{w_2^{1-p'}(\tau)}{|y-\tau|^{(1-\beta)p'}}d\tau\leq\frac{c}{|y|^{(1-\beta)p'}}\int\limits_{\frac{|y|}{4}<|\tau|<\frac{|y|}{2}}w_2^{1-p'}(\tau)d\tau \tag{4.5.2}$$

holds. Then $I_{\alpha,\beta}$ is bounded from $L^p_w(\mathbb{R}^2)$ to $L^q_v(\mathbb{R}^2)$ if and only if conditions (i)-(iv) of Theorem 4.5.4 are satisfied.

Remark 4.5.1. For example, if $1/p<\beta<1$, then the weight function $w(x,y)$ which is increasing with respect to y uniformly to $x\in\mathbb{R}$ satisfies condition (4.5.1).

Corollary 4.5.1. *Let $1<p<q<\infty$, $0<\alpha<1$, $1/p<\beta<1$. Then the two-weight inequality*

$$\bigg(\iint\limits_{\mathbb{R}^2}|(I_{\alpha,\beta}f)(x,y)|^q v(x,y)dxdy\bigg)^{1/q}\leq c\bigg(\iint\limits_{\mathbb{R}^2}|f(x,y)|^p|y|^\gamma dxdy\bigg)^{1/p}$$

holds if and only if

(i)

$$\beta p - 1 < \gamma < p - 1$$

and

(ii)

$$\sup_{\substack{a\in\mathbb{R}\\ r>0\\ t>0}} \left(\int_{a-r}^{a+r} \int_{|y|>t} \frac{v(x,y)}{|y|^{(1-\alpha)q}} dxdy \right)^{1/q} r^{\alpha-1/p} 2^{k(-\gamma/p+1/p')} < \infty.$$

To prove Theorem 4.5.4 we need some lemmas.

Lemma 4.5.1. *Let $1 < p < q < \infty$, $0 < \alpha, \beta < 1$. Suppose that $w^{1-p'} \in RD_0(y, \eta_1, \eta_2)$. Then the operator*

$$I^{(1)}_{\alpha,\beta} f(x,y) := \int_{\mathbb{R}} \int_{|\tau|\le |y|/\eta_1} f(t,\tau)|x-t|^{\alpha-1}|y-\tau|^{\beta-1} dt d\tau, \quad x, y \in \mathbb{R},$$

is bounded from $L^p_w(\mathbb{R}^2)$ to $L^q_v(\mathbb{R}^2)$ if conditions (iii) and (iv) of Theorem 4.5.4 are satisfied.

Proof. In the sequel we will use the notation:

$$\overline{v}_k(x) := \int_{\eta_1^k<|y|<\eta_1^{k+1}} \frac{v(x,y)}{|y|^{(1-\beta)q}} dy;$$

$$F_k(x) := \int_{|\tau|\le\eta_1^k} f(x,\tau)d\tau; \;\; E_k := \{y \in \mathbb{R} : \eta_1^k \le |y| \le \eta_1^{k+1}\}.$$

Let $f \ge 0$. We have that

$$J := \iint_{\mathbb{R}^2} v(x,y)\big(I^{(1)}_{\alpha,\beta} f(x,y)\big)^q dxdy$$

$$= \sum_{k\in\mathbb{Z}} \iint_{\mathbb{R}\times E_k} v(x,y) \Big(\int_{\mathbb{R}} \int_{|\tau|\le|y|/\eta_1} \frac{f(t,\tau)}{|x-t|^{1-\alpha}|y-\tau|^{1-\beta}} dt d\tau \Big)^q dxdy$$

$$\le c \sum_{k\in\mathbb{Z}} \int_{\mathbb{R}} \overline{v}_k(x) \Big(\int_{\mathbb{R}} \Big(\int_{|\tau|\le\eta_1^k} \frac{f(t,\tau)}{|x-t|^{1-\alpha}} d\tau \Big) dt \Big)^q dx$$

$$= c \sum_{k\in\mathbb{Z}} \int_{\mathbb{R}} \overline{v}_k(x) \big(I_\alpha F_k\big)^q(x) dt,$$

$$\left[\text{ by Theorem 4.1.1 }\right]$$

$$\leq c\sum_{k\in\mathbb{Z}}\left[\int_{\mathbb{R}}\left(\int_{|y|<\eta_1^k} w^{1-p'}(x,y)dy\right)^{1-p}(F_k(x))^p dx\right]^{q/p}$$

$$\leq c\left[\int_{\mathbb{R}}\sum_{k\in\mathbb{Z}}\left(\int_{|y|<\eta_1^k} w^{1-p'}(x,y)dy\right)^{1-p}\left(\sum_{j=-\infty}^{k}\int_{E_{j-1}} f(x,\tau)d\tau\right)^p dx\right]^{q/p} := J_1.$$

On the other hand,

$$\sum_{j=k}^{\infty}\left(\int_{|\tau|<\eta_1^j} w^{1-p'}(x,y)dy\right)^{1-p}\left(\sum_{i=-\infty}^{k}\int_{E_{i-1}} w^{1-p'}(x,y)dy\right)^{p-1}$$

$$=\sum_{j=k}^{\infty}\left(\int_{|\tau|<\eta_1^j} w^{1-p'}(x,y)dy\right)^{1-p}\left(\int_{|y|<\eta_1^k} w^{1-p'}(x,y)dy\right)^{p-1}$$

$$\left[\text{by the condition } w^{1-p'}\in DC_0(y,\eta_1,\eta_2)\right]$$

$$\leq\sum_{j=k}^{\infty}\eta_2^{(j-k)(1-p)}=c<\infty$$

for every $x\in\mathbb{R}$. Consequently,

$$\sup_{\substack{k\in\mathbb{Z}\\ x\in\mathbb{R}}}\sum_{j=k}^{\infty}\left(\int_{|\tau|<\eta_1^j} w^{1-p'}(x,y)dy\right)^{1-p}\left(\sum_{i=-\infty}^{k}\int_{E_{i-1}} w^{1-p'}(x,y)dy\right)^{p-1}<\infty.$$

Hence, Theorem 1.1.4 and Hölder's inequality with respect to the integral

$$\int_{E_{j-1}} f(x,\tau)d\tau$$

yield that

$$J_1\leq c\left[\int_{\mathbb{R}}\sum_{j\in\mathbb{Z}}\left(\int_{E_{j-1}} w^{1-p'}(x,y)dy\right)^{1-p}\left(\int_{E_{j-1}} f(x,y)dy\right)^p dx\right]^{q/p}$$

$$\leq c\left[\int_{\mathbb{R}}\sum_{j\in\mathbb{Z}}\left(\int_{E_{j-1}} f^p(x,y)w(x,y)dy\right)dx\right]^{q/p}=c\|f\|^q_{L^p_w(\mathbb{R}^2)}.$$

Hence we conclude that

$$J\leq c\|f\|^q_{L^p_w(\mathbb{R}^2)}.$$

The proof is complete. □

Lemma 4.5.2. *Let* $1 < p < q < \infty$, $0 < \alpha, \beta < 1$. *Suppose that* $\frac{w^{1-p'}(x,y)}{|y|^{1-\beta}} \in RD_\infty(y, \mu_1, \mu_2)$. *Then the operator*

$$I^{(2)}_{\alpha,\beta}f(x,y) = \int\limits_{\mathbb{R}} \int\limits_{|\tau|\geq \mu_1^2|y|} f(t,\tau)|\tau|^{\beta-1}|x-t|^{\alpha-1}|y-\tau|^{\beta-1}dtd\tau, \quad x,y \in \mathbb{R},$$

is bounded from $L^p_w(\mathbb{R}^2)$ *to* $L^q_v(\mathbb{R}^2)$ *if and only if* (i) and (ii) of Theorem 4.5.4 hold.

Proof. The proof of this lemma is similar to that of the previous one. In *sufficiency* Theorem 4.1.7 and the condition $\frac{w^{1-p'}(x,y)}{|y|^{1-\beta}} \in RD_\infty(y, \mu_1, \mu_2)$ are used instead of Theorem 1.1.4 and the condition $w^{1-p'} \in RD_0(y, \eta_1, \eta_2)$, respectively. The details are omitted. □

Lemma 4.5.3. *Let* $1 < p < q < \infty$, $0 < \alpha, \beta < 1$. *Suppose that* $w^{1-p'}$ *satisfies condition* (4.5.1). *Then conditions* (iii) and (iv) of Theorem 4.5.4 guarantee the boundedness of the operator

$$I^{(3)}_{\alpha,\beta}f(x,y) = \int\limits_{\mathbb{R}} \int\limits_{|y|/\eta_1<|\tau|<\mu_1^2|y|} f(t,\tau)|x-t|^{\alpha-1}|y-\tau|^{\beta-1}dtd\tau$$

from $L^p_w(\mathbb{R}^2)$ to $L^q_v(\mathbb{R}^2)$.

Proof. Suppose that $f \geq 0$. Then we have that

$$S := \iint\limits_{\mathbb{R}^2} v(x,y)\big(I^{(3)}_{\alpha,\beta}f(x,y)\big)^q dxdy$$

$$\Big[\text{ by Hölder's inequality }\Big]$$

$$\leq \iint\limits_{\mathbb{R}^2} v(x,y)\Bigg[\int\limits_{\mathbb{R}} |x-t|^{\alpha-1}\Bigg(\int\limits_{|y|/\eta_1<|\tau|<\mu_1^2|y|} f^p(t,\tau)w(t,\tau)d\tau\Bigg)^{1/p}$$

$$\times\Bigg(\int\limits_{|y|/\eta_1<|\tau|<\mu_1^2|y|} \frac{w^{1-p'}(t,\tau)d\tau}{|y-\tau|^{(1-\beta)p'}}\Bigg)^{1/p'} dt\Bigg]^q dxdy$$

$$\Big[\text{ by condition (4.5.1) }\Big]$$

$$\leq c\iint\limits_{\mathbb{R}^2} \frac{v(x,y)}{|y|^{(1-\beta)q}}\Bigg[\int\limits_{\mathbb{R}} |x-t|^{\alpha-1}\Bigg(\int\limits_{|y|/\eta_1<|\tau|<\mu_1^2|y|} f^p(t,\tau)w(t,\tau)d\tau\Bigg)^{1/p}$$

$$\times\Bigg(\int\limits_{|y|/\eta_1^2<|\tau|<|y|/\eta_1} w^{1-p'}(t,\tau)d\tau\Bigg)^{1/p'} dt\Bigg]^q dxdy$$

$$\leq c\sum_{j\in\mathbb{Z}}\int_{\mathbb{R}}\bigg(\int_{\eta_1^j<|y|<\eta_1^{j+1}}\frac{v(x,y)}{|y|^{(1-\beta)q}}dy\bigg)\bigg[\int_{\mathbb{R}}|x-t|^{\alpha-1}\bigg(\int_{\eta_1^{j-1}<|\tau|<\mu_1^2\eta_1^{j+1}}f^p(t,\tau)w(t,\tau)d\tau\bigg)^{1/p}$$

$$\times\bigg(\int_{\eta_1^{j-2}<|\tau|<\eta_1^j}w^{1-p'}(t,\tau)d\tau\bigg)^{1/p'}dt\bigg]^q dx$$

$\big[$ by Theorem 4.1.1 $\big]$

$$\leq c\sum_{j\in\mathbb{Z}}\bigg[\int_{\mathbb{R}}\bigg(\int_{\eta_1^{j-2}<|\tau|<\eta_1^j}w^{1-p'}(t,\tau)d\tau\bigg)^{1-p}\bigg(\int_{\eta_1^{j-2}<|\tau|<\eta_1^j}w^{1-p'}(t,\tau)d\tau\bigg)^{p-1}$$

$$\times\bigg(\int_{\eta_1^{j-1}<|\tau|<\mu_1^2\eta_1^{j+1}}f^p(t,\tau)w(t,\tau)d\tau\bigg)dt\bigg]^{q/p}\leq c\bigg[\iint_{\mathbb{R}^2}f^p(t,\tau)w(t,\tau)dtd\tau\bigg]^{q/p}.$$

The proof is complete. □

Proof of Theorem 4.5.4. Representing $I_{\alpha,\beta}$ as follows:

$$I_{\alpha,\beta}f=I^{(1)}_{\alpha,\beta}f+I^{(2)}_{\alpha,\beta}f+I^{(3)}_{\alpha,\beta}f \tag{4.5.3}$$

and using Lemmas 4.5.1–4.5.3 we derive sufficiency of Theorem 4.5.4.
Necessity. Suppose that $I_{\alpha,\beta}$ is bounded from $L^p_w(\mathbb{R}^2)$ to $L^q_v(\mathbb{R}^2)$. Further, let

$$f(t,\tau)=\chi_{\{|t-a|>r\}}(t)\chi_{\{|\tau|<s\}}(\tau)\frac{w^{1-p'}(t,\tau)}{|t-a|^{(1-\alpha)(p'-1)}},$$

where r,s are positive numbers and $a\in\mathbb{R}$. Then it is easy to check that

$$\|I_{\alpha,\beta}f\|^q_{L^p(\mathbb{R}^2)}\geq c\bigg(\int_{|x-a|<r}\int_{|y|>s}v(x,y)|y|^{(\beta-1)q}dxdy\bigg)$$

$$\times\bigg(\int_{|t-a|>r}\int_{|\tau|<s}\frac{w^{1-p'}(t,\tau)}{|t-a|^{1-\alpha}}dtd\tau\bigg)^q.$$

On the other hand,

$$\|f\|^q_{L^p_w(\mathbb{R}^2)}=\bigg(\int_{|t-a|>r}\int_{|\tau|<s}w^{1-p'}(t,\tau)|t-a|^{(\alpha-1)p'}dtd\tau\bigg)^{q/p}.$$

Hence we have that (iv) holds. Taking the test function

$$F(t,\tau)=\chi_{\{|t-a|<r\}}(t)\chi_{\{|\tau|<s\}}(\tau)w^{1-p'}(t,\tau),$$

and using again the two–weight inequality for the operator $I_{\alpha,\beta}$ we have that (iii) is fulfilled. The other conditions follow in a similar manner choosing the test functions:

$$g(t,\tau)=\chi_{\{|t-a|<r\}}(t)\chi_{\{|\tau|>s\}}(\tau)w^{1-p'}(t,\tau)|\tau|^{(\beta-1)(1-p')},$$

$$G(t,\tau)=\chi_{\{|t-a|>r\}}(t)\chi_{\{|\tau|>s\}}(\tau)\frac{w^{1-p'}(t,\tau)}{|t-a|^{(1-\alpha)(p'-1)}}|\tau|^{(\beta-1)(1-p')}.$$

The proof is complete. □

Lemma 4.5.4. *Let $1<p<q<\infty$, $0<\alpha,\beta<1$. Suppose that $w(x,y)=w_1(x)w_2(y)$. Then the operator $I^{(1)}_{\alpha,\beta}$ with $\eta_1=2$ (see Lemma 4.5.1) is bounded from $L^p_w(\mathbb{R}^2)$ to $L^q_v(\mathbb{R}^2)$ if conditions* (iii) and (iv) of Theorem 4.5.4 are satisfied.

Proof. Let $S:=\int_{\mathbb{R}}w_2^{1-p'}(y)dy=\infty$. Suppose that $\{x_k\}_{k=-\infty}^{+\infty}$ is a sequence of positive numbers such that

$$2^k=\int\limits_{|y|<x_k/2}w_2^{1-p'}(y)dy \tag{4.5.4}$$

holds for all $k\in\mathbb{Z}$. It is clear that $\{x_k\}$ is increasing and

$$\mathbb{R}\setminus\{0\}=\cup_{k\in\mathbb{Z}}E_k,$$

where

$$E_k:=\{y\in\mathbb{R}:x_k\le|y|<x_{k+1}\}.$$

Besides this, it is easy to verify that $2^k=\int_{G_k}w_2^{1-p'}$, where

$$G_k:=\{y\in\mathbb{R}:x_k/2\le|y|<x_{k+1}/2\}.$$

Let us denote:

$$\overline{v}_k(x):=\int\limits_{E_k}\frac{v(x,y)}{|y|^{(1-\beta)q}}dy;\;\;F_k(t):=\int\limits_{|\tau|<x_{k+1}/2}f(t,\tau)d\tau.$$

Let $f\ge0$. We have that

$$\iint\limits_{\mathbb{R}^2}v(x,y)\big(I^{(1)}_{\alpha,\beta}f(x,y)\big)^q dxdy\le c\sum_{k\in\mathbb{Z}}\int\limits_{\mathbb{R}}\overline{v}_k(x)\Big(\int\limits_{\mathbb{R}}|x-t|^{\alpha-1}F_k(t)dt\Big)^q dx=:J.$$

Further, by Theorem 4.1.1 we obtain that

$$J\le c\sum_{j\in\mathbb{Z}}\bigg[\int\limits_{\mathbb{R}}w_1(x)\Big(\int\limits_{|y|<x_j/2}w_2^{1-p'}(y)dy\Big)^{1-p}(F_j(x))^p dx\bigg]^{q/p}$$

$$\leq c\Big[\int_{\mathbb{R}} w_1(x)\sum_{j\in\mathbb{Z}}\Big(\int_{|y|<x_j/2} w_2^{1-p'}(y)dy\Big)^{1-p}\Big(\sum_{k=-\infty}^{j}\int_{G_k} f(x,\tau)d\tau\Big)^p dx\Big]^{q/p}.$$

On the other hand, (4.5.4) yields

$$\sum_{k=n}^{+\infty}\Big(\int_{|\tau|<x_k/2} w_2^{1-p'}(\tau)d\tau\Big)^{1-p}\Big(\sum_{k=-\infty}^{n}\int_{G_k} w_2^{1-p'}(\tau)d\tau\Big)^{p-1}$$

$$=\sum_{k=n}^{+\infty}\Big(\int_{|\tau|<x_k/2} w_2^{1-p'}(\tau)d\tau\Big)^{1-p}\Big(\int_{|\tau|<x_{n+1}/2} w_2^{1-p'}(y)dy\Big)^{p-1}$$

$$=\Big(\sum_{k=n}^{+\infty}2^{k(1-p)}\Big)2^{(n+1)(p-1)}=c<\infty$$

for all $n\in\mathbb{Z}$. Hence, by using Theorem 1.1.4 and Hölder's inequality with respect to the integral $\int_{G_j} f(x,\tau)d\tau$ we conclude that

$$J\leq c\Big[\int_{\mathbb{R}} w_1(x)\sum_{j\in\mathbb{Z}}\Big(\int_{G_j} w_2^{1-p'}(y)dy\Big)^{1-p}\Big(\int_{G_j} f(x,\tau)d\tau\Big)^p dx\Big]^{q/p}$$

$$\leq c\Big[\int_{\mathbb{R}} w_1(x)\sum_{j\in\mathbb{Z}}\Big(\int_{G_j} w_2(\tau)f^p(x,\tau)d\tau\Big)dx\Big]^{q/p}=c\|f\|^q_{L^p_w(\mathbb{R}^2)}.$$

If $S<\infty$, then without loss of generality we can assume that $S=1$. In this case we choose the sequence $\{x_k\}_{k=-\infty}^{0}$ for which (4.5.4) holds for all $k\leq 0$. Arguing as in the case when $S=\infty$ and using Lemma 1.1.5 instead of Lemma 1.1.4, we obtain the desired result. The proof is complete. □

The proof of the next lemma is similar to that of the previous one. The proof is omitted.

Lemma 4.5.5. *Let* $1<p<q<\infty$, $0<\alpha,\beta<1$. *Suppose that* $w(x,y)=w_1(x)w_2(y)$. *Then the operator* $I^{(2)}_{\alpha,\beta}$ *with* $\mu_1=2^{1/2}$ *(see Lemma* 4.5.2 *) is bounded from* $L^p_w(\mathbb{R}^2)$ *to* $L^q_v(\mathbb{R}^2)$ *if and only if* (i) and (ii) of Theorem 4.5.4 hold.

The following lemma can be obtained in a similar manner as Lemma 4.5.3 was derived:

Lemma 4.5.6. *Let* $1<p<q<\infty$, $0<\alpha,\beta<1$. *Suppose that* $w^{1-p'}$ *satisfies condition* (4.5.2). *Then conditions* (iii) and (iv) of Theorem 4.5.4 guarantee the boundedness of the operator $I^{(3)}_{\alpha,\beta}$ with $\eta_1=\mu_1^2=2$ (see Lemma 4.5.3) from $L^p_w(\mathbb{R}^2)$ to $L^q_v(\mathbb{R}^2)$.

Proof of Theorem 4.5.5. Using representation (4.5.3) with $\eta_1 = \mu_1^2 = 2$ and Lemmas 4.5.4–4.5.6 we have *sufficiency* of the theorem. *Necessity* follows easily choosing appropriate test functions for the two–weight inequality for $I_{\alpha,\beta}$. □

Now we investigate the two–weight problem for the operator

$$\big(I_{\alpha_1,\dots,\alpha_n}f\big)(x_1,\dots,x_n) := \underbrace{\int_{\mathbb{R}}\cdots\int_{\mathbb{R}}}_{n} \frac{f(y_1,\dots,y_n)}{\prod\limits_{i=1}^{n}|x_i-y_i|^{1-\alpha_i}}dy_1\cdots dy_n, \quad x_i\in\mathbb{R}.$$

We will need the following notation:

$$\rho_{x_1,\dots,x_{i-1},x_{i+1},\dots,x_n}(E) := \int_E \rho(x_1,\dots,x_i,\dots,x_n)dx_i,$$

where ρ is an n dimensional weight and E is an one-dimensional measurable set.

We say that an n–dimensional weight u belongs to the class $A_\infty(\mathbb{R})$ with respect to the i–th variable uniformly to other ones ($u \in A_\infty^{(x_i)}(\mathbb{R})$) if the inequality

$$\frac{u_{x_1,\dots,x_{i-1},x_{i+1},\dots,x_n}(E)}{u_{x_1,\dots,x_{i-1},x_{i+1},\dots,x_n}(I)} \le c\left(\frac{|E|}{|I|}\right)^{\delta}$$

holds for all $x_1,\dots,x_{i-1},x_{i+1},\dots,x_n\in\mathbb{R}$, all intervals $I\subset\mathbb{R}$ and all measurable sets $E\subset I$.

Theorem 4.5.1′. *Let $1<p<q<\infty$ and let $0<\alpha_1,\dots,\alpha_n<1$. Suppose that w and v are n-dimensional weights satisfying the conditions $w(x_1,\dots,x_n) = w_1(x_1)\cdots w_n(x_n)$ with $w_1^{1-p'},\dots,w_{n-1}^{1-p'} \in RD^{(d)}(\mathbb{R})$; $v\in\cap_{i=1}^{n-1}A_\infty^{(x_i)}(\mathbb{R})$. Then $I_{\alpha_1,\dots,\alpha_n}$ is bounded from $L^p_w(\mathbb{R}^n)$ to $L^q_v(\mathbb{R}^n)$ if and only if*

$$B_1^{(n)} := \sup_{\substack{a\in\mathbb{R}\\ r>0\\ I_1\subset\mathbb{R}\\ \vdots\\ I_{n-1}\subset\mathbb{R}}} |I_1|^{\alpha_1-1}\cdots|I_{n-1}|^{\alpha_{n-1}-1}\Bigg(\int_{I_1}\cdots\int_{I_{n-1}}\int_{|y_n-a|<r} w^{1-p'}(y_1,\dots,y_n)dy_1\cdots dy_n\Bigg)^{1/p'}$$

$$\times\Bigg(\int_{I_1}\cdots\int_{I_{n-1}}\int_{|y_n-a|>r}\frac{v(y_1,\dots,y_n)}{|y_n-a|^{(1-\alpha_n)q}}dy_1\cdots dy_n\Bigg)^{1/q}<\infty;$$

$$B_2^{(n)} := \sup_{\substack{a\in\mathbb{R}\\ r>0\\ I_1\subset\mathbb{R}\\ \vdots\\ I_{n-1}\subset\mathbb{R}}} |I_1|^{\alpha_1-1}\cdots|I_{n-1}|^{\alpha_{n-1}-1}\Bigg(\int_{I_1}\cdots\int_{I_{n-1}}\int_{|y_n-a|>r}\frac{w^{1-p'}(y_1,\dots,y_n)}{|y_n-a|^{(1-\alpha_n)p'}}dy_1\cdots dy_n\Bigg)^{1/p'}$$

$$\times\Bigg(\int\limits_{I_1}\cdots\int\limits_{I_{n-1}}\int\limits_{|y_n-a|<r} v(y_1,\ldots,y_n)dy_1\cdots dy_n\Bigg)^{1/q}<\infty,$$

where $I_1,\ldots,I_{n-1}$ are arbitrary bounded intervals in $\mathbb{R}$.

Proof. *Sufficiency.* Suppose that the theorem is true for the operator $I_{\alpha_2,\ldots,\alpha_n}$. Then, by the arguments used in sufficiency of Theorem 4.3.1 we can prove that the operator

$$(M_{\alpha_1}I_{\alpha_2,\ldots,\alpha_n}f)(x_1,\ldots,x_n)=\sup_{I\ni x_1}\frac{1}{|I|^{1-\alpha_1}}\underbrace{\int\limits_{\mathbb{R}}\cdots\int\limits_{\mathbb{R}}}_{n-1}\frac{f(y_1,\ldots,y_n)}{\prod_{i=2}^{n}|y_i-x_i|^{1-\alpha_i}}dy_2\cdots dy_n$$

is bounded from $L^p_w(\mathbb{R}^n)$ to $L^q_v(\mathbb{R}^n)$ if the conditions $B_1^{(n)}<\infty$ and $B_2^{(n)}<\infty$ hold. Proposition 4.1.2 completes the proof of sufficiency.

Necessity follows from choosing appropriate test functions in the two–weight inequality for $I_{\alpha_1,\ldots,\alpha_n}$. □

The proof of the next statement is similar to that of Theorem 4.5.2.

Theorem 4.5.2′. *Let $1<p<q<\infty$, $0<\alpha_1,\ldots,\alpha_n<1/p$. Suppose that the n-dimensional weight v belongs at least to any $n-1$ classes among the following ones of weights $A_\infty^{(x_1)}(\mathbb{R}),\ldots,A_\infty^{(x_n)}(\mathbb{R})$. Then the following statements are equivalent:*

(i) $I_{\alpha_1,\ldots,\alpha_n}$ *is bounded from $L^p(\mathbb{R}^n)$ to $L^q_v(\mathbb{R}^n)$;*

(ii)

$$\sup_{I_1,\ldots,I_n}\Bigg(\int\limits_{I_1}\cdots\int\limits_{I_n}v(x_1,\ldots,x_n)dx_1\cdots dx_n\Bigg)|I_1|^{q(\alpha_1-1/p)}\cdots|I_n|^{q(\alpha_n-1/p)}<\infty,$$

where the supremum is taken with respect to all bounded intervals $I_1,\ldots,I_n$ in $\mathbb{R}$.

4.6 Some Remarks

Now we give remarks regarding the results derived above:

Remark 4.6.1. It is known that $(M_\gamma\rho)^\lambda$, where ρ is a one-dimensional function and $0<\lambda<1$, belongs to the Muckenhoupt's class A_1 (see, e.g., [155]), i.e., there exists a positive constant c such that for all intervals $I\subset\mathbb{R}$ with finite length the inequality

$$1/|I|\int\limits_I(M_\gamma\rho)^\lambda\le c\operatorname*{ess\,inf}_I(M_\gamma\rho)^\lambda$$

holds. Let now $0<\alpha,\beta<1/p$.

From the latter fact and the inequalities

$$(\widetilde{M}_{\alpha,\beta}v)(x,y) \le \sup_{I\ni x} \frac{1}{|I|^{q(1/p-\alpha)}} \int_I \bigg(\sup_{J\ni y} \frac{1}{|J|^{q(1/p-\beta)}} \int_J v(t,\tau)d\tau \bigg) dt := (\widetilde{M}_\alpha v_1(\cdot,y))(x);$$

$$(\widetilde{M}_{\alpha,\beta}v)(x,y) \le \sup_{J\ni y} \frac{1}{|J|^{q(1/p-\beta)}} \int_J \bigg(\sup_{I\ni x} \frac{1}{|I|^{q(1/p-\alpha)}} \int_I v(t,\tau)dt \bigg) d\tau := (\widetilde{M}_\beta v_2(x,\cdot))(y),$$

where

$$v_1(t,y) := \sup_{J\ni y} |J|^{q(\beta-1/p)} \int_J v(t,\tau)d\tau; \quad v_2(x,\tau) := \sup_{I\ni x} |I|^{q(\alpha-1/p)} \int_I v(t,\tau)dt,$$

it follows that $(\widetilde{M}_{\alpha,\beta}v)^\lambda(\cdot,y) \in A_1(\mathbb{R})$ for every $y \in \mathbb{R}$ and $(\widetilde{M}_{\alpha,\beta}v)^\lambda(x,\cdot) \in A_1(\mathbb{R})$ for every $x \in \mathbb{R}$.

Remark 4.6.2. An analysis of the proofs of the main results enables us to conclude that all statements concerning $M_{\alpha,\beta}$, $I_{\alpha,\beta}$ and $I_\alpha J_\beta$ hold also for the following operators defined on $\mathbb{R}^n \times \mathbb{R}^m$ $(n,m > 1)$:

$$(M^{(n,m)}_{\alpha,\beta} f)(x,y) := \sup_{B_n\times B_m \ni (x,y)} \frac{1}{|B_n|^{1-\alpha/n}|B_m|^{1-\beta/m}} \int_{B_n}\int_{B_m} |f(z,s)|dzds;$$

$$(I^{(n,m)}_{\alpha,\beta} f)(x,y) := \int_{\mathbb{R}^n}\int_{\mathbb{R}^m} \frac{f(z,s)}{|x-z|^{n-\alpha}|y-s|^{m-\beta}} dzds, \quad x\in\mathbb{R}^n, y\in\mathbb{R}^m;$$

$$(I^{(n)}_\alpha J^{(m)}_\beta f)(x,y) := \int_{\mathbb{R}^n} \int_{\{\tau\in\mathbb{R}^m:|\tau|<2|y|\}} \frac{f(t,\tau)}{|x-t|^{n-\alpha}|y-\tau|^{m-\beta}} dtd\tau, \quad x\in\mathbb{R}^n, y\in\mathbb{R}^m$$

respectively, where $0<\alpha<n$, $0<\beta<m$ and B_d is a ball in $\mathbb{R}^d$.

4.7 Notes and Comments on Chapter 4

The one-weight problem for the operator $I_{\alpha,\beta}$ has been solved in [85], [86]. From the historical point of view we recall that the one-weight criteria for the Riesz potentials I_α with single kernel defined in $\mathbb{R}^n$ have been derived in [126].

The pioneering results concerning the two–weight problem for I_α have been obtained in [156], [158], while $L^p - L^q$ two–weight criteria in more transparent form have been given in [45], [46], [88] for $1 < p < q < \infty$ (see also [52], [53], [161] regarding two–weight criteria for integral transforms with positive kernels). The proof of Theorem 4.1.1 is based on the two–weight weak–type criterion for the Riesz potentials given in [156] and on more transparent one established in [45], [46] (see also [88]). These criteria avoid the concept of capacity and can be easily verified.

Earlier, a capacitary characterization of those measures which guarantee the trace inequality for the operator I_α was derived in [6], [115].

In the case $w \equiv 1$, Theorem 4.1.1 (the trace inequality) has been obtained in [4].

For the solution of the two–weight problem for fractional maximal operators M_γ we refer to [154], [155], [176], [59] (see also [53, Chapter 4]). A two–weight criterion for the strong Hardy–Littlewood maximal functions has been obtained in [155], provided that the weight on the right-hand side satisfies some additional conditions, for instance, belongs to the Muckenhoupt A_p class separately in each variable, or is a product of one–dimensional weights. For some two–weight inequalities for fractional integrals and fractional maximal functions associated to starlike sets in $\mathbb{R}^n$ we refer to [27].

A Fefferman–Stein [41] type inequality for the Riesz potentials I_α in the case $p = q$ was established by E. Sawyer (see, e.g., [5]).

Some essential properties of the Muckenhoupt A_p classes defined on rectangles has been studied in [85], [42], [43] (see also [36], [49, Chapter IV]).

We notice that the estimates similar to (4.2.1) and (4.2.2) first have been established in [41] for the classical Hardy–Littlewood maximal functions (see also [49, p. 431]). For these estimates in the case of fractional maximal functions we refer to [153]– [155].

For some properties of potentials with product kernels in unweighted case see, e.g., [151, Chapter 5].

This chapter is mainly based on the papers [93]–[96].

Chapter 5

Strong Maximal Functions and Hilbert Transforms with Product Kernels

This chapter deals mainly with two–weight inequalities for the following operators:

$$M_S f(x,y) := \sup_{\substack{x\neq x_1\\ y\neq y_1}} \frac{1}{(x-x_1)(y-y_1)} \int_{x_1}^{x}\int_{y_1}^{y} |f(s,\tau)|ds d\tau,$$

$$(M_1M_2)f(x,y) := \sup_{x\neq x_1} \frac{1}{x-x_1} \int_{x_1}^{x} \left(\sup_{y\neq y_1} \frac{1}{y-y_1}\int_{y_1}^{y} |f(s,\tau)|d\tau\right)ds,$$

where $x_1, x, y_1, y \in \mathbb{R}$;

$$\mathcal{H}f(x,y) := \lim_{\varepsilon_1,\varepsilon_2\to 0} \int_{|x-t|>\varepsilon_1}\int_{|y-\tau|>\varepsilon_2} \frac{f(t,\tau)}{(x-t)(y-\tau)} dt d\tau, \quad x,y\in\mathbb{R}.$$

In particular, we establish two–weight criteria for these operators provided that the two–dimensional weight functions are even on $\mathbb{R}$ and increasing on $\mathbb{R}_+$ in each variable for the other fixed one. The multidimensional case of this problem is also studied. Examples of appropriate weights are presented.

5.1 Single Maximal Functions

First we formulate and prove some two–weight estimates of the maximal operator

$$Mg(x) := \sup_{x\neq t} \frac{1}{x-t}\int_{t}^{x} |g(\tau)|d\tau, \quad x,t\in\mathbb{R}.$$

It is easy to check that

$$Mg(x) \approx \sup_{r>0} \frac{1}{2r} \int\limits_{x-r}^{x+r} |g(\tau)|d\tau.$$

Theorem 5.1.1. *Let $1 < p < \infty$ and let the weight functions $\sigma, u : \mathbb{R} \mapsto \mathbb{R}_+$ be even on $\mathbb{R}$ and increasing on $\mathbb{R}_+$. Then M is bounded from $L^p_u(\mathbb{R})$ to $L^p_\sigma(\mathbb{R})$ if and only if*

$$A := \sup_{t>0} \bigg(\int\limits_t^\infty \sigma(s) s^{-p} ds \bigg)^{1/p} \bigg(\int\limits_0^t u^{1-p'}(s) ds \bigg)^{1/p'} < \infty. \tag{5.1.1}$$

Moreover, there exist positive constants c_1 and c_2 depending only on p such that

$$c_1 A \le \|M\| \le c_2 A.$$

Proof. *Sufficiency.* First observe that since v and w are monotonic, condition (5.1.1) implies the following inequality

$$\sigma(2t) \le cA^p u(t). \tag{5.1.2}$$

Without loss of generality assume that σ is absolutely continuous and $\sigma(t) = \sigma(0) + \int_0^t \varphi(s)ds$, where φ is a non-negative function.

We have that

$$\|Mg\|^p_{L^p_\sigma(\mathbb{R})} = \int\limits_{\mathbb{R}} (Mg(x))^p \sigma(|x|)dx = \sigma(0) \int\limits_{\mathbb{R}} (Mg(x))^p dx$$

$$+ \int\limits_{\mathbb{R}} \bigg(\int\limits_0^{|x|} \varphi(t)dt \bigg) (Mg(x))^p dx := I_1 + I_2.$$

If $\sigma(0) = 0$, then $I_1 = 0$. If $\sigma(0) \neq 0$, then by the boundedness of M in $L^p(\mathbb{R})$ (the Hardy–Littlewood–Wiener theorem) and inequality (5.1.2) we obtain that

$$I_1 \le c\sigma(0) \int\limits_{\mathbb{R}} |g(\tau)|^p d\tau \le c \int\limits_{\mathbb{R}} \sigma(|\tau|)|g(\tau)|^p d\tau \le cA^p \int\limits_{\mathbb{R}} u(|\tau|)|g(\tau)|^p d\tau = cA^p \|g\|^p_{L^p_u(\mathbb{R})}.$$

Further, Fubini's theorem yields

$$I_2 = \int\limits_0^\infty \varphi(t) \bigg(\int\limits_{|x|>t} (Mg(x))^p dx \bigg) dt \le c \int\limits_0^\infty \varphi(t) \bigg(\int\limits_{|x|>t} (Mg_{1,t}(x))^p dx \bigg) dt$$

$$+c\int_0^\infty \varphi(t)\Big(\int_{|x|>t}(Mg_{2,t}(x))^p dx\Big)dt := I_{21}+I_{22},$$

where

$$g_{1,t}(x)=g(x)\chi_{|x|<t/2}(x);\;\; g_{2,t}(x)=g(x)\chi_{|x|\ge t/2}(x).$$

Using the boundedness of M in $L^p(\mathbb{R})$ and inequality (5.1.2) we find that

$$I_{22}\le c\int_0^\infty \varphi(t)\Big(\int_{\mathbb{R}}(Mg_{2,t}(x))^p dx\Big)dt\le c\int_0^\infty \varphi(t)\Big(\int_{|x|>t/2}|g(x)|^p\Big)dt$$

$$=c\int_{\mathbb{R}}|f(x)|^p\Big(\int_0^{2|x|}\varphi(t)dt\Big)dx\le c\int_{\mathbb{R}}|g(x)|^p\sigma(2|x|)dx\le cA^p\int_{\mathbb{R}}|g(x)|^p u(|x|)dx$$

$$=cA^p\|g\|^p_{L^p_u(\mathbb{R})}.$$

Let us take a point x so that $|x|>t$. If $|y|<t/2$, then $|y|<|x|/2$. Consequently, $|x|\le|x-y|+|x|/2$. Hence, $|x|/2\le|x-y|$. Therefore if $r<|x|/2$, we get that

$$\frac{1}{r}\int_{x-r}^{x+r}|g_{1,t}(y)|dy\le\frac{1}{r}\int_{\{|x-y|<r;|x-y|\ge|x|/2\}}|g(y)|dy=0.$$

If $r\ge|x|/2$, then

$$\frac{1}{r}\int_{x-r}^{x+r}|g_{1,t}(y)|dy\le\frac{c}{|x|}\int_{\{y:|x-y|<r;|y|\le|x|/2\}}|g(y)|dy\le\frac{c}{|x|}\int_{|y|\le|x|/2}|g(y)|dy.$$

Hence, by using Theorem 4.1.4 we have that condition (5.1.1) guarantees the estimates

$$I_{21}\le c\int_0^\infty\varphi(t)\Big(\int_{|x|>t}\Big(\frac{1}{|x|}\int_{|y|<|x|}|g(y)|dy\Big)dx\Big)^p dt$$

$$=c\int_{\mathbb{R}}\frac{1}{|x|^p}\Big(\int_{|y|<|x|}|g(y)|dy\Big)^p\Big(\int_0^{|x|}\varphi(t)dt\Big)dx$$

$$\le c\int_{\mathbb{R}}\frac{\sigma(|x|)}{|x|^p}\Big(\int_{|y|<|x|}|g(y)|dy\Big)^p dx\le cA^p\|g\|^p_{L^p_u(\mathbb{R})}.$$

Necessity. First notice that

$$Mg(x) \geq c\frac{1}{|x|}\int\limits_{|y|<|x|}|g(y)|dy,$$

where the positive constant does not depend on x. Now taking the test function

$$f(x) = u^{1-p'}(x)\chi_{\{|x|<t\}}(x)$$

in the two–weight inequality for M and noting that

$$\int\limits_{|x|<t} u^{1-p'}(s)ds < \infty$$

for every $t > 0$, we find that (5.1.1) holds.

The proof is complete. □

Theorem 5.1.2. *Let $1 < p < \infty$. Suppose that the one–dimensional weights σ and u are even on $\mathbb{R}$ and decreasing on $\mathbb{R}_+$. Then M is bounded from $L^p_u(\mathbb{R})$ to $L^p_\sigma(\mathbb{R})$ if and only if*

$$A_1 := \sup_{r>0} A_1(r) := \sup_{r>0}\bigg(\frac{1}{r}\int\limits_0^r \sigma(s)ds\bigg)^{1/p}\bigg(\frac{1}{r}\int\limits_0^r u^{1-p'}(s)ds\bigg)^{1/p'} < \infty.$$

Moreover, there exist positive constants c_1 and c_2 depending only on p such that

$$c_1A_1 \leq \|M\| \leq c_2A_1.$$

Proof. *Sufficiency.* First note that the condition $A_1 < \infty$ guarantees the pointwise inequality:

$$\sigma(r) \leq cA_1^p u(8r), \quad r > 0. \tag{5.1.3}$$

Indeed, since σ and u are decreasing,

$$A_1 \geq A_1(r) \geq \frac{1}{r}\bigg(\int\limits_0^{r/16} v(t)dt\bigg)^{1/p}\bigg(\int\limits_{r/2}^r w^{1-p'}(t)dt\bigg)^{1/p'}$$

$$\geq c\sigma^{1/p}(r/16)u^{-1/p}(r/2).$$

Hence,

$$\sigma(r/16) \leq cA_1^p u(r/2)$$

for all $r > 0$, which implies (5.1.3).

Further, we have that

$$\int_{\mathbb{R}} \sigma(x)(Mf(x))^p dx = \sum_{k\in\mathbb{Z}} \int_{2^k<|x|<2^{k+1}} \sigma(x)(Mf(x))^p dx$$

$$\leq c\Bigg[\sum_{k\in\mathbb{Z}} \int_{2^k<|x|<2^{k+1}} \sigma(x)\big[M(f\chi_{|\cdot|<2^{k-1}})(x)\big]^p dx$$

$$+c\sum_{k\in\mathbb{Z}} \int_{2^k<|x|<2^{k+1}} \sigma(x)\big[M(f\chi_{2^{k-1}|\cdot|<2^{k+2}})(x)\big]^p dx$$

$$+c\sum_{k\in\mathbb{Z}} \int_{2^k<|x|<2^{k+1}} \sigma(x)\big[M(f\chi_{|\cdot|>2^{k+2}})(x)\big]^p dx\Bigg] := c[I_1+I_2+I_3].$$

Now we claim that if $2^k < |x| < 2^{k+1}$, then

$$M(f\chi_{|\cdot|<2^{k-1}})(x) \leq \frac{c}{|x|} \int_{|y|<|x|} |f(y)|dy.$$

Indeed, if $r < 2^{k-1}$, then

$$\frac{1}{r}\int_{x-r}^{x+r} |f(y)|\chi_{|y|<2^{k-1}} dy = 0$$

because

$$|x| \leq |x-y| + |y| \leq |x-y| + 2^{k-1} \leq |x-y| + |x|/2$$

and, consequently,

$$2^{k-1} \leq |x|/2 \leq |x-y|.$$

Let $r \geq 2^{k-1}$. Then it is obvious that for x satisfying $2^k < |x| < 2^{k+1}$,

$$\frac{1}{r}\int_{x-r}^{x+r} |f(y)|\chi_{|y|<2^{k-1}} dy \leq \frac{c}{|x|} \int_{|y|<|x|} |f(y)|dy.$$

On the other hand, by (5.1.3) we see that

$$\left(\int_t^\infty \sigma(x)x^{-p}\right)^{1/p}\left(\int_0^t u^{1-p'}(x)dx\right)^{1/p'} \leq c\sigma^{1/p}(t)u^{-1/p}(t) \leq cA_1.$$

Using the latter estimates and the two–weight inequality for the Hardy transform we find that

$$I_1 \le c\int\limits_{\mathbb{R}} \frac{\sigma(x)}{|x|^p}\Big(\int\limits_{|y|<|x|}|f(y)|dy\Big)^p dx \le cA_1^p\|f\|^p_{L^p_u(\mathbb{R})}.$$

For I_2, we find that

$$I_2 \le \sum_{k\in\mathbb{Z}}\sigma(2^k)\int\limits_{2^k<|x|<2^{k+1}}\big[M(f\chi_{2^{k-1}|\cdot|<2^{k+2}})(x)\big]^p dx$$

[by using the boundedness of M in $L^p(\mathbb{R})$ and inequality (5.1.3)]

$$\le cA_1^p\sum_{k\in\mathbb{Z}}u(2^{k+2})\int\limits_{2^{k-1}<|x|<2^{k+2}}|f(x)|^p dx \le$$

[since u is decreasing]

$$c\sum_{k\in\mathbb{Z}}\int\limits_{2^{k-1}<|x|<2^{k+2}}u(x)|f(x)|^p dx \le c\|f\|^p_{L^p_u(\mathbb{R})}.$$

It remains to estimate I_3. Notice that if $|y|>2^{k+2}$ and $2^k<|x|<2^{k+1}$, then $|x-y|\ge|y|-|x|\ge 2^{k+2}-2^{k+1}=2^{k+1}$. Now observe that if $r<2^{n+1}$, then

$$\frac{1}{r}\int\limits_{x-r}^{x+r}|f(y)|\chi_{|y|>2^{k+2}}(y)dy \le \frac{1}{r}\int\limits_{\{y:|y-x|<r,|y-x|>2^{k+1}\}}|f(y)|dy = 0.$$

Let now $r\ge 2^{k+1}$. Then $r\in[2^j,2^{j+1})$ for some $j\ge k+1$. Further, if $|x-y|<r$ and $2^k<|x|<2^{k+1}$, then $|y|\le|x-y|+|x|\le r+|x|\le 2^{j+1}+2^{k+1}\le 2^{j+1}+2^j\le 2^{j+2}$. Consequently, using these estimates we obtain that

$$\frac{1}{r}\int\limits_{x-r}^{x+r}|f(y)|\chi_{|y|>2^{k+2}}(y)dy \le \frac{1}{2^j}\int\limits_{\{y:|x-y|<r,|y|>2^{k+1}\}}|f(y)|dy$$

$$\le \frac{1}{2^j}\int\limits_{2^{k+1}<|y|<2^{j+2}}|f(y)|dy = \frac{1}{2^j}\sum_{n=k+1}^{j+1}\frac{2^n}{2^n}\int\limits_{\{2^n<|y|<2^{n+1}\}}|f(y)|dy$$

$$\le \Bigg[\sup_{n\ge k+1}\frac{1}{2^n}\int\limits_{\{2^n<|y|<2^{n+1}\}}|f(y)|dy\Bigg]\frac{1}{2^j}\Bigg[\sum_{n=k+1}^{j+1}2^n\Bigg]$$

$$= c \sup_{n\geq k+1} \frac{1}{2^n} \int\limits_{2^n<|y|<2^{n+1}} |f(y)|dy.$$

Taking into account the latter estimate we conclude that

$$I_3 \leq c \sum_{k\in\mathbb{Z}} \int\limits_{2^k<|x|<2^{k+1}} \sigma(x) \Big(\sup_{n\geq k+1} \frac{1}{2^n} \int\limits_{2^n<|y|<2^{n+1}} |f(y)|dy \Big)^p dx$$

[putting the supremum outside the brackets, using the fact $\sup \leq \sum$ and observing that the supremum does not depend on x]

$$\leq c \sum_{k\in\mathbb{Z}} \sum_{n=k+1}^{\infty} \frac{1}{2^{np}} \Big(\int\limits_{2^k<|x|<2^{k+1}} \sigma(x)dx \Big) \Big(\int\limits_{2^n<|y|<2^{n+1}} |f(y)|dy \Big)^p$$

[using Hölder's inequality and change the order of summation]

$$\leq c \sum_{n\in\mathbb{Z}} \frac{1}{2^{np}} \sum_{k=-\infty}^{n-1} \Big(\int\limits_{2^k<|x|<2^{k+1}} \sigma(x)dx \Big) \Big(\int\limits_{2^n<|y|<2^{n+1}} |f(y)|^p u(y)dy \Big)$$

$$\times \Big(\int\limits_{2^n<|y|<2^{n+1}} u^{1-p'} \Big)^{p-1}$$

$$\leq c \sum_{n\in\mathbb{Z}} \frac{1}{2^{np}} \Big(\int\limits_{|x|<2^n} \sigma(x)dx \Big) \Big(\int\limits_{2^n<|y|<2^{n+1}} |f(y)|^p u(y) \Big)$$

$$\times \Big(\int\limits_{2^n<|y|<2^{n+1}} u^{1-p'} \Big)^{p-1} \leq cA_1^p \|f\|^p_{L^p_u(\mathbb{R})}.$$

Necessity. Let $f_r(y) = \chi_{(0,r)}(y) u^{1-p'}(y)$. Observe that if $x \in (0,r)$, then

$$Mf_r(x) \geq \frac{1}{2r} \int\limits_{x-r}^{x+r} u^{1-p'}(y)\chi_{(0,r)}(y)dy = \frac{1}{2r} \int\limits_0^r u^{1-p'}(x)dx.$$

Consequently,

$$\int_{\mathbb{R}} \sigma(x)(Mf_r(x))^p dx \geq \frac{c}{r^p}\Big(\int_0^r \sigma(x)dx\Big)\Big(\int_0^r u^{1-p'}(x)dx\Big)^p.$$

On the other hand,

$$\|f_r\|^p_{L^p_u(\mathbb{R})} = \int_0^r u^{1-p'}(x)dx.$$

Now the two–weight inequality for M finishes the proof of Theorem 5.1.2. □

5.2 Strong Maximal Functions

We begin this section with the following observation:

$$M_S f(x,y) \approx \sup_{\substack{r>0\\ t>0}} \frac{1}{tr}\int_{x-r}^{x+r}\int_{y-t}^{y+t} |f(s,\tau)|ds d\tau, \tag{5.2.1}$$

$$(M_1M_2)f(x,y) \approx \sup_{r>0}\frac{1}{r}\int_{x-r}^{x+r}\Big(\sup_{t>0}\frac{1}{t}\int_{y-t}^{y+t}|f(s,\tau)|d\tau\Big)ds, \tag{5.2.2}$$

where $x, y \in \mathbb{R}$ and $f \in L_{loc}(\mathbb{R}^2)$.

Relations (5.2.1) and (5.2.2) will be useful in the proofs of the main results of this chapter.

We will need the classes $DC(x)$ and $DC(y)$ of two–dimensional weights. These classes are defined in Section 1.6 (see Definition 1.6.1).

The Case of Increasing Weights

Let us introduce the following mixed–type operator

$$(MH)f(x,y) = \frac{1}{|y|}\sup_{r>0}\frac{1}{r}\int_{x-r}^{x+r}\Big(\int_{|\tau|<|y|}|f(s,\tau)|d\tau\Big)ds, \quad x,y\in\mathbb{R}.$$

Lemma 5.2.1. *Let $1<p<\infty$. Suppose that v and w are two-dimensional weights such that they are even on $\mathbb{R}^2$ and increasing on $\mathbb{R}$ in the first variable for the second fixed one and, in addition, $w^{1-p'} \in DC(y)$. Then the operator MH is bounded from $L^p_w(\mathbb{R}^2)$ to $L^p_v(\mathbb{R}^2)$ if and only if*

$$B := \sup_{\substack{a>0\\ b>0}}\Big(\int_a^\infty\int_b^\infty \frac{v(x,y)}{(xy)^p}dxdy\Big)^{1/p}\Big(\int_0^a\int_0^b w^{1-p'}(x,y)dxdy\Big)^{1/p'} < \infty. \tag{5.2.3}$$

Proof. *Sufficiency.* In the sequel we will use the notation:

$$v_1(x,y) := \frac{v(x,y)}{y^p}, \quad \tilde{v}_{1,j}(x) := \int_{2^j}^{2^{j+1}} v_1(x,y)dy;$$

$$\tilde{w}_{1,j}(x) := \Big(\int_0^{2^j} w^{1-p'}(x,\tau)d\tau\Big)^{1-p}; \quad F_j := \{y : 2^j < |y| \le 2^{j+1}\}.$$

We have that

$$I := \iint_{\mathbb{R}^2} v(x,y)\big((MH)f(x,y)\big)^p(x,y)dxdy$$

$$\le \sum_{j\in\mathbb{Z}} \iint_{\mathbb{R}\times F_j} \frac{v(x,y)}{|y|^p}\Big(\sup_{r>0}\frac{1}{r}\int_{x-r}^{x+r}\int_{|\tau|<|y|} |f(s,\tau)|dsd\tau\Big)^p dxdy$$

$$\le \sum_{j\in\mathbb{Z}} \int_{\mathbb{R}} \tilde{v}_{1,j}(x)\Big(\sup_{r>0}\frac{1}{r}\int_{x-r}^{x+r}\Big(\int_{|\tau|<2^{j+1}} |f(s,\tau)|d\tau\Big)ds\Big)^p dx.$$

Now using Theorem 5.1.1 and the fact that $\tilde{v}_{1,j}$ and $\tilde{w}_{1,j}$ are increasing weights we find that the sum does not exceed

$$B^p \sum_{j\in\mathbb{Z}} \int_{\mathbb{R}} \tilde{w}_{1,j}(x)\Big(\int_{|\tau|<2^{j+1}} |f(x,\tau)|d\tau\Big)^p dx$$

$$= B^p \int_{\mathbb{R}} \sum_{j\in\mathbb{Z}} \tilde{w}_{1,j}(x)\Big(\int_{|\tau|<2^{j+1}} |f(x,\tau)|d\tau\Big)^p dx =: I_1.$$

On the other hand,

$$\sup_{\substack{x>0\\ k\in\mathbb{Z}}} J(x,j) := \sup_{\substack{x>0\\ k\in\mathbb{Z}}} \sum_{j=k}^{\infty}\Big(\int_0^{2^j} w^{1-p'}(x,y)dy\Big)^{1-p}\Big(\sum_{j=-\infty}^{k}\int_{2^j}^{2^{j+1}} w^{1-p'}(x,y)dy\Big)^{p-1} < \infty.$$

Indeed, the condition $w^{1-p'} \in DC(y)$ and Lemma 1.6.1 yield that

$$J(x,j) = \sum_{j=k}^{\infty} \Big(\int_0^{2^j} w^{1-p'}(x,y)dy \Big)^{1-p} \Big(\int_0^{2^{k+1}} w^{1-p'}(x,y)dy \Big)^{p-1}$$

$$\leq \Big(\sum_{j=k}^{\infty} \eta^{(j-k)(1-p)} \Big) \Big(\int_0^{2^k} w^{1-p'}(x,y)dy \Big)^{1-p}$$

$$\times \Big(\int_0^{2^{k+1}} w^{1-p'}(x,y)dy \Big)^{p-1} \leq c.$$

Consequently, in virtue of Theorem 1.4.1 and Hölder's inequality we obtain that

$$I_1 \leq c \int_{\mathbb{R}} \sum_{j \in \mathbb{Z}} \tilde{w}_{1,j}(x) \Big(\sum_{k=-\infty}^{j} \int_{\{\tau: 2^k < |\tau| < 2^{k+1}\}} f(x,\tau)d\tau \Big)^p dx$$

$$\leq c \int_{\mathbb{R}} \sum_{k \in \mathbb{Z}} \Big(\int_{2^k}^{2^{k+1}} w^{1-p'}(x,\tau)d\tau \Big)^{1-p} \Big(\int_{\{\tau: 2^k < |\tau| < 2^{k+1}\}} f(x,\tau)d\tau \Big)^p dx$$

$$\leq c \int_{\mathbb{R}} \sum_{k \in \mathbb{Z}} \Big(\int_{\{\tau: 2^k < |\eta| < 2^{k+1}\}} w^{1-p'}(x,y)dy \Big)^{1-p}$$

$$\times \Big(\int_{2^k}^{2^{k+1}} w(x,\tau)|f(x,\tau)|^p d\tau \Big) \Big(\int_{2^k}^{2^{k+1}} w^{1-p'}(x,\tau)d\tau \Big)^{p-1} dx \leq c\|f\|^p_{L^p(\mathbb{R}^2)}.$$

Necessity. First we observe that the inequality

$$(MH)f(x,y) \geq c \frac{1}{|x||y|} \int_{|t|<|x|} \int_{|\tau|<|y|} |f(s,\tau)|dsd\tau$$

holds. Now applying the two–weight inequality

$$\iint_{\mathbb{R}^2} ((MH)f)^p(x,y)v(x,y)dxdy \leq \iint_{\mathbb{R}^2} |f(x,y)|^p w(x,y)dxdy$$

for the test functions $w^{1-p'}(x)\chi_{\{|x|<t\}}(x)$ we have that condition (5.2.3) holds.
The proof is complete. □

In a similar manner it follows the next lemma.

Lemma 5.2.2. *Let $1 < p < \infty$. Suppose that v is a two–dimensional weight, which is even on $\mathbb{R}$ and increasing on $\mathbb{R}$ in the first variable for the second fixed one. Let $w(x,y) = w_1(x)w_2(y)$, where w_1 and w_2 are measurable positive functions on $\mathbb{R}$ and, in addition, w_1 is increasing on $\mathbb{R}_+$. Then the operator MH is bounded from $L^p_w(\mathbb{R}^2)$ to $L^p_v(\mathbb{R}^2)$ if and only if (5.2.3) holds.*

Let us now discuss the following two–dimensional integral transform

$$(HM)f(x,y) = \frac{1}{|x|} \int\limits_{|s|<|x|} \left(\sup_{t>0} \frac{1}{t} \int\limits_{y-t}^{y+t} |f(s,\tau)| d\tau \right) ds, \quad x, y \in \mathbb{R}.$$

Lemma 5.2.3. *Let $1 < p < \infty$. Suppose that v is a two-dimensional weight, which is even on $\mathbb{R}$ and increasing on $\mathbb{R}_+$ in the second variable for the first fixed one. Further, assume that $w(x,y) = w_1(x)w_2(y)$, where w_1 and w_2 are even weights on $\mathbb{R}$ and, in addition, w_2 is increasing on $\mathbb{R}_+$. Then the operator HM is bounded from $L^p_w(\mathbb{R}^2)$ to $L^p_v(\mathbb{R}^2)$ if and only if (5.2.3) holds.*

Proof. *Sufficiency.* Suppose that $S := \int_{\mathbb{R}} w_1^{1-p'}(x)dx = \infty$. Let $\{x_k\}_{k=-\infty}^{+\infty}$ be a sequence of positive numbers for which the equality

$$2^k = \int\limits_{|x|<x_k} w_1^{1-p'}(x)dx \tag{5.2.4}$$

holds for all $k \in \mathbb{Z}$. It is clear that $\{x_k\}$ is increasing and

$$\mathbb{R} \setminus \{0\} = \cup_{k\in\mathbb{Z}} E_k,$$

where $E_k := \{x \in \mathbb{R} : x_k \le |x| \le x_{k+1}\}$. Besides this, it is easy to verify that $2^k = \int_{E_k} w_1^{1-p'}(x)dx$. Let us denote

$$\bar{v}_k(y) := \int\limits_{E_k} \frac{v(x,y)}{|x|^p} dx.$$

We have that

$$I := \iint\limits_{\mathbb{R}^2} v(x,y)\big((HM)f(x,y)\big)^p dxdy$$

$$= \sum_{k\in\mathbb{Z}} \iint\limits_{E_k\times\mathbb{R}} \frac{v(x,y)}{|x|^p} \Bigg(\int\limits_{|s|<|x|} \left(\sup_{t>0} \frac{1}{t} \int\limits_{|y-\tau|<t} |f(s,\tau)| d\tau \right) ds \Bigg)^p dxdy$$

$$\leq \sum_{k\in\mathbb{Z}} \int_{\mathbb{R}} \Big(\int_{E_k} \frac{v(x,y)}{|x|^p} dx \Big) \Big(\int_{|s|<x_{k+1}} \Big(\sup_{t>0} \frac{1}{t} \int_{|y-\tau|<t} |f(s,\tau)| d\tau \Big) ds \Big)^p dy$$

$$\leq \sum_{k\in\mathbb{Z}} \int_{\mathbb{R}} \bar{v}_k(y) \Big(\int_{|s|<x_{k+1}} [M(f(s,\cdot))(y)] ds \Big)^p dy$$

[using Minkowski's integral inequality]

$$\leq \sum_{k} \Big(\int_{|s|<x_{k+1}} \Big(\int_{\mathbb{R}} \bar{v}_k(y) \big(M(f(s,\cdot))(y)\big)^p dy \Big)^{1/p} ds \Big)^p =: I_1.$$

Moreover, notice that

$$B \geq \sup_{\substack{b>0 \\ x_k>0}} \Big(\int_b^\infty \Big(\int_{x_k}^{x_{k+1}} \frac{v(x,y)}{x^p} dx \Big) \frac{1}{y^p} dy \Big)^{1/p}$$

$$\times \Big(\int_0^b \Big(\int_0^{x_k} w_1^{1-p'}(x) dx \Big)^{(1-p)(1-p')} w_2^{1-p'}(y) dy \Big)^{1/p'}.$$

Therefore, by Theorem 5.1.1 we find that

$$I_1 \leq c \sum_{k\in\mathbb{Z}} \Big(\int_{|s|<x_{k+1}} \Big(\int_{\mathbb{R}} \Big(\int_0^{x_k} w_1^{1-p'}(\sigma) d\sigma \Big)^{1-p} |f(s,\tau)|^p w_2(\tau) d\tau \Big)^{1/p} ds \Big)^p$$

$$= c \sum_k \Big(\int_0^{x_k} w_1^{1-p'}(\sigma) d\sigma \Big)^{1-p}$$

$$\times \Big(\sum_{j=-\infty}^{k} \int_{x_j \leq |s| < x_{j+1}} \Big(\int_{\mathbb{R}} |f(s,\tau)|^p w_2(\tau) d\tau \Big)^{1/p} ds \Big)^p =: I_2.$$

It is obvious that

$$\sum_{k=n}^{+\infty} \Big(\int_0^{x_k} w_1^{1-p'}(x) dx \Big)^{1-p} \Big(\sum_{k=-\infty}^{n} \int_{x_k}^{x_{k+1}} w_1^{1-p'}(x) dx \Big)^{p-1}$$

$$= \sum_{k=n}^{+\infty} \Big(\int_0^{x_k} w_1^{1-p'}(x) dx \Big)^{1-p} \Big(\int_0^{x_{n+1}} w_1^{1-p'}(x) dx \Big)^{p-1}$$

$$= \Big(\sum_{k=n}^{+\infty} 2^{k(1-p)}\Big) 2^{(n+1)(p-1)} = c < \infty.$$

Therefore Theorem 1.1.4 yields that

$$I_2 \le c \sum_k \Big(\int_{x_k}^{x_{k+1}} w_1^{1-p'}(s)ds \Big)^{1-p}$$

$$\times \Big(\int_{x_k \le |s| < x_{k+1}} \Big(\int_{\mathbb{R}} |f(s,\tau)|^p w_2(\tau) d\tau \Big)^{1/p} ds \Big)^p$$

$\Big[$ using Hölder's inequality $\Big]$

$$\le c \sum_k \Big(\int_{x_k}^{x_{k+1}} w_1^{1-p'}(s)ds \Big)^{1-p}$$

$$\times \Big(\int_{x_k \le |s| < x_{k+1}} \int_{\mathbb{R}} |f(s,\tau)|^p w_1(s) w_2(\tau) ds d\tau \Big) \Big(\int_{x_k}^{x_{k+1}} w_1^{1-p'}(t)dt \Big)^{p-1}$$

$$= c \sum_k \int_{x_k \le |s| < x_{k+1}} \int_{\mathbb{R}} |f(s,\tau)|^p w_1(s) w_2(\tau) ds d\tau = c \|f\|^p_{L^p_w(\mathbb{R}^2)}.$$

If $S < \infty$, then without loss of generality we can assume that $S = 1$. In this case we choose the sequence $\{x_k\}_{k=-\infty}^{0}$ for which (5.2.4) holds for all $k \le 0$. Arguing as in the case $S = \infty$ and using Theorem 1.1.5 instead of Theorem 1.1.4, we obtain the desired result.

To prove *Necessity* we argue as in the case the previous lemma and we are done. □

Theorem 5.2.1. *Let $1 < p < \infty$. Suppose that the two–dimensional weight functions v and w are even on $\mathbb{R}$ and increasing on $\mathbb{R}_+$ in each variable for another fixed one and, in addition, $w^{1-p'} \in DC(x) \cap DC(y)$. Then M_S is bounded from $L^p_w(\mathbb{R}^2)$ to $L^p_v(\mathbb{R}^2)$ if and only if* (5.2.3) *holds.*

Proof. *Sufficiency.* First notice that since the weights are increasing in each variable for another fixed one, condition (5.2.3) guarantees the following estimates:

$$\infty > B \ge \sup_{\substack{a>0 \\ b>0}} \Big(\int_a^\infty \int_b^{2b} \frac{v(x,y)}{(xy)^p} dxdy \Big)^{1/p} \Big(\int_0^a \int_0^{b/4} w^{1-p'}(x,y) dxdy \Big)^{1/p'}$$

$$\geq c \sup_{\substack{a>0\\b>0}} b^{-1/p'} b^{1/p'} \bigg(\int_a^\infty \frac{v(x,b)}{x^p} dx \bigg)^{1/p} \bigg(\int_0^a w^{1-p'}(x, b/4) dx \bigg)^{1/p'}$$

$$= c \sup_{\substack{a>0\\b>0}} \bigg(\int_a^\infty \frac{v(x,b)}{x^p} dx \bigg)^{1/p} \bigg(\int_0^a w^{1-p'}(x, b/4) dx \bigg)^{1/p'}. \qquad (5.2.5)$$

By using the obvious inequality

$$f(s,\tau) = f(s,\tau)\chi_{|\tau|<|y|/2}(\tau) + f(s,\tau)\chi_{|y|/2\leq|\tau|\leq 4|y|}(\tau) + f(s,\tau)\chi_{|\tau|>4|y|}(\tau),$$

where $y \in \mathbb{R}$ we have that

$$\|M_S f\|^p_{L^p_v(\mathbb{R}^2)} \leq c\bigg[\iint_{\mathbb{R}^2} v(x,y) \big(M_S(f\chi_{|\tau|<|y|/2})\big)^p(x,y) dx dy$$

$$+ \iint_{\mathbb{R}^2} v(x,y) \big(M_S(f\chi_{|y|/2\leq|\tau|\leq 4|y|})\big)^p(x,y) dx dy$$

$$+ \iint_{\mathbb{R}^2} v(x,y) \big(M_S(f\chi_{|\tau|>4|y|})\big)^p(x,y) dx dy\bigg] = c[S_1 + S_2 + S_3].$$

Notice that if $|\tau| < |y|/2$, then $|y|/2 < |y-\tau|$. Hence, for $t < |y|/2$,

$$\frac{1}{rt} \int_{x-r}^{x+r} \int_{y-t}^{y+t} |f(s,\tau)| \chi_{\{\tau:|\tau|<|y|/2\}}(\tau) ds d\tau$$

$$\leq \frac{1}{rt} \int_{x-r}^{x+r} \bigg(\int_{\{\tau:|\tau-y|<t;|\tau-y|>|y|/2\}} |f(s,\tau)| \chi_{\{\tau:|\tau|<|y|/2\}} d\tau \bigg) ds = 0.$$

Let now $t \geq |y|/2$. Then it is easy to see that

$$\frac{1}{rt} \int_{x-r}^{x+r} \int_{y-t}^{y+t} |f(s,\tau)| \chi_{\{\tau:|\tau|<|y|/2\}} ds d\tau$$

$$\leq c \frac{1}{r} \frac{1}{|y|} \int_{x-r}^{x+r} \bigg(\int_{|\tau|<|y|/2} |f(s,\tau)| d\tau \bigg) ds$$

$$\leq c \frac{1}{|y|} \sup_{r>0} \frac{1}{r} \int_{x-r}^{x+r} \bigg(\int_{|\tau|<|y|/2} |f(s,\tau)| d\tau \bigg) ds \leq c(MH)f(x,y).$$

Now, by Lemma 5.2.1 we have that condition (5.2.3) guarantees the inequality

$$S_1 \le c\|f\|^p_{L^p_w(\mathbb{R}^2)}.$$

Let us denote

$$E_k := \{y : 2^k < |y| \le 2^{k+1}\}.$$

Further, for S_3, we find that

$$S_3 = \sum_{k\in\mathbb{Z}} \iint\limits_{\mathbb{R}^2} v(x,y)\big(M_S(f\chi_{|\tau|>4|y|})\big)^p(x,y)dxdy$$

$$\le \sum_{k\in\mathbb{Z}} \iint\limits_{\mathbb{R}\times E_k} v(x,y)\big(M_S(f\chi_{|\tau|>2^{k+2}})\big)^p(x,y)dxdy.$$

We observe also that if $|\tau| > 2^{k+2}$ and $2^k < |y| \le 2^{k+1}$, then $|y-\tau| \ge |\tau| - |y| \ge 2^{k+2} - 2^{k+1} = 2^{k+1}$. Therefore, if $t < 2^{k+1}$, $|\tau| > 2^{k+2}$ and $2^k < |y| \le 2^{k+1}$, then

$$\frac{1}{r}\int\limits_{x-r}^{x+r} \frac{1}{t} \int\limits_{\{\tau:|\tau-y|<t;|\tau|>2^{k+2}\}} |f(s,\tau)|dsd\tau$$

$$\le \frac{1}{r}\int\limits_{x-r}^{x+r} \bigg(\frac{1}{t} \int\limits_{\{\tau:|\tau-y|<t;|\tau-y|>2^{k+1}\}} |f(s,\tau)|d\tau\bigg)ds = 0.$$

Moreover, if $t \ge 2^{k+1}$, then $t \in [2^j, 2^{j+1})$ for some $j \ge k+1$. When $|y-\tau| < t$ we have that $|\tau| \le |\tau - y| + |y| \le t + |y| \le 2^{j+1} + 2^{k+1} \le 2^{j+1} + 2^j < 2^{j+2}$. Taking into account these estimates we find that for $t \ge 2^{k+1}$ and $r > 0$,

$$\frac{1}{rt}\int\limits_{x-r}^{x+r} \bigg(\int\limits_{\{\tau:|\tau-y|<t;|\tau|>2^{k+2}\}} |f(s,\tau)|d\tau\bigg)ds$$

$$\le \frac{1}{r}\frac{1}{2^j}\int\limits_{x-r}^{x+r} \bigg(\int\limits_{\{\tau:2^{k+2}<|\tau|<2^{j+2}\}} |f(s,\tau)|d\tau\bigg)ds$$

$$= \frac{1}{2^j} \int\limits_{\{\tau:2^{k+2}<|\tau|<2^{j+2}\}} \bigg(\frac{1}{r}\int\limits_{x-r}^{x+r} |f(s,\tau|ds\bigg)d\tau$$

$$\le \frac{1}{2^j}\sum_{n=k+1}^{j+1} \frac{2^n}{2^n} \int\limits_{\{\tau:2^n<|\tau|<2^{n+1}\}} \bigg(\frac{1}{r}\int\limits_{x-r}^{x+r} |f(s,\tau)|ds\bigg)d\tau$$

$$\leq \frac{1}{2^j}\Bigg[\sup_{n\geq k+1}\frac{1}{2^n}\Bigg(\int\limits_{\{\tau:2^n<|\tau|<2^{n+1}\}}\Bigg(\frac{1}{r}\int\limits_{x-r}^{x+r}|f(s,\tau)|ds\Bigg)d\tau\Bigg)\Bigg]\Bigg(\sum_{n=k+1}^{j+1}2^n\Bigg)$$

$$\leq c\Bigg[\sup_{n\geq k+1}\frac{1}{2^n}\Bigg(\int\limits_{\{\tau:2^n<|\tau|<2^{n+1}\}}\Bigg(\frac{1}{r}\int\limits_{x-r}^{x+r}|f(s,\tau)|ds\Bigg)d\tau\Bigg)\Bigg]$$

$\Big[$ changing the order of integration $\Big]$

$$= c\sup_{n\geq k+1}\frac{1}{2^n}\Bigg(\int\limits_{\{\tau:2^n<|\tau|<2^{n+1}\}}\Bigg(\frac{1}{r}\int\limits_{x-r}^{x+r}|f(s,\tau)|ds\Bigg)d\tau\Bigg)$$

$$= c\sup_{n\geq k+1}\frac{1}{2^n}\frac{1}{r}\int\limits_{x-r}^{x+r}\Bigg(\int\limits_{\{\tau:2^n<|\tau|<2^{n+1}\}}|f(s,\tau)|d\tau\Bigg)ds$$

$$\leq c\sup_{n\geq k+1}\frac{1}{2^n}\Bigg[\sup_{r>0}\frac{1}{r}\int\limits_{x-r}^{x+r}\Bigg(\int\limits_{\{\tau:2^n<|\tau|<2^{n+1}\}}|f(s,\tau)|d\tau\Bigg)\Bigg]ds.$$

Hence

$$\sup_{\substack{r>0\\t>0}}\frac{1}{rt}\int\limits_{x-r}^{x+r}\int\limits_{\{\tau:|\tau-y|<t;|\tau|>2^{k+2}\}}|f(s,\tau)|dsd\tau$$

$$\leq c\sup_{n\geq k+1}\frac{1}{2^n}\sup_{r>0}\frac{1}{r}\int\limits_{x-r}^{x+r}\int\limits_{\{\tau:2^n<|\tau|<2^{n+1}\}}|f(s,\tau)|dsd\tau.$$

Taking the inequalities derived above into account we conclude that

$$S_3\leq c\sum_k\iint\limits_{\mathbb{R}\times E_k}v(x,y)\Bigg(\sup_{n\geq k+1}\frac{1}{2^n}\sup_{r>0}\frac{1}{r}\int\limits_{x-r}^{x+r}\Bigg(\int\limits_{\{\tau:2^n<|\tau|<2^{n+1}\}}|f(s,\tau)|d\tau\Bigg)ds\Bigg)^p dxdy$$

$\Big[$putting the supremum outside the brackets $\Big]$

$$\leq c\sum_k\int\limits_{\mathbb{R}}\Bigg(\int\limits_{E_k}v(x,y)dy\Bigg)$$

$$\times\sup_{n\geq k+1}\frac{1}{2^{np}}\Bigg(\sup_{r>0}\frac{1}{r}\int\limits_{x-r}^{x+r}\Bigg(\int\limits_{\{\tau:2^n<|\tau|<2^{n+1}\}}|f(s,\tau)|d\tau\Bigg)ds\Bigg)^p dx$$

$$\Big[\text{ using the fact that } \sup \leq \sum \Big]$$

$$\leq c\sum_k \int_{\mathbb{R}} \Big(\int_{E_k} v(x,y)dy \Big)$$

$$\times \sum_{n=k+1}^{\infty} \frac{1}{2^{np}} \Big(\sup_{r>0} \frac{1}{r} \int_{x-r}^{x+r} \Big(\int_{\{\tau: 2^n<|\tau|<2^{n+1}\}} |f(s,\tau)|d\tau \Big) ds \Big)^p dx$$

$$\Big[\text{ putting the sum outside the integral}\Big]$$

$$\leq c\sum_k \sum_{n=k+1}^{\infty} \frac{1}{2^{np}} \int_{\mathbb{R}} \Big(\int_{E_k} v(x,y)dy \Big)$$

$$\times \Big(\sup_{r>0} \frac{1}{r} \int_{x-r}^{x+r} \Big(\int_{\{\tau: 2^n<|\tau|<2^{n+1}\}} |f(s,\tau)|d\tau \Big) ds \Big)^p dx$$

$\Big[$ changing the order of sums and using the condition that v is even in the second variable $\Big]$

$$= c\sum_{n\in\mathbb{Z}} \frac{1}{2^{np}} \sum_{k=-\infty}^{n-1} \int_{\mathbb{R}} \Big(\int_{2^k}^{2^{k+1}} v(x,y)dy \Big)$$

$$\times \Big(\sup_{r>0} \frac{1}{r} \int_{x-r}^{x+r} \Big(\int_{\{\tau: 2^n<|\tau|<2^{n+1}\}} |f(s,\tau)|d\tau \Big) ds \Big)^p dx$$

$\Big[$ using the fact that v in increasing in the second variable $\Big]$

$$\leq c\sum_{n\in\mathbb{Z}} \frac{1}{2^{np}} \Big(\sum_{k=-\infty}^{n-1} 2^k \Big) \int_{\mathbb{R}} v(x,2^n)$$

$$\times \Big(\sup_{r>0} \frac{1}{r} \int_{x-r}^{x+r} \Big(\int_{\{\tau: 2^n<|\tau|<2^{n+1}\}} |f(s,\tau)|d\tau \Big) ds \Big)^p dx$$

$$= c\sum_{n\in\mathbb{Z}} 2^{n(1-p)} \int_{\mathbb{R}} v(x,2^n) \Big(\sup_{r>0} \frac{1}{r} \int_{x-r}^{x+r} \Big(\int_{\{\tau: 2^n<|\tau|<2^{n+1}\}} |f(s,\tau)|d\tau \Big) ds \Big)^p dx =: I_1.$$

Note that (5.2.5) yields that

$$B \geq \sup_{\substack{a>0\\ n\in\mathbb{Z}}} \Big(\int_a^{\infty} \frac{v(x,2^n)}{x^p} dx \Big)^{1/p} \Big(\int_0^a w^{1-p'}(x,2^{n-2})dx \Big)^{1/p'}.$$

Therefore, using Theorem 5.1.1 uniformly with respect to n we find that

$$I_1 \le c \sum_{n\in\mathbb{Z}} 2^{n(1-p)} \int_{\mathbb{R}} w(x,2^{n-2}) \Big(\int_{\{\tau:2^n<|\tau|<2^{n+1}\}} |f(x,\tau)|d\tau \Big)^p dx$$

$$\le c \sum_{n\in\mathbb{Z}} 2^{n(1-p)} \int_{\mathbb{R}} w(x,2^{n-2}) \Big(\int_{\{\tau:2^n<|\tau|<2^{n+1}\}} |f(x,\tau)|d\tau \Big)^p dx$$

$$= c \sum_{n\in\mathbb{Z}} 2^{n(1-p)} \int_{\mathbb{R}} w(x,2^{n-2}) \Big(\int_{\{\tau:2^n<|\tau|<2^{n+1}\}} |f(x,\tau)|d\tau \Big)^p dx$$

[using Hölder's inequality and the fact that w is even in the second variable]

$$\le c \sum_{n\in\mathbb{Z}} 2^{n(1-p)} \int_{\mathbb{R}} w(x,2^n)$$

$$\times \Big(\int_{\{\tau:2^n<|\tau|<2^{n+1}\}} |f(x,\tau)|^p w(x,\tau)d\tau \Big) \Big(\int_{2^n}^{2^{n+1}} w^{1-p'}(x,\tau)d\tau \Big)^{p-1} dx$$

[using the fact that w is increasing on $\mathbb{R}_+$ and even on $\mathbb{R}$ in the second variable]

$$\le c \sum_{n\in\mathbb{Z}} \frac{1}{2^{n(p-1)}} \int_{\mathbb{R}} w(x,2^n)w^{-1}(x,2^n)2^{n(p-1)}$$

$$\times \Big(\int_{\{\tau:2^n<|\tau|<2^{n+1}\}} |f(x,\tau)|^p w(x,\tau)d\tau \Big) dx$$

$$= c \sum_{n\in\mathbb{Z}} \int_{\mathbb{R}} \Big(\int_{\{\tau:2^n<|\tau|<2^{n+1}\}} |f(x,\tau)|^p w(x,\tau)d\tau \Big) dx = \|f\|^p_{L^p_w(\mathbb{R}^2)}.$$

Now we estimate S_2. We have that

$$S_2 \le c \Big[\iint_{\mathbb{R}^2} v(x,y)\big(M_S(f\chi_{\{|\cdot|\le|x|/2\}}\chi_{\{|y|/2\le|\cdot|\le4|y|\}})\big)^p(x,y)dxdy$$

$$+ \iint_{\mathbb{R}^2} v(x,y)\big(M_S(f\chi_{\{|x|/2\le|\cdot|<4|x|\}}\chi_{\{|y|/2\le|\cdot|\le4|y|\}})\big)^p(x,y)dxdy$$

$$+ \iint_{\mathbb{R}^2} v(x,y)\big(M_S(f\chi_{\{|\cdot|>4|x|\}}\chi_{\{|y|/2\le|\cdot|\le4|y|\}})\big)^p(x,y)dxdy \Big] = c[S_{21}+S_{22}+S_{23}].$$

Further,

$$S_{21} \le \iint_{\mathbb{R}^2} v(x,y)\big(M_S(f\chi_{\{|\cdot|\le |x|/2\}})\big)^p(x,y)dxdy$$

$$= \iint_{\mathbb{R}^2} v(x,y)\bigg(\sup_{r,t>0}\frac{1}{rt}\int_{y-t}^{y+t}\int_{x-r}^{x+r}\chi_{\{|s|\le|x|/2\}}|f(s,\tau)|d\tau ds\bigg)^p dxdy.$$

Observe that

$$\sup_{r,t>0}\frac{1}{rt}\int_{y-t}^{y+t}\int_{x-r}^{x+r}\chi_{\{|s|\le|x|/2\}}|f(s,\tau)|d\tau ds$$

$$\le c\frac{1}{|x|}\sup_{t>0}\frac{1}{t}\int_{y-t}^{y+t}\int_{x-r}^{x+r}|f(s,\tau)|d\tau ds \le c(\overline{M}H)f(x,y),$$

where

$$(\overline{M}H)f(x,y) = c\frac{1}{|x|}\sup_{t>0}\frac{1}{t}\int_{y-t}^{y+t}\bigg(\int_{\{s:|s|<|x|/2\}}|f(s,\tau)|ds\bigg)d\tau.$$

This operator is similar to MH. The only difference is that the order of variables are interchanged. Therefore Lemma 5.2.1 guarantees the boundedness of $\overline{M}H$ provided that $w^{1-p'} \in DC(x)$.

Moreover,

$$S_{23} \le \iint_{\mathbb{R}^2} v(x,y)\big(M_S(f\chi_{\{|s|\ge 4|x|\}})\big)^p(x,y)dxdy$$

$$= \iint_{\mathbb{R}^2} v(x,y)\bigg(\sup_{\substack{r>0\\ t>0}}\frac{1}{rt}\int_{y-t}^{y+t}\int_{x-r}^{x+r}\chi_{\{|s|\ge 4|x|\}}|f(s,\tau)|d\tau ds\bigg)^p dxdy.$$

Consequently, arguing for S_{23} as in the case of the estimate of S_3, we obtain that

$$S_{23} \le c\|f\|^p_{L^p_w(\mathbb{R}^2)}.$$

Concerning S_{22} we observe that condition (5.2.3) implies that

$$v(a,b) \le cB^p w(a/4, b/4), \tag{5.2.6}$$

where the positive constant c does not depend on a and b.

Let

$$f_{k,j}(t,\tau) := f(t,\tau)\chi_{\{t:2^{k-1}<|t|<2^{k+3}\}}(t)\chi_{\{\tau:\tau 2^{j-1}<|\tau|<2^{j+3}\}}(\tau). \tag{5.2.7}$$

Using the boundedness of M_S in $L^p(\mathbb{R}^2)$ and (5.2.6) we find that

$$S_{22} \le \sum_{k\in\mathbb{Z}}\sum_{j\in\mathbb{Z}} \int\limits_{2^k<|x|<2^{k+1}} \int\limits_{2^j<|y|<2^{j+1}} v(x,y)\big(M_S f_{k,j}(x,y)\big)^p dxdy$$

$$\le \sum_{k\in\mathbb{Z}}\sum_{j\in\mathbb{Z}} v(2^{k+1},2^{j+1}) \int\limits_{2^k<|x|<2^{k+1}} \int\limits_{2^j<|y|<2^{j+1}} \big(M_S f_{k,j}(x,y)\big)^p dxdy$$

$$\le cA^p \sum_{k\in\mathbb{Z}}\sum_{j\in\mathbb{Z}} w(2^{k-1},2^{j-1}) \int\limits_{2^{k-1}<|x|<2^{k+3}} \int\limits_{2^{j-1}<|y|<2^{j+3}} |f(x,y)|^p dxdy$$

$$\le cA^p \sum_{k\in\mathbb{Z}}\sum_{j\in\mathbb{Z}} \int\limits_{2^{k-1}<|x|<2^{k+3}} \int\limits_{2^{j-1}<|y|<2^{j+3}} w(x,y)|f(x,y)|^p dxdy \le c\|f\|^p_{L^p_w(\mathbb{R}^2)}.$$

Necessity. Using the obvious inequality

$$M_s f(x,y) \ge \frac{c}{|x||y|} \int\limits_{|t|<|x|} \int\limits_{|\tau|<|y|} |f(s,\tau)|dsd\tau$$

and taking the test function

$$f(x,y) = w^{1-p'}(|x|,|y|)\chi_{\{|x|<a\}}(x)\chi_{\{|y|<b\}}(y), \quad a,b>0$$

in the two–weight inequality

$$\iint\limits_{\mathbb{R}^2} \bigg(\frac{1}{|x||y|} \int\limits_{|t|<|x|} \int\limits_{|\tau|<|y|} |f(s,\tau)|dsd\tau\bigg)^p v(|x|,|y|)dxdy \le c \iint\limits_{\mathbb{R}^2} |f(x,y)|^p w(|x|,|y|)dxdy,$$

we find that (5.2.3) holds.

The proof is complete. □

The next statement deals with the operator M_1M_2 (defined at the beginning of this chapter).

Theorem 5.2.2. *Let $1<p<\infty$. Suppose that the weight function $v:\mathbb{R}^2 \mapsto \mathbb{R}_+$ is even on $\mathbb{R}$ and increasing on $\mathbb{R}_+$ in each variable for another fixed one. Further, suppose that $w(x,y)=w_1(x)w_2(y)$, where w_1 and w_2 are increasing on $\mathbb{R}_+$ and even on $\mathbb{R}$. Then M_1M_2 is bounded from $L^p_w(\mathbb{R}^2)$ to $L^p_v(\mathbb{R}^2)$ if and only if (5.2.3) holds.*

Proof. *Sufficiency.* Splitting f with respect to the second variable, we have that

$$\|(M_1M_2)f\|^p_{L^p_v(\mathbb{R}^2)} \le c\bigg[\iint\limits_{\mathbb{R}} v(x,y)\big((M_1M_2)(f\chi_{|\tau|<|y|/2})\big)^p(x,y)dxdy$$

$$+\iint_{\mathbb{R}^2} v(x,y)\big((M_1M_2)(f\chi_{|y|/2\le|\tau|\le 4|y|})\big)^p(x,y)dxdy$$

$$+\iint_{\mathbb{R}^2} v(x,y)\big((M_1M_2)(f\chi_{|\tau|>4|y|})\big)^p(x,y)dxdy\Bigg] := c[S_1+S_2+S_3].$$

Observe that if $|\tau|<|y|/2$, then $|y|/2<|y-\tau|$. Hence for $t<|y|/2$, we get that

$$\frac{1}{t}\int_{y-t}^{y+t} |f(s,\tau)|\chi_{\{\tau:|\tau|<|y|/2\}}dsd\tau$$

$$\le \frac{1}{t}\int_{\{\tau:|\tau-y|<t;|\tau-y|>|y|/2\}} |f(s,\tau)|\chi_{\{\tau:|\tau|<|y|/2\}}d\tau = 0$$

for all s. Let now $t\ge|y|/2$. Then it is easy to see that

$$\frac{1}{l}\int_{y-t}^{y+t} |f(s,\tau)|\chi_{\{\tau:|\tau|<|y|/2\}}d\tau \le c\frac{1}{|y|}\int_{|\tau|<|y|/2} |f(s,\tau)|d\tau$$

$$\le c\frac{1}{|y|}\int_{|\tau|<|y|/2} |f(s,\tau)|d\tau$$

for all s. Therefore we find that

$$(M_1M_2)(f\chi_{|\cdot|<|y|/2})(x,y)\le c(MH)f(x,y).$$

Consequently, by Lemma 5.2.1 we have that the condition (5.2.3) guarantees the inequality:

$$S_1\le c\|f\|^p_{L^p_w(\mathbb{R}^2)}.$$

For S_2 we obtain that

$$S_2\le c\Bigg[\iint_{\mathbb{R}^2} v(x,y)\big((M_1M_2)(f\chi_{|s|<|x|/2}\chi_{|y|/2\le|\tau|<4|y|})\big)^p(x,y)dxdy$$

$$+\iint_{\mathbb{R}^2} v(x,y)\big((M_1M_2)(f\chi_{|x|/2\le|s|\le 4|x|}\chi_{|y|/2\le|\tau|<4|y|})\big)^p(x,y)dxdy$$

$$+\iint_{\mathbb{R}^2} v(x,y)\big((M_1M_2)(f\chi_{|s|>4|x|}\chi_{|y|/2\le|\tau|<4|y|})\big)^p(x,y)dxdy\Bigg] := c[S_{21}+S_{22}+S_{23}].$$

First notice that

$$(M_1M_2)\big(f\chi_{|s|<|x|/2}\chi_{|y|/2\le|\tau|<4|y|}\big)(x,y)$$

$$\le \sup_{r>0}\frac{1}{r}\int\limits_{x-r}^{x+r}\chi_{\{s:|s|<|x|/2\}}\bigg(\sup_{t>0}\frac{1}{t}\int\limits_{y-t}^{y+t}|f(s,\tau)|d\tau\bigg)ds.$$

Further, if $|s|<|x|/2$, then $|s|/2\le|x-s|$. Therefore, when $r<|x|/2$, we find that

$$\frac{1}{r}\int\limits_{x-r}^{x+r}\chi_{\{|s|<|x|/2\}}\bigg(\sup_{t>0}\frac{1}{t}\int\limits_{y-t}^{y+t}|f(s,\tau)|d\tau\bigg)ds$$

$$\le\frac{1}{r}\int\limits_{\{s:|s-x|<r,|s-x|>|x|/2\}}\bigg(\sup_{t>0}\frac{1}{t}\int\limits_{y-t}^{y+t}|f(s,\tau)|d\tau\bigg)ds=0.$$

Let now $r\ge|x|/2$. Then

$$\frac{1}{r}\int\limits_{x-r}^{x+r}\chi_{\{|s|<|x|/2\}}\bigg(\sup_{t>0}\frac{1}{t}\int\limits_{y-t}^{y+t}|f(s,\tau)|d\tau\bigg)ds$$

$$\le c\frac{1}{|x|}\int\limits_{|s|<|x|}\bigg(\sup_{t>0}\frac{1}{t}\int\limits_{y-t}^{y+t}|f(s,\tau)|d\tau\bigg)ds=(HM)f(x,y).$$

By Lemma 5.2.3 we have that

$$S_{21}\le c\|f\|^p_{L^p_w(\mathbb{R}^2)}.$$

To estimate S_{22} first observe that the condition (5.2.3) implies (5.2.6).

Applying notation (5.2.7), inequality (5.2.6) and boundedness of M_1M_2 in $L^p(\mathbb{R}^2)$ we obtain that

$$S_{22}=\sum_{k\in\mathbb{Z}}\sum_{j\in\mathbb{Z}}\int\limits_{2^k<|x|<2^{k+1}}\int\limits_{2^j<|y|<2^{j+1}}v(x,y)\big((M_1M_2)f_{k,j}(x,y)\big)^p dxdy$$

$$\le\sum_{k\in\mathbb{Z}}\sum_{j\in\mathbb{Z}}v(2^{k+1},2^{j+1})\int\limits_{2^k<|x|<2^{k+1}}\int\limits_{2^j<|y|<2^{j+1}}((M_1M_2)f_{k,j}(x,y))^p dxdy$$

$$\le cA^p\sum_{k\in\mathbb{Z}}\sum_{j\in\mathbb{Z}}w_1(2^{k-1})w_2(2^{j-1})\int\limits_{2^{k-1}<|x|<2^{k+3}}\int\limits_{2^{j-1}<|y|<2^{j+3}}|f(x,y)|^p dxdy$$

$$\leq cA^p \sum_{k\in\mathbb{Z}} \sum_{j\in\mathbb{Z}} \int\limits_{2^{k-1}<|x|<2^{k+1}} \int\limits_{2^{j-1}<|y|<2^{j+1}} w_1(x)w_2(y)|f(x,y)|^p dxdy \leq c\|f\|^p_{L^p_w(\mathbb{R}^2)}.$$

Let us denote $E_i =: \{x : 2^i < |x| \leq 2^{i+1}\}$. For S_{23} we have that

$$S_{23} = \iint\limits_{\mathbb{R}^2} v(x,y)\big((M_1M_2)(f\chi_{|s|>4|x|}\chi_{|y|/2\leq|\tau|\leq 4|y|})\big)^p(x,y)dxdy$$

$$\leq \sum_{i\in\mathbb{Z}} \iint\limits_{E_i\times\mathbb{R}} v(x,y)\big((M_1M_2)(f\chi_{|s|>4|x|}\chi_{|y|/2\leq|\tau|\leq 4|y|})\big)^p(x,y)dxdy.$$

Observe that if $|s| > 2^{i+2}$ and $2^i < |x| \leq 2^{i+1}$, then $|x-s| \geq |s| - |x| \geq 2^{i+2} - 2^{i+1} = 2^{i+1}$. Hence, if $r < 2^{i+1}$, $|s| > 2^{i+2}$ and $2^i < |x| \leq 2^{i+1}$, then

$$\frac{1}{r} \int\limits_{\{s:|s-x|<r;|s|>2^{i+2}\}} M\big(f(s,\cdot)\chi_{\{|y|/2\leq|\cdot|\leq 4|y|\}}\big)(y)ds = 0.$$

Therefore we discuss the case when $r \geq 2^{i+1}$. For such r, we have that there is an integer j, $j \geq i+1$ such that $r \in [2^j, 2^{j+1})$. Further, if $|x-s| < r$, we have $|s| \leq |x| + r \leq 2^{j+1} + 2^{i+1} \leq 2^{j+1} + 2^j \leq 2^{j+2}$. Taking into account these estimates we conclude that if $r \geq 2^{i+1}$, then

$$\frac{1}{r} \int\limits_{\{s:|s-x|<r;|s|>2^{i+2}\}} M\big(f(s,\cdot)\chi_{\{|y|/2\leq|\cdot|\leq 4|y|\}}\big)(y)ds$$

$$\leq \frac{c}{2^j} \int\limits_{\{s:2^{i+2}<|s|<2^{j+2}\}} M\big(f(s,\cdot)\chi_{\{|y|/2\leq|\cdot|\leq 4|y|\}}\big)(y)ds$$

$$\leq \frac{c}{2^j} \sum_{m=i+1}^{j+1} \int\limits_{\{s:2^m<|s|<2^{m+1}\}} M\big(f(s,\cdot)\chi_{\{|y|/2\leq|\cdot|\leq 4|y|\}}\big)(y)ds$$

$$\leq \frac{c}{2^j}\Bigg(\sup_{m\geq i+1} \frac{1}{2^m} \int\limits_{\{s:2^m<|s|<2^{m+1}\}} M\big(f(s,\cdot)\chi_{\{|y|/2\leq|\cdot|\leq 4|y|\}}\big)(y)ds\Bigg)\Bigg(\sum_{m=i+1}^{j+1} 2^m\Bigg)$$

$$\leq c \sup_{m\geq i+1} \frac{1}{2^m} \int\limits_{\{s:2^m<|s|<2^{m+1}\}} M\big(f(s,\cdot)\chi_{\{|y|/2\leq|\cdot|\leq 4|y|\}}\big)(y)ds.$$

It follows that

$$\sup_{r>0} \frac{1}{r} \int\limits_{\{s:|s-x|<r;|s|>2^{i+2}\}} M\big(f(s,\cdot)\chi_{\{|y|/2\leq|\cdot|\leq 4|y|\}}\big)(y)ds$$

$$\leq c \sup_{m\geq i+1} \frac{1}{2^m} \int\limits_{\{s:2^m<|s|<2^{m+1}\}} M\big(f(s,\cdot)\chi_{\{|y|/2\leq|\cdot|\leq 4|y|\}}\big)(y)ds.$$

From this estimate we obtain that

$$S_{23} \leq c \sum_i \sum_k \iint\limits_{E_i\times E_k} v(x,y)$$

$$\times\Bigg(\sup_{m\geq i+1} \frac{1}{2^m} \int\limits_{\{s:2^m<|\tau|<2^{m+1}\}} M\big(f(s,\cdot)\chi_{\{|y|/2\leq|\cdot|\leq 4|y|\}}\big)(y)ds\Bigg)^p dxdy$$

$$\Big[\text{ putting the supremum outside the brackets }\Big]$$

$$= \sum_i \sum_k \int\limits_{E_i}\int\limits_{E_k} v(x,y)$$

$$\times \sup_{m\geq i+1} \frac{1}{2^{mp}} \Bigg(\int\limits_{\{s:2^m<|\tau|<2^{m+1}\}} M\big(f(s,\cdot)\chi_{\{|y|/2\leq|\cdot|\leq 4|y|\}}\big)(y)ds\Bigg)^p dxdy$$

$$\Big[\text{ observing that the inner integral does not depend on } x \text{ }\Big]$$

$$= \sum_i \sum_k \int\limits_{E_k} \Bigg(\int\limits_{E_i} v(x,y)dx\Bigg)$$

$$\times \sup_{m\geq i+1} \frac{1}{2^{mp}} \Bigg(\int\limits_{\{s:2^m<|\tau|<2^{m+1}\}} M\big(f(s,\cdot)\chi_{\{|y|/2\leq|\cdot|\leq 4|y|\}}\big)(y)ds\Bigg)^p dy$$

$$\Bigg[\text{ using the fact that } v \text{ is even on } \mathbb{R}, \text{ increasing on } \mathbb{R}_+$$

$$\text{in both variables and the inequality } \sup \leq \sum \Bigg]$$

$$\leq \sum_i 2^i \sum_k v(2^{i+1}, 2^{k+1})$$

$$\times \sum_{m\geq i+1} \frac{1}{2^{mp}} \int\limits_{E_k} \Bigg(\int\limits_{\{s:2^m<|\tau|<2^{m+1}\}} M\big(f(s,\cdot)\chi_{\{|y|/2\leq|\cdot|\leq 4|y|\}}\big)(y)ds\Bigg)^p dy$$

$$\Big[\text{ changing the order of sums }\Big]$$

$$\leq c \sum_k \sum_i 2^i v(2^{i+1}, 2^{k+1})$$

$$\times \sum_{m\geq i+1} \frac{1}{2^{mp}} \int\limits_{E_k} \Bigg(\int\limits_{\{s:2^m<|\tau|<2^{m+1}\}} M\big(f(s,\cdot)\chi_{\{|y|/\leq|\cdot|\leq 4|y|\}}\big)(y)ds\Bigg)^p dy$$

$$\Big[\text{changing again the order of sums and taking into account that}$$

$$v(2^{i+1}, 2^{k+1}) \le v(2^m, 2^{k+1})\Big]$$

$$\le c \sum_k \sum_m \frac{v(2^m, 2^{k+1})}{2^{mp}} \Big(\sum_{i=-\infty}^{m-1} 2^i \Big)$$

$$\times \int_{E_k} \Big(\int_{\{s:2^m<|\tau|<2^{m+1}\}} M\big(f(s,\cdot)\chi_{\{|y|/2\le|\cdot|\le 4|y|\}}\big)(y)ds\Big)^p dy$$

$$\Big[\text{taking into account that} \sum_{i=-\infty}^{m-1} 2^i = c2^m$$

$$\text{and using Hölder's inequality in the inner integral}\Big]$$

$$\le c \sum_k \sum_m \frac{v(2^m, 2^{k+1})}{2^{m(p-1)}} 2^{m(p-1)}$$

$$\times \int_{E_k} \Big(\int_{\{s:2^m<|\tau|<2^{m+1}\}} M\big(f(s,\cdot)\chi_{\{|y|/2\le|\cdot|\le 4|y|\}}\big)(y)^p ds\Big) dy \le$$

$$\Big[\text{changing the order of sums}\Big]$$

$$c \sum_k \sum_m v(2^m, 2^{k+1}) \Big(\int_{\{s:2^m<|\tau|<2^{m+1}\}} \Big(\int_{E_k} \Big[M\big(f(s,\cdot)\chi_{\{|y|/2\le|\cdot|\le 4|y|\}}\big)(y)\Big]^p dy\Big) ds$$

$$\Big[\text{using the boundedness of the maximal operator } M \text{ in the space } L^p(\mathbb{R})\Big]$$

$$\le c \sum_k \sum_m v(2^m, 2^{k+1})$$

$$\times \Big(\int_{\{s:2^m<|s|<2^{m+1}\}} \int_{\{y:2^{k-1}<|y|<2^{k+3}\}} |f(s,y)|^p dy\Big) ds$$

$$\Big[\text{using the inequality } v(2^m, 2^{k+1}) \le cw_1(2^m)w_2(2^{k-1})$$

$$\text{and the fact that } w_1 \text{ and } w_2 \text{ are increasing}\Big]$$

$$\le c \sum_k \sum_m \int_{\{s:2^m<|s|<2^{m+1}\}} \int_{\{y:2^{k-1}<|y|<2^{k+3}\}} w_1(s)w_2(y)|f(s,y)|^p dyds = \|f\|^p_{L^p_w(\mathbb{R}^2)}.$$

For S_3 we have that

$$S_3 \le c\Big[\iint_{\mathbb{R}^2} v(x,y)\big((M_1M_2)(f\chi_{|s|<|x|/2}\chi_{|\tau|>4|y|})\big)^p(x,y)dxdy$$

$$+\int_{\mathbb{R}}\int_{\mathbb{R}} v(x,y)\big((M_1M_2)(f\chi_{|x|/2\le|s|\le4|x|}\chi_{|\tau|>4|y|})\big)^p(x,y)dxdy$$

$$\left.+\int_{\mathbb{R}}\int_{\mathbb{R}} v(x,y)\big((M_1M_2)(f\chi_{|s|>4|x|}\chi_{|\tau|>4|y|})\big)^p(x,y)dxdy\right] := c[S_{31}+S_{32}+S_{33}].$$

Observe that

$$(M_1M_2)(f\chi_{|s|<|x|/2}\chi_{|\tau|>4|y|})(x,y)dxdy \le c(HM)f(x,y).$$

Therefore, by Lemma 5.2.3 we find that

$$S_{31}\le c\|f\|^p_{L^p_w(\mathbb{R}^2)}.$$

To estimate S_{32} we notice that if $2^k\le|y|<2^{k+1}$, then

$$(M_1M_2)(f\chi_{|x|/2\le|s|<4|x|}\chi_{|\tau|>4|y|})(x,y)dxdy$$

$$\le \sup_{r>0}\frac{1}{r}\int_{x-r}^{x+r}\chi_{\{|x|/2\le|s|\le4|x|\}}\bigg(\sup_{n\ge k+1}\frac{1}{2^n}\int_{2^n<|\tau|<2^{n+1}}|f(s,\tau)|d\tau\bigg)ds.$$

Hence, it yields that

$$S_{32}\le c\sum_{i\in\mathbb{Z}}\sum_{k\in\mathbb{Z}}\iint_{E_i\times E_k} v(x,y)\bigg(\sup_{r>0}\frac{1}{r}\int_{x-r}^{x+r}\chi_{\{s:2^{i-1}\le|s|\le2^{i+3}\}}(s)$$

$$\times\bigg(\sup_{n\ge k+1}\frac{1}{2^n}\int_{2^n<|\tau|<2^{n+1}}|f(s,\tau)|d\tau\bigg)ds\bigg)^p dxdy$$

[observing that the inner expression does not depend on y]

$$= c\sum_{i\in\mathbb{Z}}\sum_{k\in\mathbb{Z}}\int_{E_i}\bigg(\int_{E_k} v(x,y)dy\bigg)\bigg(\sup_{r>0}\frac{1}{r}\int_{x-r}^{x+r}\chi_{\{s:2^{i-1}\le|s|<2^{i+3}\}}$$

$$\times\bigg(\sup_{n\ge k+1}\frac{1}{2^n}\int_{2^n<|\tau|<2^{n+1}}|f(s,\tau)|d\tau\bigg)ds\bigg)^p dx$$

[using the fact that v is even and increasing in each variable]

$$\le c\sum_{i\in\mathbb{Z}}\sum_{k\in\mathbb{Z}} v(2^{i+1},2^{k+1})2^k\int_{\mathbb{R}}\bigg(\sup_{r>0}\frac{1}{r}\int_{x-r}^{x+r}\chi_{\{s:2^{i-1}\le|s|<2^{i+3}\}}$$

$$\times\Big(\sup_{n\ge k+1}\frac{1}{2^n}\int\limits_{2^n<|\tau|<2^{n+1}}|f(s,\tau)|d\tau\Big)ds\Big)^p dx$$

$\Big[$ using the boundedness of M in $L^p(\mathbb{R})$ $\Big]$

$$\le\sum_{i\in\mathbb{Z}}\sum_{k\in\mathbb{Z}}v(2^{i+1},2^{k+1})2^k\int\limits_{2^{i-1}<|x|<2^{i+3}}\Big(\sup_{n\ge k+1}\frac{1}{2^n}\int\limits_{2^n<|\tau|<2^{n+1}}|f(x,\tau)|d\tau\Big)^p dx$$

$\Big[$putting the supremum outside the brackets and using the inequality $\sup\le\sum$$\Big]$

$$\le\sum_{i\in\mathbb{Z}}\sum_{k\in\mathbb{Z}}v(2^{i+1},2^{k+1})2^k\int\limits_{2^{i-1}<|x|<2^{i+3}}\Big[\sup_{n\ge k+1}\frac{1}{2^{np}}\Big(\int\limits_{2^n<|\tau|<2^{n+1}}|f(x,\tau)|d\tau\Big)^p\Big]dx$$

$$\le\sum_{i\in\mathbb{Z}}\sum_{k\in\mathbb{Z}}v(2^{i+1},2^{k+1})2^k\sum_{n=k+1}^{\infty}\frac{1}{2^{np}}\int\limits_{2^{i-1}<|x|<2^{i+3}}\Big(\int\limits_{2^n<|\tau|<2^{n+1}}|f(x,\tau)|d\tau\Big)^p dx$$

$\Big[$ changing the order of the summation and using the inequality

$v(2^{j+1},2^{k+1})\le v(2^{j+1},2^n)$ $\Big]$

$$=\sum_{i\in\mathbb{Z}}\sum_{n\in\mathbb{Z}}\frac{v(2^{i+1},2^n)}{2^{np}}\Big(\sum_{k=-\infty}^{n-1}2^k\Big)\int\limits_{2^{i-1}<|x|<2^{i+3}}\Big(\int\limits_{2^n<|\tau|<2^{n+1}}|f(x,\tau)|d\tau\Big)^p dx$$

$\Big[$ using Hölder's inequality $\Big]$

$$\le\sum_{i\in\mathbb{Z}}\sum_{n\in\mathbb{Z}}v(2^{i+1},2^n)\int\limits_{2^{i-1}<|x|<2^{i+3}}\int\limits_{2^n<|\tau|<2^{n+1}}|f(x,\tau)|^p d\tau dx$$

$\Big[$ using the inequality $v(2^{i+1},2^n)\le cw_1(2^{j-1})w_2(2^n)$ $\Big]$

$$\le\sum_{i\in\mathbb{Z}}\sum_{n\in\mathbb{Z}}\int\limits_{2^{i-1}<|x|<2^{i+3}}\int\limits_{2^n<|\tau|<2^{n+1}}w_1(x)w_2(\tau)|f(x,\tau)|^p d\tau dx\le\|f\|^p_{L^p_w(\mathbb{R}^2)}.$$

It remains to estimate S_{33}. First notice that if $2^k\le|y|<2^{k+1}$ and $2^i\le|x|<2^{i+1}$, then

$$(M_1M_2)(f\chi_{|s|>4|x|}\chi_{|\tau|>4|y|})(x,y)dxdy$$

$$\le c\sup_{m\ge i+1}\frac{1}{2^m}\int\limits_{2^i<|s|<2^{i+1}}\Big(\sup_{n\ge k+1}\frac{1}{2^n}\int\limits_{2^n<|\tau|<2^{n+1}}|f(s,\tau)|d\tau\Big)ds.$$

Further, we have that

$$S_{33}\le c\sum_{i\in\mathbb{Z}}\sum_{k\in\mathbb{Z}}\iint\limits_{E_i\times E_k}v(x,y)\Big(\sup_{m\ge i+1}\frac{1}{2^m}\int\limits_{2^m\le|s|<2^{m+1}}$$

$$\times\Big(\sup_{n\geq k+1}\frac{1}{2^n}\int\limits_{2^n<|\tau|<2^{n+1}}|f(s,\tau)|d\tau\Big)ds\Big)^p dxdy$$

[putting the supremum outside the brackets, using the fact that the expression in the integral does not depend on x and y and the inequality $\sup\leq\sum$]

$$\leq c\sum_{i\in\mathbb{Z}}\sum_{k\in\mathbb{Z}}\Big(\iint\limits_{E_i\times E_k}v(x,y)dxdy\Big)$$

$$\times\sup_{m\geq i+1}\frac{1}{2^{mp}}\Big(\int\limits_{2^m\leq|s|<2^{m+1}}\Big(\sup_{n\geq k+1}\frac{1}{2^n}\int\limits_{2^n<|\tau|<2^{n+1}}|f(s,\tau)|d\tau\Big)ds\Big)^p$$

$$\leq\sum_{i\in\mathbb{Z}}\sum_{k\in\mathbb{Z}}\Big(\int\limits_{E_i}\int\limits_{E_k}v(x,y)dxdy\Big)$$

$$\times\sum_{m=i+1}^{\infty}\frac{1}{2^{mp}}\Big(\int\limits_{2^m\leq|s|<2^{m+1}}\Big(\sup_{n\geq k+1}\frac{1}{2^n}\int\limits_{2^n<|\tau|<2^{n+1}}|f(s,\tau)|d\tau\Big)ds\Big)^p$$

[using the fact that v is even and increasing in each variable and Hölder's inequality]

$$\leq c\sum_{i\in\mathbb{Z}}2^i\sum_{k\in\mathbb{Z}}v(2^{i+1},2^{k+1})2^k$$

$$\times\sum_{m=i+1}^{\infty}\frac{1}{2^{mp}}\Big(\int\limits_{2^m\leq|s|<2^{m+1}}\Big(\sup_{n\geq k+1}\frac{1}{2^n}\int\limits_{2^n<|\tau|<2^{n+1}}|f(s,\tau)|d\tau\Big)ds\Big)^p$$

$$\leq c\sum_{i\in\mathbb{Z}}2^i\sum_{k\in\mathbb{Z}}v(2^{i+1},2^{k+1})2^k\sum_{m=i+1}^{\infty}2^{-m}$$

$$\times\int\limits_{2^m\leq|s|<2^{m+1}}\Big(\sup_{n\geq k+1}\frac{1}{2^n}\int\limits_{2^n<|\tau|<2^{n+1}}|f(s,\tau)|d\tau\Big)^p ds$$

[changing the order of the sums and using the monotonicity of v]

$$=c\sum_{k\in\mathbb{Z}}2^k\sum_{m\in\mathbb{Z}}\frac{v(2^m,2^{k+1})}{2^m}\Big(\sum_{i=-\infty}^{m-1}2^i\Big)$$

$$\times\int\limits_{2^m\leq|s|<2^{m+1}}\Big(\sup_{n\geq k+1}\frac{1}{2^n}\int\limits_{2^n<|\tau|<2^{n+1}}|f(s,\tau)|d\tau\Big)^p ds$$

$$= c\sum_{k\in\mathbb{Z}} 2^k \sum_{m\in\mathbb{Z}} v(2^m, 2^{k+1})$$

$$\times \int\limits_{2^m\le|s|<2^{m+1}} \Big(\sup_{n\ge k+1} \frac{1}{2^n} \int\limits_{2^n<|\tau|<2^{n+1}} |f(s,\tau)|d\tau \Big)^p ds$$

[putting the supremum outside the brackets and using the inequality $\sup \le \sum$]

$$\le c\sum_{k\in\mathbb{Z}} 2^k \sum_{m\in\mathbb{Z}} v(2^m, 2^{k+1}) \sum_{n=k+1}^{\infty} \frac{1}{2^{np}} \int\limits_{2^m\le|s|<2^{m+1}} \Big(\int\limits_{2^n<|\tau|<2^{n+1}} |f(s,\tau)|d\tau \Big)^p ds$$

[changing the order of sums, using the monotonicity of v and Hölder's inequality]

$$\le c\sum_{m\in\mathbb{Z}}\sum_{n\in\mathbb{Z}} \frac{v(2^m,2^n)}{2^n} \Big(\sum_{k=-\infty}^{n-1} 2^k \Big) \int\limits_{2^m\le|s|<2^{m+1}} \int\limits_{2^n<|\tau|<2^{n+1}} |f(s,\tau)|^p d\tau ds$$

[using the inequality $v(2^n, 2^m) \le c w_1(2^n) w_2(2^m)$ and the fact that w_1 and w_2 are even and increasing]

$$\le c\sum_{m\in\mathbb{Z}}\sum_{n\in\mathbb{Z}} w_1(2^m) w_2(2^n) \int\limits_{2^m\le|s|<2^{m+1}} \int\limits_{2^n<|\tau|<2^{n+1}} |f(s,\tau)|^p d\tau ds \le c\|f\|^p_{L^p_w(\mathbb{R}^2)}.$$

Necessity. It is easy to check that the inequality

$$(M_1M_2)f(x,y) \ge c\frac{1}{|x||y|} \int\limits_{|s|<|x|} \int\limits_{|\tau|<|y|} |f(s,\tau)|dsd\tau$$

holds. Substituting now the test functions

$$f(x,y) = w^{1-p'}(x,y)\chi_{\{|x|<a\}}(x)\chi_{\{|y|<b\}}(y), \quad a,b>0$$

in the two–weight inequality for M_1M_2 we find that (5.2.3) holds.
The proof is complete. □

Notice that the next inequality holds

$$M_S f(x,y) \le (M_1M_2)f(x,y).$$

Here M_S and M_1M_2 are the operators defined at the beginning of this chapter. Applying now the previous theorem we have the next statement for which the proof is omitted.

Theorem 5.2.3. *Let $1<p<\infty$. Suppose that the two–dimensional weight v is even on $\mathbb{R}$ and increasing on $\mathbb{R}_+$ in each variable for fixed another one. Suppose also that w satisfies the condition $w(x,y) = w_1(x)w_2(y)$, where w_1 and w_2 are even on $\mathbb{R}$ and increasing on $\mathbb{R}_+$. Then M_S is bounded from $L^p_w(\mathbb{R}^2)$ to $L^p_v(\mathbb{R}^2)$ if and only if* (5.2.3) *holds.*

The Case of Decreasing Weights

Now we study the case of decreasing weights. In this case we will see that the criteria for the boundedness of strong and iterated maximal operators are different from those for increasing weights.

We begin with the following lemma:

Lemma 5.2.4. *Let $1 < p < \infty$. Suppose that v and w are two–dimensional weight functions such that they are even on $\mathbb{R}$ and decreasing on $\mathbb{R}_+$ in the first variable for the second fixed one, and, in addition, assume that $w^{1-p'}(x,y) \in DC(y)$. Then the operator MH is bounded from $L^p_w(\mathbb{R}^2)$ to $L^p_v(\mathbb{R}^2)$ if and only if*

$$\widetilde{B} := \sup_{a,b>0} \left(\frac{1}{a} \int_0^a \int_b^\infty \frac{v(x,y)}{y^p} dxdy \right)^{1/p} \left(\frac{1}{a} \int_0^a \int_0^b w^{1-p'}(x,y) dxdy \right)^{1/p'} < \infty \qquad (5.2.8)$$

(*here MH is the operator defined at the beginning of Section* 5.2).

Proof. *Sufficiency.* In the proof we will use the notation:

$$v_1(x,y) := \frac{v(x,y)}{y^p}, \quad \tilde{v}_{1,j}(x) := \int_{2^j}^{2^{j+1}} v_1(x,y)dy;$$

$$\tilde{w}_{1,j}(x) := \left(\int_0^{2^j} w^{1-p'}(x,\tau)d\tau \right)^{1-p}, \quad F_j := \{y : 2^j < |y| < 2^{j+1}\}.$$

We have that

$$I := \iint_{\mathbb{R}^2} v(x,y)\big((MH)f(x,y)\big)^p(x,y)dxdy$$

$$\leq \sum_{j\in\mathbb{Z}} \iint_{\mathbb{R}\times F_j} \frac{v(x,y)}{|y|^p} \left(\sup_{r>0} \frac{1}{r} \int_{x-r}^{x+r} \int_{|\tau|<|y|} |f(s,\tau)|dsd\tau \right)^p dxdy$$

$$\leq \sum_{j\in\mathbb{Z}} \int_{\mathbb{R}} \tilde{v}_{1,j}(x) \left(\sup_{r>0} \frac{1}{r} \int_{x-r}^{x+r} \left(\int_{|\tau|<2^{j+1}} |f(s,\tau)|d\tau \right) ds \right)^p dx$$

[using Theorem 5.1.2 and the fact that $\tilde{v}_{1,j}$ and $\tilde{w}_{1,j}$ are decreasing]

$$\leq c \sum_{j\in\mathbb{Z}} \int_{\mathbb{R}} \tilde{w}_{1,j}(x) \left(\int_{|\tau|<2^{j+1}} |f(x,\tau)|d\tau \right)^p dx$$

$$= c \int\limits_{\mathbb{R}} \sum_{j\in\mathbb{Z}} \tilde{w}_{1,j}(x) \bigg(\int\limits_{|\tau|<2^{j+1}} |f(x,\tau)| d\tau \bigg)^p dx := I_1.$$

On the other hand, it yields that

$$\sup_{\substack{x>0\\k\in\mathbb{Z}}} J(x,k) := \sup_{\substack{x>0\\k\in\mathbb{Z}}} \sum_{j=k}^{\infty} \bigg(\int\limits_0^{2^j} w^{1-p'}(x,y)dy \bigg)^{1-p}$$

$$\times \bigg(\sum_{j=-\infty}^{k} \int\limits_{2^j}^{2^{j+1}} w^{1-p'}(x,y)dy \bigg)^{p-1} < \infty.$$

Indeed, the doubling condition for $w^{1-p'}$ with respect to second variable and Lemma 1.6.1 imply that

$$J(x,k) = \sum_{j=k}^{\infty} \bigg(\int\limits_0^{2^j} w^{1-p'}(x,y)dy \bigg)^{1-p}$$

$$\times \bigg(\int\limits_0^{2^{k+1}} w^{1-p'}(x,y)dy \bigg)^{p-1} \leq \bigg(\sum_{j=k}^{\infty} \eta^{(j-k)(1-p)} \bigg)$$

$$\times \bigg(\int\limits_0^{2^k} w^{1-p'}(x,y)dy \bigg)^{1-p} \bigg(\int\limits_0^{2^{k+1}} w^{1-p'}(x,y)dy \bigg)^{p-1} \leq c.$$

Consequently, in virtue of Theorem 1.1.4 and Hölder's inequality we find that

$$I_1 \leq \int\limits_{\mathbb{R}} \sum_{j\in\mathbb{Z}} \tilde{w}_{1,j}(x) \bigg(\sum_{k=-\infty}^{j} \int\limits_{\{\tau:2^k<|\tau|<2^{k+1}\}} f(x,\tau)d\tau \bigg)^p dx$$

$$\leq c \int\limits_{\mathbb{R}} \sum_{k\in\mathbb{Z}} \bigg(\int\limits_{2^k}^{2^{k+1}} w^{1-p'}(x,\tau)d\tau \bigg)^{1-p} \bigg(\int\limits_{\{\tau:2^k<|\tau|<2^{k+1}\}} f(x,\tau)d\tau \bigg)^p dx$$

$$\leq c \int\limits_{\mathbb{R}} \sum_{k\in\mathbb{Z}} \bigg(\int\limits_{\{\tau:2^k<|\eta|<2^{k+1}\}} w^{1-p'}(x,y)dy \bigg)^{1-p}$$

$$\times \bigg(\int\limits_{2^k}^{2^{k+1}} w(x,\tau) f^p(x,\tau)d\tau \bigg) \bigg(\int\limits_{2^k}^{2^{k+1}} w^{1-p'}(x,\tau)d\tau \bigg)^{p-1} dx \leq c\|f\|^p_{L^p(\mathbb{R}^2)}.$$

Necessity follows easily if we substitute the test function

$$f(x,y) = w^{1-p'}(x,y)\chi_{\{|x|<a\}}(x)\chi_{\{|y|<b\}}(y)$$

in the two–weight inequality

$$\iint_{\mathbb{R}^2} ((MH)f(x,y))^p v(x,y)dxdy \le c \iint_{\mathbb{R}^2} |f(x,y)|^p w(x,y)dxdy$$

and make some simple calculations. The proof is complete. □

Analogously, it follows the next statement:

Lemma 5.2.5. *Let $1 < p < \infty$. Suppose that v is a two-dimensional weight, which is decreasing on $\mathbb{R}_+$ and even on $\mathbb{R}$ in the first variable for the the second fixed one. Let another two–dimensional weight w be of the form $w(x,y) = w_1(x)w_2(y)$, where w_1 is decreasing on $\mathbb{R}_+$ and even on $\mathbb{R}$. Then the operator MH is bounded from $L^p_w(\mathbb{R}^2)$ to $L^p_v(\mathbb{R}^2)$ if and only if* (5.2.8) *holds.*

The following lemma will be also useful for us.

Lemma 5.2.6. *Let $1 < p < \infty$. Suppose that a two–dimensional weight v is even on $\mathbb{R}$ and decreasing on $\mathbb{R}_+$ in the second variable for the first fixed one. Suppose also that $w(x,y) = w_1(x)w_2(y)$, where w_2 is even on $\mathbb{R}$ and decreasing on $\mathbb{R}_+$. Then the operator HM is bounded from $L^p_w(\mathbb{R}^2)$ to $L^p_v(\mathbb{R}^2)$ if and only if*

$$\overline{B} := \sup_{\substack{a>0\\ b>0}} \left(\frac{1}{b}\int_a^\infty x^{-p}\left(\int_0^b v(x,y)dy\right)dx\right)^{1/p} \left(\frac{1}{b}\int_0^b \left(\int_0^a w^{1-p'}(x,y)dy\right)dx\right)^{1/p'} < \infty.$$

Proof. *Necessity.* follows in the same way as in the proof of the previous theorem. *Sufficiency.* First suppose that $S := \int_{\mathbb{R}} w_1^{1-p'}(x)dx = \infty$. Let $\{x_k\}_{k=-\infty}^{+\infty}$ be a sequence of positive numbers for which the equality

$$2^k = \int_{|x|<x_k} w_1^{1-p'}(x)dx$$

holds for all $k \in \mathbb{Z}$. It is clear that $\{x_k\}$ is increasing and

$$\mathbb{R} \setminus \{0\} = \cup_{k\in\mathbb{Z}} E_k,$$

where $E_k := \{x \in \mathbb{R} : x_k \le |x| < x_{k+1}\}$. Besides this, it is easy to verify that $2^k = \int_{E_k} w_1^{1-p'}(x)dx$.

Let

$$\bar{v}_k(y) := \int_{E_k} \frac{v(x,y)}{|x|^p}dx.$$

We have that

$$I := \iint_{\mathbb{R}^2} v(x,y)\big((HM)f(x,y)\big)^p dxdy$$

$$= \sum_{k\in\mathbb{Z}} \iint_{E_k\times\mathbb{R}} \frac{v(x,y)}{|x|^p}\Bigg(\int_{|s|<|x|} \Bigg(\sup_{t>0}\frac{1}{t}\int_{|y-\tau|<t} |f(s,\tau)|d\tau\Bigg)ds\Bigg)^p dxdy$$

$$\leq \sum_{k\in\mathbb{Z}} \int_{\mathbb{R}} \Bigg(\int_{E_k} \frac{v(x,y)}{|x|^p}dx\Bigg)\Bigg(\int_{|s|<x_{k+1}} \Bigg(\sup_{t>0}\frac{1}{t}\int_{|y-\tau|<t} |f(s,\tau)|d\tau\Bigg)ds\Bigg)^p dy$$

$$\leq \sum_{k\in\mathbb{Z}} \int_{\mathbb{R}} \bar{v}_k(y)\Bigg(\int_{|s|<x_{k+1}} M(f(s,\cdot))(y)ds\Bigg)^p dy$$

[using generalized Minkowski's inequality]

$$\leq \sum_{k}\Bigg(\int_{|s|<x_{k+1}} \Bigg(\int_{\mathbb{R}} \bar{v}_k(y)\Big(M(f(s,\cdot))(y)\Big)^p dy\Bigg)^{1/p} ds\Bigg)^p := I_1.$$

Notice that

$$\overline{B} \geq \sup_{b,x_k>0}\Bigg(\frac{1}{b}\int_0^b \Bigg(\int_{x_k}^{x_{k+1}} \frac{v(x,y)}{x^p}dx\Bigg)dy\Bigg)^{1/p}$$

$$\times\Bigg(\frac{1}{b}\int_0^b \Bigg(\int_0^{x_k} w_1^{1-p'}(x)dx\Bigg)^{\frac{1-p'}{1-p'}} w_2^{1-p'}(y)dy\Bigg)^{1/p'}.$$

Hence, by using Theorem 5.1.2 we find that

$$I_1 \leq c\sum_{k\in\mathbb{Z}}\Bigg(\int_{|s|<x_{k+1}} \Bigg(\int_{\mathbb{R}} \Bigg(\int_0^{x_k} w_1^{1-p'}(\sigma)d\sigma\Bigg)^{1-p} |f(s,\tau)|^p w_2(\tau)d\tau\Bigg)^{1/p} ds\Bigg)^p$$

$$= c\sum_{k} \Bigg(\int_0^{x_k} w_1^{1-p'}(\sigma)d\sigma\Bigg)^{1-p}$$

$$\times\Bigg(\sum_{j=-\infty}^{k} \int_{x_j\leq|s|<x_{j+1}} \Bigg(\int_{\mathbb{R}} |f(s,\tau)|^p w_2(\tau)d\tau\Bigg)^{1/p} ds\Bigg)^p := I_2.$$

Further, it is obvious that

$$\sum_{k=n}^{+\infty}\Big(\int_0^{x_k} w_1^{1-p'}(x)dx\Big)^{1-p}\Big(\sum_{k=-\infty}^{n}\int_{x_k}^{x_{k+1}} w_1^{1-p'}(x)dx\Big)^{p-1}$$

$$=\sum_{k=n}^{+\infty}\Big(\int_0^{x_k} w_1^{1-p'}(x)dx\Big)^{1-p}\Big(\int_0^{x_{n+1}} w_1^{1-p'}(x)dx\Big)^{p-1}$$

$$=\Big(\sum_{k=n}^{+\infty}2^{k(1-p)}\Big)2^{(n+1)(p-1)}=c<\infty.$$

Therefore, Theorem 1.1.4 yields

$$I_2\le c\sum_k\Big(\int_{x_k}^{x_{k+1}} w_1^{1-p'}(s)ds\Big)^{1-p}$$

$$\times\Big(\int_{x_k\le|s|<x_{k+1}}\Big(\int_{\mathbb{R}}|f(s,\tau)|^p w_2(\tau)d\tau\Big)^{1/p}ds\Big)^p$$

$$\Big[\text{ using Hölder's inequality }\Big]$$

$$\le c\sum_k\Big(\int_{x_k}^{x_{k+1}} w_1^{1-p'}(s)ds\Big)^{1-p}$$

$$\times\Big(\int_{x_k\le|s|<x_{k+1}}\int_{\mathbb{R}}|f(s,\tau)|^p w_2(\tau)w_1(s)d\tau ds\Big)\Big(\int_{x_k}^{x_{k+1}} w_1^{1-p'}(t)dt\Big)^{p-1}$$

$$=c\sum_k\int_{x_k\le|s|<x_{k+1}}\int_{\mathbb{R}}|f(s,\tau)|^p w_2(\tau)w_1(s)d\tau ds=c\|f\|^p_{L^p_w(\mathbb{R}^2)}.$$

The proof in the case $S<\infty$ follows in the same way as in the previous case. The details are omitted.

The proof is complete. □

Theorem 5.2.4. *Let $1<p<\infty$. Suppose that the two-dimensional weights v and w are even on $\mathbb{R}$ and decreasing on $\mathbb{R}_+$ in each variable for another fixed one. Assume, in addition, that $w^{1-p'}(x,y)$ belongs to $DC(x)\cap DC(y)$. Then M_S is bounded from $L^p_w(\mathbb{R}^2)$ to $L^p_v(\mathbb{R}^2)$ if and only if*

$$B_1:=\sup_{a,b>0}B_1(a,b):=\sup_{a,b>0}\Big(\frac{1}{ab}\int_0^a\int_0^b v(x,y)dxdy\Big)^{1/p}$$

$$\times\left(\frac{1}{ab}\int_0^a\int_0^b w^{1-p'}(x,y)dxdy\right)^{1/p'} < \infty. \tag{5.2.9}$$

Proof. *Sufficiency.* Notice that condition (5.2.9) implies the following estimates:

$$v(a,b) \le cB_1^p w(8a,8b), \quad a,b>0; \tag{5.2.10}$$

$$\sup_{\substack{a>0\\b>0}}\left(\frac{1}{a}\int_0^a\int_b^\infty \frac{v(x,y)}{y^p}dxdy\right)^{1/p}\left(\frac{1}{a}\int_0^a\int_0^b w^{1-p'}(x,y)dxdy\right)^{1/p'} \le cB_1; \tag{5.2.11}$$

$$\sup_{\substack{a>0\\b>0}}\left(\frac{1}{b}\int_a^\infty\int_0^b \frac{v(x,y)}{x^p}dxdy\right)^{1/p}\left(\frac{1}{b}\int_0^a\int_0^b w^{1-p'}(x,y)dxdy\right)^{1/p'} \le cB_1. \tag{5.2.12}$$

Indeed, to prove (5.2.10) we observe that monotonicity of the weights v and w yields

$$\infty > B_1(a,b) \ge c\left(\frac{1}{ab}\int_0^{a/16}\int_0^{b/16} v(x,y)dxdy\right)^{1/p}\left(\frac{1}{ab}\int_{a/2}^a\int_{b/2}^b w^{1-p'}(x,y)dxdy\right)^{1/p'}$$

$$\ge cv^{1/p}(a/16,b/16)w^{-1/p}(a/2,b/2).$$

Observe now that for any positive a and b, we have that

$$\left(\frac{1}{a}\int_0^a\int_b^\infty \frac{v(x,y)}{y^p}dxdy\right)^{1/p}\left(\frac{1}{a}\int_0^a\int_0^b w^{1-p'}(x,y)dxdy\right)^{1/p'}$$

$$\le c\left(\frac{1}{a}\int_0^a v(x,b)dx\right)^{1/p}\left(\frac{1}{a}\int_0^a w^{1-p'}(x,b)dx\right)^{1/p'}$$

$$\le c\left(\frac{1}{ab}\int_0^a\int_0^{2b} v(x,y)dxdy\right)^{1/p}\left(\frac{1}{ab}\int_0^a\int_0^{2b} w^{1-p'}(x,y)dxdy\right)^{1/p'} \le cB_1.$$

Thus (5.2.11) holds. Analogously, it follows (5.2.12).

Further, splitting f into three parts with respect to the second variable we find that

$$\|M_S f\|^p_{L^p_v(\mathbb{R}^2)} \le c\left[\iint_{\mathbb{R}^2} v(x,y)\big(M_S(f\chi_{|\cdot|<|y|/2})\big)^p(x,y)dxdy\right.$$

$$+\iint\limits_{\mathbb{R}^2} v(x,y)\big(M_S(f\chi_{|y|/2\leq|\cdot|\leq 4|y|})\big)^p(x,y)dxdy$$

$$+\iint\limits_{\mathbb{R}^2} v(x,y)\big(M_S(f\chi_{|\cdot|>4|y|})\big)^p(x,y)dxdy\Bigg] := c[S_1+S_2+S_3].$$

Suppose that $|\tau|<|y|/2$. Then $|y|/2<|y-\tau|$. Therefore if $t<|y|/2$, then it yields that

$$\frac{1}{rt}\int\limits_{x-r}^{x+r}\int\limits_{y-t}^{y+t}|f(s,\tau)|\chi_{\{\tau:|\tau|<|y|/2\}}(\tau)dsd\tau$$

$$\leq \frac{1}{rt}\int\limits_{x-r}^{x+r}\Bigg(\int\limits_{\{\tau:|\tau-y|<t;|\tau-y|>|y|/2\}}|f(s,\tau)|\chi_{\{\tau:|\tau|<|y|/2\}}(\tau)d\tau\Bigg)ds=0.$$

Let now $t\geq|y|/2$. Then it is easy to see that

$$\frac{1}{rt}\int\limits_{x-r}^{x+r}\int\limits_{y-t}^{y+t}|f(s,\tau)|\chi_{\{\tau:|\tau|<|y|/2\}}(\tau)dsd\tau\leq c\frac{1}{r}\frac{1}{|y|}\int\limits_{x-r}^{x+r}\Bigg(\int\limits_{|\tau|<|y|/2}|f(s,\tau)|d\tau\Bigg)ds$$

$$\leq c\frac{1}{|y|}\sup_{r>0}\frac{1}{r}\int\limits_{x-r}^{x+r}\Bigg(\int\limits_{|\tau|<|y|/2}|f(s,\tau)|d\tau\Bigg)ds\leq c(MH)f(x,y).$$

Now, by using (5.2.11) and Lemma 5.2.4 we conclude that

$$S_1\leq c\|f\|^p_{L^p_w(\mathbb{R}^2)}.$$

To estimate S_3, first we split f into three parts with respect to the first variable. Then

$$M_S(f\chi_{|\tau|>4|y|})(x,y)=M_S(f\chi_{|s|<|x|/2}\chi_{|\tau|>4|y|})(x,y)$$

$$+M_S(f\chi_{|x|/2<|s|<4|x|}\chi_{|\tau|>4|y|})(x,y)+M_S(f\chi_{|s|>4|x|}\chi_{|\tau|>4|y|})(x,y)$$

$$:=M_S^{(3,1)}f(x,y)+M_S^{(3,2)}f(x,y)+M_S^{(3,3)}f(x,y).$$

It is obvious that

$$M_S^{(3,1)}f(x,y)\leq\frac{c}{|x|}\sup_{t>0}\frac{1}{t}\int\limits_{y-t}^{y+t}\Bigg(\int\limits_{|s|<|x|/2}|F(\tau,s)|ds\Bigg)d\tau,$$

where $F(\tau, s) := f(s, \tau)$. Now we observe that by (5.2.12) it follows that

$$\left(\frac{1}{b}\int_0^b\left(\int_a^\infty \frac{V(y,x)}{x^p}dx\right)dy\right)^{1/p}\left(\frac{1}{b}\int_0^b\left(\int_0^a W^{1-p'}(y,x)dx\right)dy\right)^{1/p'}$$

$$=\left(\frac{1}{b}\int_a^\infty x^{-p}\left(\int_0^b v(x,y)dy\right)dx\right)^{1/p}\left(\frac{1}{b}\int_0^a\left(\int_0^b w^{1-p'}(x,y)dy\right)dx\right)^{1/p'} \le cB_1,$$

where $V(y,x) := v(x,y)$, $W(y,x) := w(x,y)$.

Hence, by using Lemma 5.2.4 (interchanging the order of x and y) we conclude that

$$S_{3,1} := \iint_{\mathbb{R}^2} v(x,y)(M_S^{(3,1)})^p(x,y)dxdy \le c\|f\|^p_{L^p_w(\mathbb{R}^2)}.$$

Further, by the arguments used in Theorem 5.2.1 we have that

$$M_S^{(3,2)}f(x,y) \le c \sup_{n\ge k+1}\frac{1}{2^n}\sup_{r>0}\frac{1}{r}\int_{x-r}^{x+r}\chi_{\{|x|/2<|s|<4|x|\}}(s)$$

$$\times\left(\int_{\{2^n<|\tau|<2^{n+1}\}}|f(s,\tau)|d\tau\right)ds$$

for some integer k.

Putting the supremum outside the brackets, using the obvious inequality $\sup \le \sum$, the fact that v and w are decreasing, the boundedness of M in $L^p(\mathbb{R})$ (Hardy–Littlewood– Wiener theorem) we find that

$$S_{3,2} := \sum_{i\in\mathbb{Z}}\sum_{k\in\mathbb{Z}}\iint_{E_i\times E_k} v(x,y)\sup_{n\ge k+1}\frac{1}{2^{np}}\left(\sup\frac{1}{r}\int_{x-r}^{x+r}\chi_{\{2^{i-1}<|s|<2^{i+2}\}}(s)\right.$$

$$\left.\times\left(\int_{2^n<|\tau|<2^{n+1}}|f(s,\tau)|d\tau\right)ds\right)^p dxdy$$

$$\le \sum_{i\in\mathbb{Z}}\sum_{k\in\mathbb{Z}}\int_{E_i}\left(\int_{E_k}v(x,y)dy\right)\sum_{n=k+1}^{\infty}\frac{1}{2^{np}}\left(\sup_{r>0}\frac{1}{r}\int_{x-r}^{x+r}\chi_{\{2^{i-1}<|s|<2^{i+2}\}}(s)\right.$$

$$\left.\times\left(\int_{2^n<|\tau|<2^{n+1}}|f(s,\tau)|d\tau\right)ds\right)^p dx$$

$$\leq \sum_{i\in\mathbb{Z}}\sum_{k\in\mathbb{Z}} 2^k \sum_{n=k+1}^{\infty} \frac{1}{2^{np}} v(2^i,2^k) \int_{E_i} \Bigg(\sup_{r>0} \frac{1}{r} \int_{x-r}^{x+r} \chi_{\{2^{i-1}<|s|<2^{i+2}\}}(s)$$

$$\times \Bigg(\int_{2^n<|\tau|<2^{n+1}} |f(s,\tau)|d\tau \Bigg) ds \Bigg)^p dx$$

$$\leq \sum_{i\in\mathbb{Z}}\sum_{k\in\mathbb{Z}} 2^k \sum_{n=k+1}^{\infty} \frac{1}{2^{np}} v(2^i,2^k) \int_{2^i<|x|<2^{i+2}} \Bigg(\int_{2^n<|\tau|<2^{n+1}} |f(x,\tau)|d\tau \Bigg)^p dx$$

$$= \sum_{i\in\mathbb{Z}}\sum_{n\in\mathbb{Z}} \frac{1}{2^{np}} \sum_{k=-\infty}^{n-1} \int_{2^i<|x|<2^{i+2}} \Bigg(\int_{2^{k-1}<|y|<2^k} v(2^i,y)dy \Bigg)$$

$$\times \Bigg(\int_{2^n<|\tau|<2^{n+1}} |f(x,\tau)|d\tau \Bigg)^p dx$$

$$= \sum_{i\in\mathbb{Z}}\sum_{n\in\mathbb{Z}} \frac{1}{2^{np}} \int_{2^i<|x|<2^{i+2}} \Bigg(\int_{|y|<2^{n-1}} v(2^i,y)dy \Bigg) \Bigg(\int_{2^n<|\tau|<2^{n+1}} |f(x,\tau)|d\tau \Bigg)^p dx$$

[using Hölder's inequality]

$$\leq \sum_{i\in\mathbb{Z}}\sum_{n\in\mathbb{Z}} \frac{1}{2^{np}} \int_{2^i<|x|<2^{i+2}} \Bigg(\int_{|y|<2^{n-1}} v(2^i,y)dy \Bigg)$$

$$\times \Bigg(\int_{2^n<|\tau|<2^{n+1}} |f(x,\tau)|^p w(2^{i+2},\tau)d\tau \Bigg)$$

$$\times \Bigg(\int_{2^n<|\tau|<2^{n+1}} w^{1-p'}(2^{i+2},\tau)d\tau \Bigg)^{p-1} dx := S.$$

In addition, using the fact that v and w are decreasing we have that

$$\frac{1}{2^{np}} \Bigg(\int_{|y|<2^{n-1}} v(2^i,y)dy \Bigg) \Bigg(\int_{2^n<|\tau|<2^{n+1}} w^{1-p'}(2^{i+2},\tau)d\tau \Bigg)^{p-1} \leq cB_1^p$$

Hence

$$S \le c \sum_{i\in\mathbb{Z}} \sum_{n\in\mathbb{Z}} \int\limits_{2^{i-1}<|x|<2^{i+2}} \int\limits_{2^n<|\tau|<2^{n+1}} w(2^{i+2},\tau)|f(x,\tau)|^p d\tau dx$$

$$\le \sum_{i\in\mathbb{Z}} \sum_{n\in\mathbb{Z}} \int\limits_{2^{i-1}<|x|<2^{i+2}} \int\limits_{2^n<|\tau|<2^{n+1}} w(x,\tau)|f(x,\tau)|^p d\tau dx \le c\|f\|^p_{L^p_w(\mathbb{R}^2)}.$$

Now we estimate $M_S^{(3,3)}$. We claim that

$$M_S^{(3,3)} f(x,y) \le c \sup_{\substack{n_1\ge i+1\\ n_2\ge k+1}} \frac{1}{2^{n_1}2^{n_2}} \int\limits_{\{s:2^{n_1}<|s|<2^{n_1+1}\}} \int\limits_{\{\tau:2^{n_2}<|\tau|<2^{n_2+1}\}} |f(s,\tau)| ds d\tau$$

when $2^i < |x| < 2^{i+1}$ and $2^k < |y| < 2^{k+1}$. Indeed, if $t < 2^{k+1}$, then for $2^k < |y| < 2^{k+1}$, we find that

$$\frac{1}{rt} \int\limits_{x-r}^{x+r} \int\limits_{y-t}^{y+t} |f(s,\tau)| \chi_{|s|>4|x|}(s) \chi_{|\tau|>4|y|}(\tau) ds d\tau$$

$$\le \frac{1}{r} \int\limits_{x-r}^{x+r} \frac{1}{t} \Bigg(\int\limits_{\{|y-\tau|<t,|y-\tau|>2^{k+1}\}} |f(s,\tau)| d\tau \Bigg) ds = 0.$$

By using the same arguments we obtain that if $r < 2^{i+1}$ and $2^i < |x| < 2^{i+1}$, then

$$\frac{1}{rt} \int\limits_{x-r}^{x+r} \int\limits_{y-t}^{y+t} |f(s,\tau)| \chi_{|s|>4|x|}(s) \chi_{|\tau|>4|y|}(\tau) ds d\tau = 0.$$

It remains to consider the case when $r > 2^{i+1}$ and $t > 2^{k+1}$. In this case $r \in [2^n, 2^{n+1})$ and $t \in [2^j, 2^{j+1})$ for some $n \ge i+1$ and $j \ge k+1$. If $|y-\tau| < t$, then for $2^k < |y| < 2^{k+1}$, we have that $|\tau| \le |y-\tau| + |y| \le t + |y| \le 2^{j+1} + 2^{k+1} \le 2^{j+1} + 2^j \le 2^{j+2}$. Analogously, if $|x-s| < r$, then for $2^i < |x| < 2^{i+1}$ and $|s| > 2^{i+2}$ it yields that $|s| \le |x-s| + |x| \le r + |x| \le 2^{n+1} + 2^{i+1} \le 2^{n+2}$. Using this observation we find that

$$\frac{1}{rt} \int\limits_{x-r}^{x+r} \int\limits_{y-t}^{y+t} |f(s,\tau)| ds d\tau$$

$$\le \frac{1}{2^j 2^n} \int\limits_{\{s:2^{i+2}<|s|<2^{n+2}\}} \int\limits_{\{\tau:2^{k+2}<|\tau|<2^{j+2}\}} |f(s,\tau)| ds d\tau$$

$$\leq \frac{1}{2^j 2^n} \sum_{n_1=i+1}^{n+1} \sum_{n_2=k+1}^{j+1} \int\limits_{\{s:2^{n_1}<|s|<2^{n_1+1}\}} \int\limits_{\{\tau:2^{n_2}<|\tau|<2^{n_2+1}\}} |f(s,\tau)|dsd\tau$$

$$\leq \frac{1}{2^j 2^n}\Bigg(\sup_{n_1\geq i+1} \sup_{n_2>k+1} \frac{1}{2^{n_1}2^{n_2}} \int\limits_{\{s:2^{n_1}<|s|<2^{n_1+1}\}} \int\limits_{\{\tau:2^{n_2}<|\tau|<2^{n_2+1}\}} |f(s,\tau)|dsd\tau\Bigg)$$

$$\times\Big(\sum_{n_1=j+1}^{n+1} 2^{n_1}\Big)\Big(\sum_{n_2=k+1}^{j+1} 2^{n_2}\Big)$$

$$= c \sup_{n_1\geq i+1} \sup_{n_2>k+1} \frac{1}{2^{n_1}2^{n_2}} \int\limits_{\{s:2^{n_1}<|s|<2^{n_1+1}\}} \int\limits_{\{\tau:2^{n_2}<|\tau|<2^{n_2+1}\}} |f(s,\tau)|dsd\tau.$$

Taking these estimates into account we obtain that

$$S_{3,3} \leq c \sum_{i\in\mathbb{Z}} \sum_{k\in\mathbb{Z}} \int_{2^i}^{2^{i+1}} \int_{2^k}^{2^{k+1}} v(x,y)$$

$$\times\Bigg(\int\limits_{\{s:2^{n_1}<|s|<2^{n_1+1}\}} \int\limits_{\{\tau:2^{n_2}<|\tau|<2^{n_2+1}\}} |f(s,\tau)|dsd\tau\Bigg)^p dxdy$$

[putting the supremum outside the brackets]

$$= \sum_{i\in\mathbb{Z}} \sum_{k\in\mathbb{Z}} \int_{2^i}^{2^{i+1}} \int_{2^k}^{2^{k+1}} v(x,y) \sup_{n_1\geq i+1} \sup_{n_2>k+1} \frac{1}{2^{n_1 p}2^{n_2 p}}$$

$$\times\Bigg(\int\limits_{\{s:2^{n_1}<|s|<2^{n_1+1}\}} \int\limits_{\{\tau:2^{n_2}<|\tau|<2^{n_2+1}\}} |f(s,\tau)|dsd\tau\Bigg)^p dxdy$$

[using the inequality $\sup \leq \sum$]

$$\leq \sum_{i\in\mathbb{Z}} \sum_{k\in\mathbb{Z}} \int_{2^i}^{2^{i+1}} \int_{2^k}^{2^{k+1}} v(x,y) \sum_{n_1=i+1}^{\infty} \sum_{n_2=k+1}^{\infty} \frac{1}{2^{n_1 p}2^{n_2 p}}$$

$$\Bigg(\int\limits_{\{s:2^{n_1}<|s|<2^{n_1+1}\}} \int\limits_{\{\tau:2^{n_2}<|\tau|<2^{n_2+1}\}} |f(s,\tau)|dsd\tau\Bigg)^p dxdy$$

$$= \sum_{i\in\mathbb{Z}} \sum_{k\in\mathbb{Z}} \sum_{n_1=i+1}^{\infty} \sum_{n_2=k+1}^{\infty} \bigg(\int_{2^i}^{2^{i+1}} \int_{2^k}^{2^{k+1}} v(x,y)dxdy \bigg)$$

$$\times \frac{1}{2^{n_1 p} 2^{n_2 p}} \bigg(\int_{\{s:2^{n_1}<|s|<2^{n_1+1}\}} \int_{\{\tau:2^{n_2}<|\tau|<2^{n_2+1}\}} |f(s,\tau)| ds d\tau \bigg)^p$$

[changing the order of sums twice]

$$\leq \sum_{n_1\in\mathbb{Z}} \sum_{n_2\in\mathbb{Z}} \frac{1}{2^{n_1 p} 2^{n_2 p}} \sum_{i=-\infty}^{n_1-1} \sum_{k=-\infty}^{n_2-1} \bigg(\int_{2^i}^{2^{i+1}} \int_{2^k}^{2^{k+1}} v(x,y)dxdy \bigg)$$

$$\times \bigg(\int_{\{s:2^{n_1}<|s|<2^{n_1+1}\}} \int_{\{\tau:2^{n_2}<|\tau|<2^{n_2+1}\}} |f(s,\tau)| ds d\tau \bigg)^p$$

$$= \sum_{n_1\in\mathbb{Z}} \sum_{n_2\in\mathbb{Z}} \bigg(\frac{1}{2^{n_1 p} 2^{n_2 p}} \int_0^{2^{n_1-1}} \int_0^{2^{n_2-1}} v(x,y)dxdy \bigg)$$

$$\times \bigg(\int_{\{s:2^{n_1}<|s|<2^{n_1+1}\}} \int_{\{\tau:2^{n_2}<|\tau|<2^{n_2+1}\}} |f(s,\tau)| ds d\tau \bigg)^p$$

[using Hölder's inequality]

$$\leq \sum_{n_1\in\mathbb{Z}} \sum_{n_2\in\mathbb{Z}} \bigg(\frac{1}{2^{n_1 p} 2^{n_2 p}} \int_0^{2^{n_1-1}} \int_0^{2^{n_2-1}} v(x,y)dxdy \bigg)$$

$$\times \bigg(\int_{\{s:2^{n_1}<|s|<2^{n_1+1}\}} \int_{\{\tau:2^{n_2}<|\tau|<2^{n_2+1}\}} |f(s,\tau)|^p w(s,\tau) ds d\tau \bigg)$$

$$\times \bigg(\int_{\{s:2^{n_1}<|s|<2^{n_1+1}\}} \int_{\{\tau:2^{n_2}<|\tau|<2^{n_2+1}\}} w^{1-p'}(s,\tau) ds d\tau \bigg)^{p-1} \leq c B_1^p \|f\|^p_{L^p_w(\mathbb{R}^2)}.$$

It remains to estimate S_2. Let us denote

$$M_S^{(2)} f(x,y) := \sup_{\substack{r>0\\t>0}} \frac{1}{rt} \int_{x-r}^{x+r} \int_{y-t}^{y+t} \chi_{\{\tau:|y|/2<|\tau|<4|y|\}}(\tau) |f(s,\tau)| ds d\tau.$$

Then it is easy to see that the next estimate holds:

$$M_S^{(2)} f(x,y) \leq \sup_{\substack{r>0\\t>0}} \frac{1}{rt} \int_{x-r}^{x+r} \int_{y-t}^{y+t} \chi_{s:|s|<|x|/2\}}(s) \chi_{\{\tau:|y|/2<|\tau|<4|y|\}}(\tau) |f(s,\tau)| ds d\tau$$

$$+\sup_{\substack{r>0\\t>0}}\frac{1}{rt}\int\limits_{x-r}^{x+r}\int\limits_{y-t}^{y+t}\chi_{s:|x|/2\le|s|\le4|x|\}}(s)\chi_{\{\tau:|y|/2<|\tau|<4|y|\}}(\tau)|f(s,\tau)|dsd\tau$$

$$+\sup_{\substack{r>0\\t>0}}\frac{1}{rt}\int\limits_{x-r}^{x+r}\int\limits_{y-t}^{y+t}\chi_{s:|s|>4|x|\}}(s)\chi_{\{\tau:|y|/2<|\tau|<4|y|\}}(\tau)|f(s,\tau)|dsd\tau$$

$$:=M_S^{(2,1)}f(x,y)+M_S^{(2,2)}f(x,y)+M_S^{(2,3)}f(x,y).$$

By the arguments used above but interchanging the order of the variables x and y we observe that

$$M^{2,1}f(x,y)\le c\big(\overline{M}H\big)f(x,y),$$

where

$$(\overline{M}H)f(x,y)=\frac{1}{|x|}\sup_{t>0}\frac{1}{t}\int\limits_{y-t}^{y+t}\Bigg(\int\limits_{\{s:|s|<|x|/2\}}|f(s,\tau)|ds\Bigg)d\tau.$$

Consequently, using the notation

$$V(y,x):=v(x,y),\;\; W(y,x):=w(x,y),\;\; F(\tau,s):=f(s,\tau)$$

we find that

$$S_{2,1}:=\iint\limits_{\mathbb{R}^2}v(x,y)\big((\overline{M}H)f(x,y)\big)^pdxdy$$

$$=\iint\limits_{\mathbb{R}^2}V(y,x)\Bigg(\frac{1}{|x|}\sup_{t>0}\frac{1}{t}\int\limits_{y-t}^{y+t}\Bigg(\int\limits_{\{s:|s|<|x|/2\}}|F(\tau,s)|ds\Bigg)d\tau\Bigg)^pdydx$$

[Using Lemma 5.2.4 and (5.2.12)]

$$\le\iint\limits_{\mathbb{R}}W(y,x)|F(y,x)|^pdydx=\|f\|^p_{L^p_w(\mathbb{R}^2)}.$$

Let us denote

$$f_{k,j}(x,y):=f(x,y)\chi_{\{2^{k-1}<|x|<2^{k+3}\}}(x)\chi_{\{\tau 2^{j-1}<|y|<2^{j+3}\}}(y).$$

Now (5.2.10), the boundedness of M_S in $L^p(\mathbb{R}^2)$ and the fact that the weights v and w are decreasing yield that

$$S_{2,2}:=\iint\limits_{\mathbb{R}}v(x,y)(M_S^{(2,2)}f(x,y))^pdxdy$$

$$\leq \sum_{k\in\mathbb{Z}}\sum_{j\in\mathbb{Z}} \int\limits_{2^k<|x|<2^{k+1}} \int\limits_{2^j<|y|<2^{j+1}} v(x,y)\big|M_S f_{k,j}(x,y)\big|^p dxdy$$

$$\leq \sum_{k\in\mathbb{Z}}\sum_{j\in\mathbb{Z}} v(2^k,2^j) \int\limits_{2^k<|x|<2^{k+1}} \int\limits_{2^j<|y|<2^{j+1}} \big|M_S f_{k,j}(x,y)\big|^p dxdy$$

$$\leq cB_1^p \sum_{k\in\mathbb{Z}}\sum_{j\in\mathbb{Z}} w(2^{k+2},2^{j+2}) \int\limits_{2^{k-1}<|x|<2^{k+3}} \int\limits_{2^{j-1}<|y|<2^{j+3}} |f(x,y)|^p dxdy$$

$$\leq cB_1^p \sum_{k\in\mathbb{Z}}\sum_{j\in\mathbb{Z}} \int\limits_{2^{k-1}<|x|<2^{k+3}} \int\limits_{2^{j-1}<|y|<2^{j+3}} w(x,y)|f(x,y)|^p dxdy \leq cB_1^p\|f\|^p_{L^p_w(\mathbb{R}^2)}.$$

The estimate of the integral

$$S_{2,3} := \iint\limits_{\mathbb{R}} v(x,y)(M_S^{(2,3)} f(x,y))^p dxdy$$

is similar to that of $S_{3,2}$. The only difference is that the order of the variables x and y are interchanged. Therefore using the arguments for the estimate of $S_{3,2}$ we conclude that

$$S_{2,3} \leq \|f\|^p_{L^p_w(\mathbb{R}^2)}.$$

Hence, sufficiency is proved.

To show *necessity* we take the test functions

$$f_{a,b}(x,y) = \chi_{(0,a)\times(0,b)}(x,y)w^{1-p'}(x,y)$$

in the two–weight inequality for M_S and observe that if $x \in (0,a)$ and $y \in (0,b)$, then

$$M_S f(x,y) \geq \frac{1}{ab}\int\limits_{x-a}^{x+a}\int\limits_{y-b}^{y+b} w^{1-p'}(s,\tau)dsd\tau = \frac{1}{ab}\int\limits_0^a\int\limits_0^b w^{1-p'}(s,\tau)dsd\tau.$$

The proof is complete. □

Theorem 5.2.5. *Let $1 < p < \infty$. Suppose that the two-dimensional weight v is even on $\mathbb{R}$ and decreasing on $\mathbb{R}_+$ in each variable for another fixed one. Assume also that another two–dimensional weight w has the form $w(x,y) = w_1(x)w_2(y)$, where w_1 and w_2 are decreasing on $\mathbb{R}_+$ and even on $\mathbb{R}$. Then M_1M_2 is bounded from $L^p_w(\mathbb{R}^2)$ to $L^p_v(\mathbb{R}^2)$ if and only if (5.2.9) holds.*

Proof. *Sufficiency.* Splitting f into three parts with respect to the second variable we have that

$$\|(M_1M_2)f\|^p_{L^p_v(\mathbb{R}^2)} \le c\Big[\iint\limits_{\mathbb{R}^2} v(x,y)\big((M_1M_2)^{(1)}f\big)^p(x,y)dxdy$$

$$+\iint\limits_{\mathbb{R}^2} v(x,y)\big((M_1M_2)^{(2)}f\big)^p(x,y)dxdy$$

$$+\int\limits_{\mathbb{R}^2} v(x,y)\big((M_1M_2)^{(3)}f\big)^p(x,y)dxdy\Big] := c[S_1+S_2+S_3],$$

where we used the notation:

$$(M_1M_2)^{(1)}f(x,y) := (M_1M_2)(f\chi_{|\cdot|<|y|/2})(x,y);$$

$$(M_1M_2)^{(2)}f(x,y) := (M_1M_2)(f\chi_{|y|/2\le|\cdot|\le 4|y|})(x,y);$$

$$(M_1M_2)^{(3)}f(x,y) := (M_1M_2)(f\chi_{|\cdot|>4|y|})(x,y).$$

We have already seen in the proof of Theorem 5.2.2 that the following estimate holds:

$$(M_1M_2)^{(1)}f(x,y) \le c(MH)f(x,y).$$

Hence, Lemma 5.2.5 and (5.2.11) yield that

$$S_1 \le c\|f\|^p_{L^p_w(\mathbb{R}^2)}.$$

For $(M_1M_2)^{(2)}$ we obtain that

$$(M_1M_2)^{(2)}f(x,y) \le \sup_{r>0}\frac{1}{r}\int\limits_{x-r}^{x+r}\chi_{\{|s|<|x|/2\}}(s)$$

$$\times\Big(\sup_{t>0}\frac{1}{t}\int\limits_{y-t}^{y+t}|f(s,\tau)|\chi_{\{|y|/2<|\tau|<4y|\}}(\tau)(s,\tau)d\tau\Big)ds$$

$$+\sup_{r>0}\frac{1}{r}\int\limits_{x-r}^{x+r}\chi_{\{|x|/2<|s|<4|x|\}}(s)\Big(\sup_{t>0}\frac{1}{t}\int\limits_{y-t}^{y+t}|f(s,\tau)|\chi_{|y|/2<|\tau|<4y|}(\tau)(s,\tau)d\tau\Big)ds$$

$$+\sup_{r>0}\frac{1}{r}\int\limits_{x-r}^{x+r}\chi_{\{|s|>4|x|\}}(s)\Big(\sup_{t>0}\frac{1}{t}\int\limits_{y-t}^{y+t}|f(s,\tau)|\chi_{|y|/2<|\tau|<4y|}(\tau)(s,\tau)d\tau\Big)ds$$

$$=: (M_1M_2)^{(2,1)}f(x,y) + (M_1M_2)^{(2,2)}f(x,y) + (M_1M_2)^{(2,3)}f(x,y).$$

Arguing as in the proof of Theorem 5.2.2 we find that

$$(M_1M_2)^{(2,1)}f(x,y) \le c(HM)f(x,y)$$

and, consequently, by Lemma 5.2.6 we obtain that

$$S_{2,1} := \iint\limits_{\mathbb{R}^2} v(x,y)\big((M_1M_2)^{(2,1)}f(x,y)\big)^p dxdy \le c\|f\|^p_{L^p_w(\mathbb{R}^2}.$$

The estimate

$$S_{2,2} := \iint\limits_{\mathbb{R}^2} v(x,y)\big((M_1M_2)^{(2,2)}f(x,y)\big)^p dxdy \le c\|f\|^p_{L^p_w(\mathbb{R}^2)}$$

is clear because of (5.2.12), the fact that v and w are decreasing weights and the boundedness of M_1M_2 in $L^p(\mathbb{R}^2)$ (see also the proof of Theorem 5.2.5 for the details).

Further, using the inequality

$$(M_1M_2)^{(2,3)}f(x,y) \le c \sup_{m\ge i+1} \frac{1}{2^m} \int\limits_{2^m<|s|<2^{m+1}} M(f(s,\cdot)\chi_{|y|/2<|\cdot|<4|y|})(y)ds,$$

which holds for x satisfying the condition $2^i < |x| < 2^{i+1}$, we have that

$$S_{2,3} := \iint\limits_{\mathbb{R}^2} v(x,y)\big((M_1M_2)^{(2,3)}f(x,y)\big)^p dxdy$$

$$\big[\text{ putting the supremum outside the integral }\big]$$

$$\le c\sum_{i\in\mathbb{Z}}\sum_{k\in\mathbb{Z}} \int\limits_{2^i<|x|<2^{i+1}} \int\limits_{2^k<|y|<2^{k+1}} v(x,y)$$

$$\times \sup_{m\ge i+1} \frac{1}{2^{mp}}\Bigg(\int\limits_{2^m<|s|<2^{m+1}} M(f(s,\cdot)\chi_{|y|/2<|\cdot|<4|y|})(y)ds\Bigg)^p dxdy$$

$$\big[\text{ using the inequality} \sup \le \textstyle\sum \big]$$

$$\le c\sum_{i\in\mathbb{Z}}\sum_{k\in\mathbb{Z}}\sum_{m=i+1}^{\infty} \frac{1}{2^{mp}} \int\limits_{2^i<|x|<2^{i+1}} \Bigg(\int\limits_{2^k<|y|<2^{k+1}} v(x,y)dx\Bigg)$$

$$\times\Big(\int\limits_{2^m<|s|<2^{m+1}} M(f(s,\cdot)\chi_{|y|/2<|\cdot|<4|y|})(y)ds\Big)^p dy$$

[using the fact that v is decreasing and even in each variable]

$$\leq c\sum_{i\in\mathbb{Z}}2^i\sum_{k\in\mathbb{Z}}\sum_{m=i+1}^{\infty}\frac{1}{2^{mp}}v(2^i,2^k)$$

$$\times\int\limits_{2^k<|y|<2^{k+1}}\Big(\int\limits_{2^m<|s|<2^{m+1}} M(f(s,\cdot)\chi_{|y|/2<|\cdot|<4|y|})(y)ds\Big)^p dy$$

[using Hölder's inequality in the inner integral]

$$\leq c\sum_{i\in\mathbb{Z}}2^i\sum_{k\in\mathbb{Z}}\sum_{m=i+1}^{\infty}\frac{1}{2^{mp}}v(2^i,2^k)$$

$$\times\int\limits_{2^k<|y|<2^{k+1}}\Big(\int\limits_{2^m<|s|<2^{m+1}} (M(f(s,\cdot)\chi_{|y|/2<|\cdot|<4|y|})(y))^p w(s,2^{k+3})ds\Big)dy$$

$$\times\Big(\int\limits_{2^m}^{2^{m+1}} w^{1-p'}(s,2^{k+3})ds\Big)^{p-1}$$

[changing the order of sums and integration]

$$= c\sum_{k\in\mathbb{Z}}\sum_{m\in\mathbb{Z}}2^{-mp}\Big(\sum_{i=-\infty}^{m-1}2^i v(2^i,2^k)\Big)\Big(\int\limits_{2^m}^{2^{m+1}} w^{1-p'}(s,2^{k+3})ds\Big)^{p-1}$$

$$\times\int\limits_{2^m<|s|<2^{m+1}} w(s,2^{k+3})\Big(\int\limits_{2^k<|y|<2^{k+1}} (M(f(s,\cdot)\chi_{|y|/2<|\cdot|<4|y|})(y))^p dy\Big)ds$$

[using again the fact that v is decreasing]

$$\leq c\sum_{k\in\mathbb{Z}}\sum_{m\in\mathbb{Z}}2^{-mp}\Big(\int\limits_{0}^{2^{m-1}} v(x,2^k)dx\Big)\Big(\int\limits_{2^m}^{2^{m+1}} w^{1-p'}(s,2^{k+3})ds\Big)^{p-1}$$

$$\times \int\limits_{2^m<|s|<2^{m+1}} w(s,2^{k+3})\Big(\int\limits_{2^k<|y|<2^{k+1}} (M(f(s,\cdot)\chi_{|y|/2<|\cdot|<4|y|})(y))^p dy\Big) ds$$

$\big[$ using the boundedness of M in $L^p(\mathbb{R})$ $\big]$

$$\leq c\sum_{k\in\mathbb{Z}}\sum_{m\in\mathbb{Z}} 2^{-mp}\Big(\int\limits_0^{2^{m-1}} v(x,2^k)dx\Big)\Big(\int\limits_{2^m}^{2^{m+1}} w^{1-p'}(s,2^{k+3})ds\Big)^{p-1}$$

$$\times \int\limits_{2^m<|s|<2^{m+1}} w(s,2^{k+3}) \int\limits_{2^{k-1}<|y|<2^{k+3}} |f(s,y)|^p ds dy := T.$$

Observe now that the condition (5.2.9) implies the estimate

$$2^{-mp}\Big(\int\limits_0^{2^{m-1}} v(x,2^k)dx\Big)\Big(\int\limits_{2^m}^{2^{m+1}} w^{1-p'}(s,2^{k+3})ds\Big)^{p-1} \leq cB_1^p.$$

Hence,

$$T\leq c\sum_{k\in\mathbb{Z}}\sum_{m\in\mathbb{Z}} \int\limits_{2^m<|s|<2^{m+1}} \int\limits_{2^{k-1}<|y|<2^{k+3}} |f(s,y)|^p w(s,2^{k+3}) ds dy$$

$\big[$ using the fact that w is decreasing in the second variable $\big]$

$$\leq c\sum_{k\in\mathbb{Z}}\sum_{m\in\mathbb{Z}} \int\limits_{2^m<|s|<2^{m+1}} \int\limits_{2^{k-1}<|y|<2^{k+3}} |(f(s,y)|^p w(s,y) ds dy \leq c\|f\|^p_{L^p_w(\mathbb{R}^2)}.$$

Moreover, by using the estimates

$$(M_1M_2)^{(3)}f(x,y)$$

$$\leq \sup_{r>0}\frac{1}{r}\int\limits_{x-r}^{x+r}\chi_{\{|s|<|x|/2\}}(s)\Big(\sup_{t>0}\frac{1}{t}\int\limits_{y-t}^{y+t}|f(s,\tau)|\chi_{\{|\tau|>4|y|\}}(\tau)d\tau\Big)ds$$

$$+\sup_{r>0}\frac{1}{r}\int\limits_{x-r}^{x+r}\chi_{\{|x|/2\leq|s|\leq4|x|\}}(s)\Big(\sup_{t>0}\frac{1}{t}\int\limits_{y-t}^{y+t}|f(s,\tau)|\chi_{\{|\tau|>4|y|\}}(\tau)d\tau\Big)ds$$

$$+\sup_{r>0}\frac{1}{r}\int\limits_{x-r}^{x+r}\chi_{\{|s|>4|x|\}}(s)\Big(\sup_{t>0}\frac{1}{t}\int\limits_{y-t}^{y+t}|f(s,\tau)|\chi_{\{|\tau|>4|y|\}}(\tau)d\tau\Big)ds$$

$$=: (M_1M_2)^{(3,1)}f(x,y) + (M_1M_2)^{(3,2)}f(x,y) + (M_1M_2)^{(3,3)}f(x,y),$$

we find that

$$S_3 \le c\Big[\iint\limits_{\mathbb{R}^2} v(x,y)\big((M_1M_2)^{(3,1)}f(x,y)\big)^p dxdy + \iint\limits_{\mathbb{R}^2} v(x,y)\big((M_1M_2)^{(3,2)}f(x,y)\big)^p dxdy$$
$$+ \iint\limits_{\mathbb{R}^2} v(x,y)\big((M_1M_2)^{(3,3)}f(x,y)\big)^p dxdy\Big] =: c[S_{3,1} + S_{3,2} + S_{3,3}].$$

By the inequality

$$(M_1M_2)^{(3,1)}f(x,y) \le c(HM)f(x,y), \quad x,y \in \mathbb{R},$$

Lemma 5.2.6 and (5.2.12) we obtain that

$$S_{3,1} \le c\|f\|^p_{L^p_w(\mathbb{R})}.$$

Next we estimate $S_{3,2}$:

$$S_{3,2} := \iint\limits_{\mathbb{R}^2} v(x,y)\big((M_1M_2)^{(3,2)}f(x,y)\big)^p dxdy$$
$$\le c\sum_{i\in\mathbb{Z}}\sum_{k\in\mathbb{Z}}\int\limits_{2^i<|x|<2^{i+1}}\Bigg(\int\limits_{2^k<|y|<2^{k+1}} v(x,y)dy\Bigg)$$
$$\times\Bigg(\sup_{r>0}\frac{1}{r}\int\limits_{x-r}^{x+r}\chi_{\{|x|/2<|s|<4|x|\}}(s)\Bigg(\sup_{n\ge k+1}\frac{1}{2^n}\int\limits_{2^n<|\tau|<2^{n+1}}|f(s,\tau)|d\tau\Bigg)ds\Bigg)^p dx$$

$\big[$ using the fact that v is decreasing in each variable$\big]$

$$\le c\sum_{i\in\mathbb{Z}}\sum_{k\in\mathbb{Z}}\int\limits_{2^i<|x|<2^{i+1}} v(2^i,2^k)2^k \int\limits_{2^i<|x|<2^{i+1}}\Bigg(\sup_{r>0}\frac{1}{r}\int\limits_{x-r}^{x+r}\chi_{|x|/2<|s|<4|x|}(s)$$
$$\times\Bigg(\sup_{n\ge k+1}\frac{1}{2^n}\int\limits_{2^n<|\tau|<2^{n+1}}|f(s,\tau)|d\tau\Bigg)ds\Bigg)^p dx$$

$\big[$ using the boundedness of M in $L^p(\mathbb{R})\big]$

$$\le c\sum_{i\in\mathbb{Z}}\sum_{k\in\mathbb{Z}}\int\limits_{2^i<|x|<2^{i+1}} v(2^i,2^k)2^k$$

$$\times \int\limits_{2^{i-1}<|x|<2^{i+3}} \left(\sup_{n\geq k+1} \frac{1}{2^n} \int\limits_{2^n<|\tau|<2^{n+1}} |f(x,\tau)|d\tau \right)^p dx$$

[putting the supremum outside the brackets and using the inequality sup $\leq \sum$]

$$\leq c\sum_{i\in\mathbb{Z}}\sum_{k\in\mathbb{Z}} v(2^i,2^k)2^k \sum_{n=k+1}^{\infty} \frac{1}{2^{np}} \int\limits_{2^{i-1}<|x|<2^{i+3}} \left(\int\limits_{2^n<|\tau|<2^{n+1}} |f(x,\tau)|d\tau \right)^p dx$$

[using Hölder's inequality]

$$\leq c\sum_{i\in\mathbb{Z}}\sum_{k\in\mathbb{Z}} v(2^i,2^k)2^k \sum_{n=k+1}^{\infty} \frac{1}{2^{np}} \Bigg(\int\limits_{2^{i-1}<|x|<2^{i+3}} \Bigg(\int\limits_{2^n<|\tau|<2^{n+1}} w(2^{i+3},\tau)$$

$$\times |f(x,\tau)|^p d\tau dx \Bigg) \Bigg(\int\limits_{2^n<|\tau|<2^{n+1}} w^{1-p'}(2^{i+3},\tau)d\tau \Bigg)^{p-1} =$$

[changing the order of sums]

$$c\sum_{i\in\mathbb{Z}}\sum_{n\in\mathbb{Z}} \frac{1}{2^{np}} \sum_{k=-\infty}^{n-1} v(2^i,2^k)2^k \Bigg(\int\limits_{2^{i-1}<|x|<2^{i+3}} \int\limits_{2^n<|\tau|<2^{n+1}} w(x,\tau)$$

$$\times |f(x,\tau)|^p dx d\tau \Bigg) \Bigg(\int\limits_{2^n<|\tau|<2^{n+1}} w^{1-p'}(2^{i+3},\tau)d\tau \Bigg)^{p-1}$$

[using the fact that v is decreasing in the second variable and using (5.2.12)]

$$\leq c\sum_{i\in\mathbb{Z}}\sum_{n\in\mathbb{Z}} \frac{1}{2^{np}} \Bigg(\int_0^{2^{n-1}} v(2^i,\tau)d\tau \Bigg) \Bigg(\int\limits_{2^n<|\tau|<2^{n+1}} w^{1-p'}(2^{i+3},\tau)d\tau \Bigg)^{p-1}$$

$$\times \Bigg(\int\limits_{2^{i-1}<|x|<2^{i+3}} \int\limits_{2^n<|\tau|<2^{n+1}} w(x,\tau)|f(x,\tau)|^p dx d\tau \Bigg) \leq c\|f\|^p_{L^p_w(\mathbb{R}^2)}.$$

Further, taking into account the inequality

$$(M_1M_2)^{(3,3)}f(x,y) \leq c \sup_{m\geq i+1} \frac{1}{2^m} \int\limits_{2^m\leq|s|<2^{m+1}} \left(\sup_{n\geq k+1} \frac{1}{2^n} \int\limits_{2^n<|\tau|<2^{n+1}} |f(s,\tau)|d\tau \right) ds,$$

which holds for x and y satisfying the conditions $2^i < |x| < 2^{i+1}$ and $2^k < |y| < 2^{k+1}$, we obtain that

$$S_{3,3} := \iint\limits_{\mathbb{R}^2} v(x,y)\big((M_1M_2)^{(3,3)}f(x,y)\big)^p dx dy$$

$$\leq c \sum_{i\in\mathbb{Z}} \sum_{k\in\mathbb{Z}} \int\limits_{2^i<|x|<2^{i+1}} \int\limits_{2^k<|y|<2^{k+1}} v(x,y)dxdy$$

$$\times \Bigg(\sup_{m\geq i+1} \frac{1}{2^m} \Bigg(\int\limits_{2^m}^{2^{m+1}} \sup_{n\geq k+1} \frac{1}{2^n} \int\limits_{2^n<|\tau|<2^{n+1}} |f(s,\tau)|d\tau \Bigg) ds \Bigg)^p$$

[applying the fact that v is decreasing in both variables

and the inequality $\sup \leq \sum$]

$$\leq c \sum_{i\in\mathbb{Z}} \sum_{k\in\mathbb{Z}} v(2^i,2^k)2^i2^k \sum_{m=i+1}^{\infty} \frac{1}{2^{mp}}$$

$$\Bigg(\int\limits_{2^m<|s|<2^{m+1}} \Bigg(\sup_{n\geq k+1} \frac{1}{2^n} \int\limits_{2^n<|\tau|<2^{n+1}} |f(s,\tau)|d\tau \Bigg) ds \Bigg)^p$$

[using Hölder's inequality in the outer integral]

$$\leq c \sum_{i\in\mathbb{Z}} \sum_{k\in\mathbb{Z}} v(2^i,2^k)2^i2^k$$

$$\times \sum_{m=i+1}^{\infty} 2^{-m} \Bigg(\int\limits_{2^m<|s|<2^{m+1}} \Bigg(\sup_{n\geq k+1} \frac{1}{2^n} \Bigg(\int\limits_{2^n<|\tau|<2^{n+1}} |f(s,\tau)|d\tau \Bigg)^p ds \Bigg)$$

[putting the supremum outside the brackets, using the inequality $\sup \leq \sum$

and Hölder's inequality]

$$\leq c \sum_{i\in\mathbb{Z}} \sum_{k\in\mathbb{Z}} v(2^i,2^k)2^i2^k \sum_{m=i+1}^{\infty} 2^{-m} \sum_{n=k+1}^{\infty} \frac{1}{2^{np}} \times$$

$$\int\limits_{2^m<|s|<2^{m+1}} \Bigg(\int\limits_{2^n<|\tau|<2^{n+1}} |f(s,\tau)|^p w(s,\tau)d\tau \Bigg)$$

$$\times \Bigg(\int\limits_{2^n<|\tau|<2^{n+1}} w^{1-p'}(s,\tau)d\tau \Bigg)^{p-1} ds$$

[applying the fact that w is decreasing in the first variable]

$$\leq c \sum_{i\in\mathbb{Z}} \sum_{k\in\mathbb{Z}} v(2^i,2^k)2^i2^k \sum_{m=i+1}^{\infty} 2^{-m} \sum_{n=k+1}^{\infty} \frac{1}{2^{np}}$$

$$\times\left(\int\limits_{2^m<|s|<2^{m+1}}\int\limits_{2^n<|\tau|<2^{n+1}} |f(s,\tau)|^p w(s,\tau)d\tau ds\right)$$

$$\times\left(\int\limits_{2^n<|\tau|<2^{n+1}} w^{1-p'}(2^{m+1},\tau)d\tau\right)^{p-1}$$

[changing the order of sums twice]

$$= c\sum_{m\in\mathbb{Z}} 2^{-m}\sum_{n\in\mathbb{Z}}\frac{1}{2^{np}}\left(\int\limits_{2^m<|s|<2^{m+1}}\int\limits_{2^n<|\tau|<2^{n+1}} |f(s,\tau)|^p w(s,\tau)d\tau ds\right)\times$$

$$\left(\int\limits_{2^n<|\tau|<2^{n+1}} w^{1-p'}(2^{m+1},\tau)d\tau\right)^{p-1}\sum_{i=-\infty}^{m-1}\sum_{k=-\infty}^{n-1} v(2^i,2^k)2^i2^k \le$$

[using the fact that v is decreasing in each variable]

$$c\sum_{m\in\mathbb{Z}}\sum_{n\in\mathbb{Z}} 2^{-mp}2^{-np}\left(\int\limits_{2^m<|s|<2^{m+1}}\int\limits_{2^n<|\tau|<2^{n+1}} |f(s,\tau)|^p w(s,\tau)d\tau ds\right)$$

$$\times\left(\int\limits_{2^{m+1}<|s|<2^{m+2}}\int\limits_{2^n<|\tau|<2^{n+1}} w^{1-p'}(s,\tau)dsd\tau\right)^{p-1}$$

$$\times\left(\int\limits_{|s|<2^{m-1}}\int\limits_{|\tau|<2^{n-1}} v(x,y)dxdy\right) \le$$

[using condition (5.1.9)]

$$cB_1^p\sum_m\sum_n \int\limits_{2^m<|s|<2^{m+1}}\int\limits_{2^n<|\tau|<2^{n+1}} |f(s,\tau)|^p w(s,\tau)d\tau ds = c\|f\|^p_{L^p_w(\mathbb{R}^2)}.$$

Necessity follows from obvious estimate $(M_1M_2)f(x,y) \ge M_S f(x,y)$ and Theorem 5.2.4. The proof is complete. □

Theorem 5.2.5 yields the following statement:

Theorem 5.2.6. *Let $1 < p < \infty$. Suppose that the two-dimensional weight v is even on $\mathbb{R}$ and decreasing on $\mathbb{R}_+$ in each variable for another fixed one. Suppose also that another two-dimensional weight w has the form $w(x,y) = w_1(x)w_2(y)$, where w_1 and w_2 are decreasing on $\mathbb{R}_+(0,\infty)$ and even on $\mathbb{R}$. Then M_S is bounded from $L^p_w(\mathbb{R}^2)$ to $L^p_v(\mathbb{R}^2)$ if and only if (5.2.9) holds.*

The Case of Mixed–type Weights

The next statements follows in the same manner as the previous ones; therefore the proofs are omitted.

Theorem 5.2.7. *Let $1 < p < \infty$. Suppose that the two-dimensional weights v and w are even on $\mathbb{R}$. Further, let these weights be increasing on $\mathbb{R}_+$ in the first variable for the second fixed one and decreasing on $\mathbb{R}_+$ in the second variable for the first fixed one. In addition, assume that $w^{1-p'} \in DC(x) \cap DC(y)$. Then M_S is bounded from $L^p_w(\mathbb{R}^2)$ to $L^p_v(\mathbb{R}^2)$ if and only if* (5.2.12) *holds.*

Theorem 5.2.8. *Let $1 < p < \infty$. Suppose that the two-dimensional weight v is even on $\mathbb{R}$. Suppose also that v is increasing on $\mathbb{R}_+$ in the first variable for the second fixed one and decreasing on $\mathbb{R}_+$ in the second variable for the first fixed one. Assume that w is another two–dimensional weight which has the form $w(x,y) = w_1(x)w_2(y)$, where w_1 is even on $\mathbb{R}$ and increasing on $\mathbb{R}_+$; w_2 is even on $\mathbb{R}$ and decreasing on $\mathbb{R}_+$. Then the operators M_S and M_1M_2 are bounded from $L^p_w(\mathbb{R}^2)$ to $L^p_v(\mathbb{R}^2)$ if and only if* (5.2.12) *holds.*

5.3 Two–weight Estimates for Hilbert Transforms with Single Kernel

Let $\mathbb{H}$ be the one–dimensional Hilbert transform given by

$$\mathbb{H}f(x) = \lim_{\varepsilon \to 0} \int\limits_{|x-t|>\varepsilon} \frac{f(t)}{x-t} dt.$$

For the following statements we refer to [37] (see also [127], [39], p. 553).

Theorem 5.3.1. *Let $1 < p < \infty$ and let the one–dimensional weights σ and u be even on $\mathbb{R}$ and increasing on $\mathbb{R}_+$. Then $\mathbb{H}$ is bounded from $L^p_u(\mathbb{R})$ to $L^p_\sigma(\mathbb{R})$ if and only if* (5.1.1) *holds. Moreover, there exist positive constants c_1 and c_2 depending only on p such that $c_1 A \le \|\mathbb{H}\| \le c_2 A$.*

Theorem 5.3.2. *Let $1 < p < \infty$ and let the one-dimensional weights σ and u be even on $\mathbb{R}$ and decreasing on $\mathbb{R}_+$. Then the Hilbert transform $\mathbb{H}$ is bounded from $L^p_u(\mathbb{R})$ to $L^p_\sigma(\mathbb{R})$ if and only if*

$$\overline{A} := \sup_{t>0} \left(\int\limits_0^t \sigma(x)dx \right)^{1/p} \left(\int\limits_t^\infty u^{1-p'}(x) x^{-p'} dx \right)^{1/p'} < \infty.$$

Moreover, there exist positive constants c_1 and c_2 depending only on p such that $c_1 \overline{A}_1 \le \|\mathbb{H}\| \le c_2 \overline{A}_1$.

5.4 Hilbert Transforms with Product Kernels

Before formulating the results of this section observe that by iteration we have the boundedness of the Hilbert transform with product kernels $\mathcal{H}$ in $L^p(\mathbb{R}^2)$.

The Case of Increasing Weights

Theorem 5.4.1. *Let $1 < p < \infty$. Suppose that the two-dimensional weights v and w are even on $\mathbb{R}$ and increasing on $\mathbb{R}_+$ in each variable for another fixed one. Further, assume that $w(x,y) = w_1(x)w_2(y)$. Then the two–weight inequality*

$$\iint\limits_{\mathbb{R}^2} |\mathcal{H}f(x,y)|^p v(x,y)dxdy \le c \iint\limits_{\mathbb{R}^2} |f(x,y)|^p w(x,y)dxdy \tag{5.4.1}$$

holds if and only if condition (5.2.3) *is fulfilled.*

To prove the theorem we need to discuss the following mixed–type operators:

$$\mathcal{H}^{(1)}f(x,y) := \int\limits_{|t|<|x|} \left| \lim_{\varepsilon\to 0} \int\limits_{|y-\tau|>\varepsilon} \frac{f(t,\tau)}{y-\tau} d\tau \right| dt;$$

$$\mathcal{H}^{(2)}f(x,y) := \int\limits_{|t|>|x|} \left| \lim_{\varepsilon\to 0} \int\limits_{|y-\tau|>\varepsilon} \frac{f(t,\tau)}{y-\tau} d\tau \right| dt.$$

For these operators we present the following two auxiliary results of independent interest:

Lemma 5.4.1. *Let $1 < p < \infty$. Suppose that the two-dimensional weight U is even on $\mathbb{R}$ and increasing on $\mathbb{R}_+$ in the second variable for the first fixed one. Suppose also that another two–dimensional weight W has the form $W(x,y) = W_1(x)W_2(y)$, where W_2 is even on $\mathbb{R}$ and increasing on $\mathbb{R}_+$. Then the operator $\mathcal{H}^{(1)}$ is bounded from $L^p_W(\mathbb{R}^2)$ to $L^p_U(\mathbb{R}^2)$ if and only if*

$$D_1 := \sup_{\substack{a>0\\ b>0}} \left(\int_a^\infty \int_b^\infty \frac{U(x,y)}{y^p} dxdy \right)^{1/p} \left(\int_0^a W_1^{1-p'}(x)dx \right)^{1/p'} \left(\int_0^b W_2^{1-p'}(y)dy \right)^{1/p'} < \infty.$$

Proof. First suppose that $S := \int_{\mathbb{R}} W_1^{1-p'}(x)dx = \infty$. Let $\{x_k\}_{k=-\infty}^{+\infty}$ be a sequence of positive numbers for which the equality

$$2^k = \int\limits_{|x|<x_k} W_1^{1-p'}(x)dx \tag{5.4.2}$$

holds for all $k\in\mathbb{Z}$. It is clear that $\{x_k\}$ is increasing and

$$\mathbb{R}\setminus\{0\}=\cup_{k\in\mathbb{Z}}E_k,$$

where $E_k =: \{x\in\mathbb{R}: x_k\le |x|<x_{k+1}\}$. Moreover, it is easy to verify that $2^k=\int_{E_k}W_1^{1-p'}$. Let us introduce the notation

$$\overline{U}_k(y):=\int\limits_{E_k}U(x,y)dx.$$

We have that

$$I:=\iint\limits_{\mathbb{R}^2}U(x,y)(\mathcal{H}^{(1)}f(x,y))^p dxdy$$

$$=\sum_{k\in\mathbb{Z}}\iint\limits_{E_k\times\mathbb{R}}U(x,y)\Bigg(\int\limits_{|t|<|x|}\Bigg|\lim_{\varepsilon\to 0}\int\limits_{|y-\tau|>\varepsilon}\frac{f(t,\tau)}{y-\tau}d\tau\Bigg|dt\Bigg)^p dxdy$$

$$\le\sum_{k\in\mathbb{Z}}\int\limits_{\mathbb{R}}\Bigg(\int\limits_{E_k}U(x,y)dx\Bigg)\Bigg(\int\limits_{|t|<x_{k+1}}\Bigg|\lim_{\varepsilon\to 0}\int\limits_{|y-\tau|>\varepsilon}\frac{f(t,\tau)}{y-\tau}d\tau\Bigg|dt\Bigg)^p dy$$

$$\le\sum_{k\in\mathbb{Z}}\int\limits_{\mathbb{R}}\overline{U}_k(y)\Bigg(\int\limits_{|t|<x_{k+1}}|H(f(t,\cdot))(y)|dt\Bigg)^p dy$$

[applying generalized Minkowski's inequality]

$$\le\sum_{k}\Bigg(\int\limits_{|t|<x_{k+1}}\Bigg(\int\limits_{\mathbb{R}}\overline{U}_k(y)|H(f(t,\cdot))(y)|^p dy\Bigg)^{1/p}dt\Bigg)^p:=I_1.$$

Notice that

$$D_1\ge\sup_{\substack{b>0\\ x_k>0}}\Bigg(\int\limits_b^\infty\Bigg(\int\limits_{x_k}^{x_{k+1}}U(x,y)dx\Bigg)\frac{1}{y^p}dy\Bigg)^{1/p}$$

$$\times\Bigg(\int\limits_0^b\Bigg(\int\limits_0^{x_k}W_1^{1-p'}(x)dx\Bigg)^{\frac{1-p'}{1-p'}}W_2^{1-p'}(y)dy\Bigg)^{1/p'}.$$

Therefore, using Theorem 5.3.1 we find that

$$I_1\le c\sum_{k\in\mathbb{Z}}\Bigg(\int\limits_{|t|<x_{k+1}}\Bigg(\int\limits_{\mathbb{R}}\Bigg(\int\limits_0^{x_k}W_1^{1-p'}(s)ds\Bigg)^{1-p}|f(t,\tau)|^p W_2(\tau)d\tau\Bigg)^{1/p}dt\Bigg)^p$$

$$=c\sum_k\Big(\int_0^{x_k}W_1^{1-p'}(s)ds\Big)^{1-p}\Big(\sum_{j=-\infty}^{k}\int_{x_j\le|t|<x_{j+1}}\Big(\int_{\mathbb{R}}|f(t,\tau)|^pW_2(\tau)d\tau\Big)^{1/p}dt\Big)^p:=I_2.$$

It is easy to check that

$$\sum_{k=n}^{+\infty}\Big(\int_0^{x_k}W_1^{1-p'}(x)dx\Big)^{1-p}\Big(\sum_{k=-\infty}^{n}\int_{x_k}^{x_{k+1}}W_1^{1-p'}(x)dx\Big)^{p-1}$$

$$=\sum_{k=n}^{+\infty}\Big(\int_0^{x_k}W_1^{1-p'}(x)dx\Big)^{1-p}\Big(\int_0^{x_{n+1}}W_1^{1-p'}(x)dx\Big)^{p-1}$$

$$=\Big(\sum_{k=n}^{+\infty}2^{k(1-p)}\Big)2^{(n+1)(p-1)}=c<\infty.$$

Hence, Theorem 1.1.4 yields that

$$I_2\le c\sum_k\Big(\int_{x_k}^{x_{k+1}}W_1^{1-p'}(s)ds\Big)^{1-p}\Big(\int_{x_k\le|t|<x_{k+1}}\Big(\int_{\mathbb{R}}|f(t,\tau)|^pW_2(\tau)d\tau\Big)^{1/p}dt\Big)^p$$

[using Hölder's inequality]

$$\le c\sum_k\Big(\int_{x_k}^{x_{k+1}}W_1^{1-p'}(s)ds\Big)^{1-p}$$

$$\times\Big(\int_{x_k\le|s|<x_{k+1}}\int_{\mathbb{R}}|f(s,\tau)|^pW_2(\tau)W_1(s)d\tau ds\Big)\Big(\int_{x_k}^{x_{k+1}}W_1^{1-p'}(t)dt\Big)^{p-1}$$

$$=c\sum_k\int_{x_k\le|s|<x_{k+1}}\int_{\mathbb{R}}|f(s,\tau)|^pW_2(\tau)W_1(s)d\tau ds=c\|f\|^p_{L^p_W(\mathbb{R}^2)}.$$

If $S<\infty$, then without loss of generality we can assume that $S=1$. In this case we choose the sequence $\{x_k\}_{k=-\infty}^0$ for which (5.4.2) holds for all $k\le 0$. Arguing as in the case $S=\infty$ and using Theorem 1.1.5 instead of Theorem 1.1.4, we obtain the desired result.

Necessity. Substituting the test function

$$f(x,y)=\chi_{(0,a)\times(0,b)}(x,y)W^{1-p'}(x,y)$$

in the inequality

$$\iint\limits_{\mathbb{R}^2} |\mathcal{H}^{(1)} f(x,y)|^p U(x,y)dxdy \le c \iint\limits_{\mathbb{R}^2} |f(x,y)|^p W(x,y)dxdy$$

and making some simple estimates we derive the condition $D_1 < \infty$.

The proof is complete. □

Lemma 5.4.2. *Let $1 < p < \infty$. Suppose that a two-dimensional weight U is even on $\mathbb{R}$ and increasing on $\mathbb{R}_+$ in the second variable for the the first fixed one. Assume also that $W(x,y) = W_1(x)W_2(y)$, where W_2 is even on $\mathbb{R}$ and increasing on $\mathbb{R}_+$. Then the operator $\mathcal{H}^{(2)}$ is bounded from $L^p_W(\mathbb{R}^2)$ to $L^p_U(\mathbb{R}^2)$ if and only if*

$$D_2 := \sup_{a,b>0} \Big(\int_0^a \int_b^\infty \frac{U(x,y)}{y^p} dxdy \Big)^{1/p} \Big(\int_a^\infty W_1^{1-p'}(x)dx \Big)^{1/p'} \Big(\int_0^b W_2^{1-p'}(y)dy \Big)^{1/p'} < \infty.$$

Proof. First suppose that $S := \int_{\mathbb{R}} W_1^{1-p'}(x)dx = \infty$. Let $\{x_k\}_{k=-\infty}^{+\infty}$ be the sequence of positive numbers for which the equality

$$2^{-k} = \int\limits_{|x|>x_k} W_1^{1-p'}(x)dx \tag{5.4.3}$$

holds for all $k \in \mathbb{Z}$. It is clear that $\{x_k\}$ is increasing and

$$\mathbb{R} \setminus \{0\} = \cup_{k\in\mathbb{Z}} F_k,$$

where $F_k := \{x \in \mathbb{R} : x_k \le |x| < x_{k+1}\}$. Moreover, it is easy to verify that

$$2^{-k-1} = \int\limits_{|x|>x_k} W_1^{1-p'}(x)dx - \int\limits_{|x|>x_{k+1}} W_1^{1-p'}(x)dx = \int\limits_{F_k} W_1^{1-p'}(x)dx.$$

Using the notation

$$\widetilde{U}_k(y) := \int\limits_{F_k} U(x,y)dx$$

and generalized Minkowski's inequality we have that

$$J := \iint\limits_{\mathbb{R}^2} U(x,y)(\mathcal{H}^{(2)} f(x,y))^p dxdy$$

$$= \sum_{k\in\mathbb{Z}} \iint\limits_{F_k\times\mathbb{R}} U(x,y) \Big(\int\limits_{|t|>|x|} \Big| \lim_{\varepsilon\to 0} \int\limits_{|y-\tau|>\varepsilon} \frac{f(t,\tau)}{(y-\tau)} d\tau \Big| dt \Big)^p dxdy$$

$$\leq \sum_{k\in\mathbb{Z}} \int_{\mathbb{R}} \Big(\int_{F_k} U(x,y)dx \Big) \Big(\int_{|t|>x_k} \Big| \lim_{\varepsilon\to 0} \int_{|y-\tau|>\varepsilon} \frac{f(t,\tau)}{(y-\tau)} d\tau \Big| dt \Big)^p dy$$

$$\leq \sum_{k\in\mathbb{Z}} \int_{\mathbb{R}} \widetilde{U}_k(y) \Big(\int_{|t|<x_k} |H(f(t,(\cdot))(y)|dt \Big)^p dy$$

$$\leq \sum_k \Big(\int_{|t|>x_k} \Big(\int_{\mathbb{R}} \widetilde{U}_k(y)|H(f(t,\cdot)(y)|^p dy \Big)^{1/p} dt \Big)^p := J_1.$$

Further, we find that

$$D_2 \geq \sup_{\substack{b>0\\ x_k>0}} \Big(\int_b^\infty \Big(\int_0^{x_k} U(x,y)dx \Big) \frac{1}{y^p} dy \Big)^{1/p}$$

$$\times \Big(\int_0^b \Big(\int_{x_k}^{x_{k+1}} W_1^{1-p'}(x)dx \Big)^{\frac{1-p'}{1-p'}} W_2^{1-p'}(y)dy \Big)^{1/p'}.$$

Consequently, Theorem 5.3.1 yields that

$$J_1 \leq c \sum_k \Big(\int_{|t|>x_k} \Big(\int_{\mathbb{R}} \Big(\int_{x_k}^{x_{k+1}} W_1^{1-p'}(s)ds \Big)^{1-p} |f(t,\tau)|^p W_2(\tau)d\tau \Big)^{1/p} dt \Big)^p$$

$$= \sum_k \Big(\int_{x_k}^{x_{k+1}} W_1^{1-p'}(s)ds \Big)^{1-p}$$

$$\times \Big(\sum_{j=k}^{\infty} \int_{x_j<|t|\leq x_{j+1}} \Big(\int_{\mathbb{R}} |f(t,\tau)|^p W_2(\tau)d\tau \Big)^{1/p} dt \Big)^p := J_2.$$

Further, it is easy to verify that

$$\sum_{k=-\infty}^{n} \Big(\int_{x_k}^{x_{k+1}} W_1^{1-p'}(x)dx \Big)^{1-p} \Big(\sum_{k=n}^{\infty} \int_{x_k}^{x_{k+1}} W_1^{1-p'}(x)dx \Big)^{p-1}$$

$$= \sum_{k=-\infty}^{n} \Big(\int_{x_k}^{x_{k+1}} W_1^{1-p'}(y)dy \Big)^{1-p} \Big(\int_{x_n}^{\infty} W_1^{1-p'}(x)dx \Big)^{p-1}$$

$$= \Big(\sum_{k=-\infty}^{n} 2^{-(k-1)(1-p)} \Big) 2^{-n(p-1)} = c2^{-n(p-1)} \sum_{k=-n}^{\infty} 2^{k(1-p)} = c < \infty.$$

Hence, by Theorem 4.1.8 we have that

$$J_2 \le c \sum_k \Big(\int_{x_k}^{x_{k+1}} W_1^{1-p'}(s)ds \Big)^{1-p}$$

$$\times \Big(\int_{x_k \le |t| < x_{k+1}} \Big(\int_{\mathbb{R}} |f(t,\tau)|^p W_2(\tau) d\tau \Big)^{1/p} dt \Big)^p$$

$$\Big[\text{ using Hölder's inequality } \Big]$$

$$\le c \sum_k \Big(\int_{x_k}^{x_{k+1}} W_1^{1-p'}(s)ds \Big)^{1-p}$$

$$\times \int_{x_k \le |t| < x_{k+1}} \int_{\mathbb{R}} |f(t,\tau)|^p W_2(\tau) W_1(t) d\tau dt \Big(\int_{x_k}^{x_{k+1}} W_1^{1-p'}(t)dt \Big)^{1-p}$$

$$= c \sum_k \int_{x_k \le |t| < x_{k+1}} \int_{\mathbb{R}} |f(t,\tau)|^p W_2(\tau) W_1(t) d\tau dt = c \|f\|^p_{L^p_w(\mathbb{R}^2)}.$$

If $S < \infty$, then without loss of generality we can assume that $S = 1$. In this case we choose the sequence $\{x_k\}_{k=0}^{\infty}$ for which (5.4.3) holds for all $k \ge 0$. Arguing as in the case $S = \infty$ and using Theorem 1.1.5 instead Theorem 1.1.4 we obtain the desired result.

The proof is complete. □

Now we are ready to prove Theorem 5.4.1.

Proof of Theorem 5.4.1. *Sufficiency.* Splitting f into three parts with respect to the first variable we find that

$$\|\mathcal{H}f\|^p_{L^p_v(\mathbb{R}^2)} = \sum_{k\in\mathbb{Z}} \int_{2^k<|x|<2^{k+1}} \int_{\mathbb{R}} v(x,y) |\mathcal{H}f(x,y)|^p dxdy$$

$$\le c \Big[\sum_{k\in\mathbb{Z}} \int_{2^k<|x|<2^{k+1}} \int_{\mathbb{R}} v(x,y) |\mathcal{H}f_{1,k}(x,y)|^p dxdy$$

$$+ \sum_{k\in\mathbb{Z}} \int_{2^k<|x|<2^{k+1}} \int_{\mathbb{R}} v(x,y) |\mathcal{H}f_{2,k}(x,y)|^p dxdy$$

$$+ \sum_{k\in\mathbb{Z}} \int_{2^k<|x|<2^{k+1}} \int_{\mathbb{R}} v(x,y) |\mathcal{H}f_{3,k}(x,y)|^p dxdy \Big] =: c[S_1 + S_1 + S_3],$$

where

$$f_{1,k}(t,\tau) := \chi_{\{t:|t|<2^{k-1}\}}(t) f(t,\tau);$$

$$f_{2,k}(t,\tau) := \chi_{\{t:2^{k-1}\le|t|\le 2^{k+2}\}}(t) f(t,\tau);$$

$$f_{3,k}(t,\tau) := \chi_{\{t:|t|>2^{k+2}\}}(t) f(t,\tau).$$

It is easy to check that if $|t| < 2^{k-1}$ and $2^k < |x| < 2^{k+1}$, then $|x| \le |x-t|+|t| \le |x-t| + |x|/2$. Hence $|x|/2 \le |x-t|$. Therefore, using the formula

$$\mathcal{H}f(x,y) = \mathbb{H}(\mathbb{H}(f(\cdot,y))(x))$$

we find that

$$S_1 = \sum_{k\in\mathbb{Z}} \int\limits_{2^k<|x|<2^{k+1}} \int\limits_{\mathbb{R}} v(x,y) \Bigg| \lim_{\varepsilon_1\to 0} \int\limits_{|x-t|>\varepsilon_1} (x-t)^{-1} \chi_{\{t:|t|<2^{k-1}\}}(t)$$

$$\times \Bigg(\lim_{\varepsilon_2\to 0} \int\limits_{|y-\tau|>\varepsilon_2} \frac{f(t,\tau)}{y-\tau} d\tau \Bigg) dt \Bigg|^p dxdy \le c \sum_{k\in\mathbb{Z}} \int\limits_{2^k<|x|<2^{k+1}} \int\limits_{\mathbb{R}} v(x,y)|x|^{-p}$$

$$\times \Bigg(\int\limits_{|t|<|x|} \Bigg| \lim_{\varepsilon_2\to 0} \int\limits_{|y-\tau|>\varepsilon_2} \frac{f(t,\tau)}{y-\tau} d\tau \Bigg| dt \Bigg)^p dxdy$$

$$\le c \iint\limits_{\mathbb{R}^2} v(x,y)|x|^{-p} \Bigg(\int\limits_{|t|<|x|} \Bigg| \lim_{\varepsilon_2\to 0} \int\limits_{|y-\tau|>\varepsilon_2} \frac{f(t,\tau)}{y-\tau} d\tau \Bigg| dt \Bigg)^p dxdy$$

[applying Lemma 5.4.1 to the weight $u(x,y) = v(x,y)|x|^{-p}$]

$$\le c \|f\|^p_{L^p(\mathbb{R}^2)}.$$

Observe now that, taking Lemma 5.4.2 into account, the condition

$$\overline{D}_1 := \sup_{\substack{a>0\\ b>0}} \Bigg(\int_0^a \int_b^\infty \frac{v(x,y)}{y^p} dxdy \Bigg)^{1/p} \Bigg(\int_a^\infty \frac{w_1^{1-p'}(x)}{x^{p'}} dx \Bigg)^{1/p'}$$

$$\times \Bigg(\int_0^b w_2^{1-p'}(y) dy \Bigg)^{1/p'} < \infty \tag{5.4.4}$$

guarantees the estimates

$$S_3 \le \sum_{k\in\mathbb{Z}} \int\limits_{2^k<|x|<2^{k+1}} \int\limits_{\mathbb{R}} v(x,y)$$

$$\times\Big|\lim_{\varepsilon_1\to 0}\int\limits_{|x-t|>\varepsilon_1}(x-t)^{-1}\chi_{\{t:|t|>2^{k+2}\}}(t)\Big(\lim_{\varepsilon_2\to 0}\int\limits_{|y-\tau|>\varepsilon_2}\frac{f(t,\tau)}{y-\tau}d\tau\Big)dt\Big|^p dxdy$$

$$\leq c\sum_{k\in\mathbb{Z}}\int\limits_{2^k<|x|<2^{k+1}}\int\limits_{\mathbb{R}}v(x,y)\Big(\int\limits_{|t|>|x|}|t|^{-1}\Big|\lim_{\varepsilon_2\to 0}\int\limits_{|y-\tau|>\varepsilon_2}\frac{f(t,\tau)}{y-\tau}d\tau\Big|dt\Big)dxdy$$

$$\leq c\iint\limits_{\mathbb{R}^2}v(x,y)\Big(\int\limits_{|t|>|x|}|t|^{-1}\Big|\lim_{\varepsilon_2\to 0}\int\limits_{|y-\tau|>\varepsilon_2}\frac{f(t,\tau)}{y-\tau}d\tau\Big|dt\Big)^p dxdy\leq c\|f\|^p_{L^p(\mathbb{R}^2)}.$$

Splitting the integral with respect to τ we have for S_2 that

$$S_2=\sum_{k\in\mathbb{Z}}\int\limits_{2^k<|x|<2^{k+1}}\int\limits_{\mathbb{R}}v(x,y)\Big|\lim_{\varepsilon_2\to 0}\int\limits_{|y-\tau|>\varepsilon_2}(y-\tau)^{-1}$$

$$\times\Big(\lim_{\varepsilon_1\to 0}\int\limits_{|x-t|>\varepsilon_1}\frac{f(t,\tau)\chi_{\{t:2^{k-1}\leq|t|\leq 2^{k+2}\}}(t)}{x-t}dt\Big)d\tau\Big|^p dxdy$$

$$\leq c\Big[\sum_{k\in\mathbb{Z}}\sum_{j\in\mathbb{Z}}\int\limits_{2^k<|x|<2^{k+1}}\int\limits_{2^j<|y|<2^{j+1}}v(x,y)\Big|\lim_{\varepsilon_2\to 0}\int\limits_{|y-\tau|>\varepsilon_2}(y-\tau)^{-1}$$

$$\times\chi_{\{\tau:|\tau|<2^{j-1}\}}(\tau)\Big(\lim_{\varepsilon_1\to 0}\int\limits_{|x-t|>\varepsilon_1}\frac{f(t,\tau)\chi_{\{t:2^{k-1}\leq|t|\leq 2^{k+2}\}}(t)}{x-t}dt\Big)d\tau\Big|^p dxdy$$

$$+c\sum_{k\in\mathbb{Z}}\sum_{j\in\mathbb{Z}}\int\limits_{2^k<|x|<2^{k+1}}\int\limits_{2^j<|y|<2^{j+1}}v(x,y)\Big|\lim_{\varepsilon_2\to 0}\int\limits_{|y-\tau|>\varepsilon_2}(y-\tau)^{-1}$$

$$\times\chi_{\{\tau:2^{j-1}\leq|\tau|\leq 2^{j+2}\}}(\tau)\Big(\lim_{\varepsilon_1\to 0}\int\limits_{|x-t|>\varepsilon_1}\frac{f(t,\tau)\chi_{\{t:2^{k-1}\leq|t|\leq 2^{k+2}\}}(t)}{x-t}dt\Big)d\tau\Big|^p dxdy$$

$$+c\sum_{k\in\mathbb{Z}}\sum_{j\in\mathbb{Z}}\int\limits_{2^k<|x|<2^{k+1}}\int\limits_{2^j<|y|<2^{j+1}}v(x,y)\Big|\lim_{\varepsilon_2\to 0}\int\limits_{|y-\tau|>\varepsilon_2}(y-\tau)^{-1}\chi_{\{\tau:|\tau|>2^{j+2}\}}(\tau)$$

$$\times\Big(\lim_{\varepsilon_1\to 0}\int\limits_{|x-t|>\varepsilon_1}\frac{f(t,\tau)\chi_{\{t:2^{k-1}\leq|t|\leq 2^{k+2}\}}(t)}{x-t}dt\Big)d\tau\Big|^p dxdy\Big]$$

$$=:c[S_{2,1}+S_{2,2}+S_{2,3}].$$

If $2^j<|y|<2^{j+1}$ and $|\tau|<2^{j-1}$, then $|y|/2\leq|y-\tau|$. Using Lemma 5.4.1 (uniformly to k), but replacing the variables x and y, we find that

$$S_{2,1}\leq c\sum_{k\in\mathbb{Z}}\sum_{j\in\mathbb{Z}}\int\limits_{2^k<|x|<2^{k+2}}\int\limits_{2^j<|y|<2^{j+1}}v(x,y)|y|^{-p}$$

$$\times\Big(\int_{|\tau|<|y|}\Big|\lim_{\varepsilon_1\to 0}\int_{|x-t|>\varepsilon_1}\frac{f(t,\tau)\chi_{\{2^{k-1}\le|t|\le 2^{k+2}\}}(t)}{x-t}dt\Big|d\tau\Big)^p dxdy$$

$$\le c\sum_{k\in\mathbb{Z}}\int_{2^k<|x|<2^{k+1}}\int_{\mathbb{R}}v(x,y)|y|^{-p}$$

$$\times\Big(\int_{|\tau|<|y|}\Big|\lim_{\varepsilon_1\to 0}\int_{|x-t|>\varepsilon_1}\frac{f(t,\tau)\chi_{\{t:2^{k-1}\le|t|\le 2^{k+2}\}}(t)}{x-t}dt\Big|d\tau\Big)^p dxdy$$

$$\le c\sum_{k\in\mathbb{Z}}\int_{2^{k-1}<|x|<2^{k+2}}\int_{\mathbb{R}}|f(x,y)|^p w(x,y)dxdy = c\iint_{\mathbb{R}^2}|f(x,y)|^p w(x,y)dxdy.$$

Arguing as in the case of the estimate of $S_{2,1}$ but using Lemma 5.4.2 instead of Lemma 5.4.1 (again replacing the variables x and y) we find that the condition

$$\widetilde{D}_1 := \sup_{\substack{a>0\\ b>0}}\Big(\int_a^\infty\int_0^b\frac{v(x,y)}{x^p}dxdy\Big)^{1/p}\Big(\int_0^a\int_b^\infty\frac{w_1^{1-p'}(x,y)}{y^{p'}}dxdy\Big)^{1/p'}<\infty \tag{5.4.5}$$

guarantees the inequality

$$S_{2,3}\le c\|f\|^p_{L^p_w(\mathbb{R}^2)}.$$

To estimate $S_{2,2}$ recall that the condition (5.2.3) implies the pointwise inequality (see (5.2.6))

$$v(4a,4b)\le cB^p w(a,b)$$

for all t and τ, where c depends only on p. Therefore, using the boundedness of $\mathcal{H}$ in $L^p(\mathbb{R}^2)$, the fact that w_1, w_2 and v are increasing in each variable for another fixed one and the notation

$$f_{k,j}(t,\tau) := f(x,y)\chi_{\{t:2^{k-1}\le|t|\le 2^{k+2}\}}(t)\chi_{\{\tau:2^{j-1}\le|\tau|\le 2^{j+2}\}}(\tau),$$

we find that

$$S_{2,2}=\sum_{k\in\mathbb{Z}}\sum_{j\in\mathbb{Z}}\int_{2^k<|x|<2^{k+1}}\int_{2^j<|y|<2^{j+1}}v(x,y)|\mathcal{H}f_{k,j}(x,y)|^p dxdy$$

$$\le\sum_{k\in\mathbb{Z}}\sum_{j\in\mathbb{Z}}v(2^{k+1},2^{j+1})\int_{2^k<|x|<2^{k+1}}\int_{2^j<|y|<2^{j+1}}|\mathcal{H}f_{k,j}(x,y)|^p dxdy$$

$$\le cB^p\sum_{k\in\mathbb{Z}}\sum_{j\in\mathbb{Z}}w_1(2^{k-1})w_2(2^{j-1})\int_{2^{k-1}\le|x|\le 2^{k+2}}\int_{2^{j-1}\le|y|\le 2^{j+2}}|f(x,y)|^p dxdy$$

$$\leq cB^p \sum_{k\in\mathbb{Z}} \sum_{j\in\mathbb{Z}} \int\limits_{2^{k-1}\leq|x|\leq 2^{k+1}} \int\limits_{2^{j-1}\leq|y|\leq 2^{j+1}} w_1(x)w_2(y)|f(x,y)|^p dxdy \leq c\|f\|^p_{L^p_w(\mathbb{R}^2)}.$$

It remains to show that (5.2.3) implies conditions (5.4.4), (5.4.5). For example, let us show that (5.2.3) implies (5.4.5). Indeed, since the weights are increasing in each variable we have

$$\Big(\int_a^\infty\int_0^b \frac{v(x,y)}{x^p}dxdy\Big)^{1/p}\Big(\int_0^a\int_b^\infty \frac{w^{1-p'}(x,y)}{y^{p'}}dxdy\Big)^{1/p'}$$

$$\leq cb^{1/p}b^{-1/p}\Big(\int_a^\infty \frac{v(x,b)}{x^p}dx\Big)^{1/p}\Big(\int_0^a w^{1-p'}(x,b)dx\Big)^{1/p'}$$

$$\leq cb^{1/p'}b^{-1/p'}\Big(\int_a^\infty \frac{v(x,b)}{x^p}dx\Big)^{1/p}\Big(\int_0^a w^{1-p'}(x,b)dx\Big)^{1/p'}$$

$$\leq c\Big(\int_a^\infty\int_b^\infty \frac{v(x,y)}{(xy)^p}dxdy\Big)^{1/p}\Big(\int_0^a\int_0^b w^{1-p'}(x,y)dxdy\Big)^{1/p'},$$

where the positive constant depends only on p. Consequently $\widetilde{D}_1 \leq cB$. Analogously, it follows that $\overline{D}_1 \leq cB$.

Necessity. First we show that

$$I(a) := \iint\limits_{[-a,a]^2} w^{1-p'}(x,y)dxdy < \infty$$

for all $a > 0$.

Indeed, if $I(a) = \infty$ for some $a > 0$, then there exists $g \in L^p([-a,a])^2$ such that

$$\int\limits_{[-a,a]^2} g(x,y)w^{-1/p}(x,y)dxdy = \infty.$$

Let us assume that $f_a(x,y) = g(x,y)w^{-1/p}(x,y)\chi_{[-a,a]^2}(x,y)$. Then we find that

$$\|\mathcal{H}f_a\|_{L^p_v(\mathbb{R}^2)} \geq \|\chi_{(\mathbb{R}\setminus[-a,a])^2}\mathcal{H}f_t\|_{L^p_v(\mathbb{R}^2)}$$

$$\geq c_1\Big(\int\limits_{|x|>a}\int\limits_{|y|>t} v(x,y)(|x||y|)^{-p}dx\Big)^{1/p}\iint\limits_{[-a,a]^2} g(x,y)w^{-1/p}(x,y)dxdy = \infty.$$

On the other hand,

$$\|f_a\|_{L^p_w(\mathbb{R})} = \iint\limits_{[-a,a]^2} g(x,y)dxdy < \infty.$$

Thus, by (5.4.1) we conclude that $I(a) < \infty$. This contradiction proves what we have claimed.

To show that condition (5.2.3) holds, we observe that for the test functions

$$f_{a,b}(x,y) = w^{1-p'}(x,y)\chi_{[-a,a]}(x)\chi_{[-b,b]}(x),$$

we have that

$$\|\mathcal{H}f_{a,b}\|_{L^p_v(\mathbb{R}^2)} \geq \|\chi_{\mathbb{R}^2\setminus([-a,a]\times[-b,b])}Hf_{a,b}\|_{L^p_v(\mathbb{R}^2)}$$

$$\geq c_1\Big(\int\limits_{|x|>a}\int\limits_{|y|>b} v(x,y)(x,y)^{-p}dxdy\Big)^{1/p}\int\limits_{|x|<a}\int\limits_{|y|<b} w^{1-p'}(x,y)dxdy,$$

while

$$\|f_{a,b}\|_{L^p_w(\mathbb{R}^2)} = \iint\limits_{[-a,a]\times[-b,b]} w^{1-p'}(x,y)dxdy < \infty.$$

Finally, an application of inequality (5.4.1) completes the proof. □

The Case of Other Type of Weights

The following statements can be proved in the same way as Theorem 5.4.1; therefore we omit the proofs.

Theorem 5.4.2. *Let $1 < p < \infty$. Suppose that thc two-dimensional weights v and w are even on $\mathbb{R}$ and decreasing on $\mathbb{R}_+$ in each variable for another fixed one. Suppose also that $w(x,y) = w_1(x)w_2(y)$. Then inequality (5.4.1) holds if and only if*

$$B_2 := \sup_{\substack{a>0\\b>0}}\Big(\int\limits_0^a\int\limits_0^b v(x,y)dxdy\Big)^{1/p}$$

$$\times\Big(\int\limits_a^\infty w_1^{1-p'}(x)x^{-p'}dx\Big)^{1/p'}\Big(\int\limits_b^\infty w_2^{1-p'}(y)y^{-p'}dy\Big)^{1/p'} < \infty.$$

Theorem 5.4.3. *Let $1 < p < \infty$. Suppose that the two-dimensional weights v and w are even on $\mathbb{R}$. Further assume that these weights are increasing on $\mathbb{R}_+$ in the first variable for the second fixed one and decreasing on $\mathbb{R}_+$ in the second variable for* *the first fixed one. In addition, suppose that $w(x,y) = w_1(x)w_2(y)$. Then inequality* (5.4.1) *holds if and only if*

$$B_3 := \sup_{\substack{a>0 \\ b>0}} \bigg(\int_a^\infty \int_0^b v(x,y) x^{-p} dx dy \bigg)^{1/p}$$

$$\times \bigg(\int_0^a w_1^{1-p'}(x) dx \bigg)^{1/p'} \bigg(\int_b^\infty w_2^{1-p'}(y) y^{-p'} dy \bigg)^{1/p'} < \infty.$$

The n–dimensional Case

The next statements concerns with the two–weight inequality for $n-$ dimensional Hilbert transform with product kernels:

$$(\mathcal{H}^{(n)} f)(x_1, \dots, x_n) := \lim_{\varepsilon_1, \dots, \varepsilon_n \to 0} \int\limits_{|x_1 - t_1| > \varepsilon_1} \cdots \int\limits_{|y_n - t_n| > \varepsilon_n} \frac{f(t_1, \dots, t_n)}{\prod\limits_{i=1}^{n} (x_i - t_i)} dt_1 \cdots dt_n,$$

where $x_1, \dots, x_n \in \mathbb{R}$.

Theorem 5.4.1$'$. *Let $1 < p < \infty$. Suppose that the n–dimensional weights v and w are even on $\mathbb{R}$ and increasing on $\mathbb{R}_+$ in each variable for the other fixed ones. Further, assume that $w(x_1, \dots, x_n) = w_1(x) \cdots w_n(y)$. Then the two–weight inequality*

$$\int\limits_{\mathbb{R}^n} \Big| (\mathcal{H}^{(n)} f)(x_1, \dots, x_n) \Big|^p v(x_1, \dots, x_n) dx_1 \cdots dx_n$$

$$\leq c \int\limits_{\mathbb{R}^n} |f(x_1, \dots, x_n)|^p w(x_1, \dots, x_n) dx_1 \cdots dx_n \tag{5.4.6}$$

holds if and only if the condition

$$B^{(n)} := \sup_{\substack{a_1>0 \\ \vdots \\ a_n>0}} \bigg(\int_{a_1}^\infty \cdots \int_{a_n}^\infty \frac{v(x_1, \dots, x_n)}{(x_1 \cdots x_n)^p} dx_1 \cdots dx_n \bigg)^{1/p}$$

$$\times \bigg(\int_0^{a_1} \cdots \int_0^{a_n} w^{1-p'}(x_1, \dots, x_n) dx_1 \cdots dx_n \bigg)^{1/p'} < \infty$$

is satisfied.

Proof. *Sufficiency* follows in the same way as in the proof of Theorem 5.4.1 using two–weight criteria for the operators

$$(\mathcal{H}_n^{(1)}f)(x_1,\ldots,x_n)=\int\limits_{|t_1|<|x_1|}\cdots\int\limits_{|t_{n-1}|<|x_{n-1}|}\left|\lim_{\varepsilon\to 0}\int\limits_{|x_n-t_n|>\varepsilon}\frac{f(t_1,\ldots,t_n)}{x_n-t_n}dt_n\right|dt_1\cdots dt_{n-1},$$

$$(\mathcal{H}_n^{(2)}f)(x_1,\ldots,x_n)=\int\limits_{|t_1|>|x_1|}\cdots\int\limits_{|t_{n-1}|>|x_{n-1}|}\left|\lim_{\varepsilon\to 0}\int\limits_{|x_n-t_n|>\varepsilon}\frac{f(t_1,\ldots,t_n)}{x_n-t_n}dt_n\right|dt_1\cdots dt_{n-1}$$

and the boundedness of the Hilbert transforms with product kernels in unweighted Lebesgue spaces. There arise also other mixed– type operators, two–weight criteria of which can be deduces to those for $\mathcal{H}_n^{(1)}$ and $\mathcal{H}_n^{(2)}$.

Necessity can be easily proved substituting the test functions

$$f_{a_1,\ldots,a_n}=\chi_{[-a_1,a_1]}(x_1)\cdots\chi_{[-a_n,a_n]}(x_n)w^{1-p'}(x_1,\ldots,x_n),\quad a_i>0,\ i=1,\cdots,n,$$

in the two–weight inequality for $\mathcal{H}^{(n)}$.

The proof is complete. □

The proof of the next statement is similar to that of Theorem 5.4.1$'$. Details are omitted.

Theorem 5.4.2$'$. *Let $1<p<\infty$. Suppose that the n–dimensional weights v and w are even on $\mathbb{R}$ and decreasing on $\mathbb{R}_+$ in each variable for the other fixed ones. Suppose also that $w(x_1,\ldots,x_n)=w_1(x_1)\cdots w_n(x_n)$. Then inequality (5.4.6) holds if and only if the condition*

$$\overline{B}^{(n)}:=\sup_{\substack{a_1>0\\ \vdots\\ a_n>0}}\left(\int\limits_0^{a_1}\cdots\int\limits_0^{a_n}v(x_1,\ldots,x_n)dx_1\cdots dx_n\right)^{1/p}$$

$$\times\left(\int\limits_{a_n}^{\infty}\cdots\int\limits_{a_n}^{\infty}w^{1-p'}(x_1,\ldots,x_n)(x_1\cdots x_n)^{-p'}dx_1\cdots dx_n\right)^{1/p'}<\infty$$

is satisfied.

5.5 Examples

Now we give some examples of weights for which the two–weight inequality holds for the strong and iterated maximal functions, and the Hilbert transforms with product kernels.

Example 5.5.1. Let $2 < p < \infty$, $v(x,y) = (|x|+|y|)^{-\alpha}(|xy|)^p$ and $w(x,y) = (|x|+|y|)^{\beta}$, where $\alpha = 2p - \beta$, $p \le \beta < 2(p-1)$. Then by Theorem 5.2.1 we have that M_S is bounded from $L^p_w(\mathbb{R}^2)$ to $L^p_v(\mathbb{R}^2)$.

Example 5.5.2. Let $1 < p < \infty$ and $A = \min\{e^{-\frac{\beta}{p-1}}, e^{-\frac{\gamma}{p-1}}\}$, where $\beta = \gamma - 2p$, $\gamma > 2p - 1$. Suppose that

$$v(x,y) = \begin{cases} |xy|^{p-1} \ln^{\beta} \frac{2A}{|xy|}, & |x|, |y| < \sqrt{A} \\ (\ln^{\beta} 2) A^{p-1-\lambda} |xy|^{\lambda}, & \text{therwise}, \end{cases}$$

$$w(x,y) = \begin{cases} |xy|^{p-1} \ln^{\gamma} \frac{2A}{|xy|}, & |x|, |y| < \sqrt{A} \\ (\ln^{\gamma} 2) A^{p-1-\lambda} |xy|^{\lambda}, & \text{therwise}, \end{cases}$$

where $-1 < \lambda < p - 1$. Then the operator M_S is bounded from $L^p_w(\mathbb{R}^2)$ to $L^p_v(\mathbb{R}^2)$.

Example 5.5.3. Let $1 < p < \infty$ and let

$$v(x,y) = \begin{cases} |xy|^{p-1} \ln \frac{4\beta}{|xy|}, & |x|, |y| \le \min\left\{1, \frac{2}{e^{\frac{\beta}{2(p-1)}}}\right\} \\ (\ln^{\beta} 4) |xy|^{\lambda}, & \text{otherwise}, \end{cases}$$

$$w(x,y) = \begin{cases} |x|^{p-1} |y|^{\eta} \ln^{\gamma} \frac{2}{|x|}, & |x|, |y| \le 1, \\ (\ln^{\gamma} 2) |xy|^{\lambda}, & \text{otherwise}, \end{cases}$$

where $\beta > -1$, $\gamma > p$, $0 < \eta < p - 1$, $\beta = \gamma - p - 1$. Then it is easy to verify that the pair of weights (v, w) satisfies the conditions of Theorem 5.2.2 and 5.4.1 and, consequently, the operators M_S, $M_1 M_2$ and $\mathcal{H}$ are bounded from $L^p_w(\mathbb{R}^2)$ to $L^p_v(\mathbb{R}^2)$.

5.6 Applications to the Fourier Multipliers

Two–weighted criteria for potentials and Hilbert transforms with product kernels given in this and the previous chapters enable us to derive two–weighted estimates for Fourier multipliers in multidimensional Triebel-Lizorkin spaces. We emphasize that the stated conditions are not only sufficient but also necessary for a whole class of multipliers under consideration.

Let $S(\mathbb{R}^n)$ be the Schwartz space of rapidly decreasing functions. For $\varphi \in S(\mathbb{R}^n)$, the Fourier transform $\widehat{\varphi}$ is defined by

$$\widehat{\varphi}(\lambda) = (2\pi)^{-\frac{n}{2}} \int\limits_{\mathbb{R}^n} \varphi(x) e^{-i\lambda x} dx.$$

For the Fourier transform and its inverse, the notation $F(\varphi)$ and $F^{-1}(\varphi)$, respectively, will also be used.

Let S' be the space of tempered distributions, i.e., a space of linear bounded functionals over $S(\mathbb{R}^n)$. In the sequel the Fourier transform in the framework of the theory of S'-distributions will be considered.

Let $m = (m_1, m_2, \dots, m_n)$, $m_j \in \mathbb{Z}$, $j = 1, \dots, n$. Let Q_m be a subset of $\mathbb{R}^n$ defined by

$$2^{m_j} < |\lambda_j| \le 2^{m_{j+1}}; \quad j = 1, 2, \dots, n.$$

For $\varphi \in S(\mathbb{R}^n)$, we have that

$$\varphi(x) = \sum_m \varphi_m(x),$$

where

$$\varphi_m(x) = (2\pi)^{-\frac{n}{2}} \int\limits_{Q_m} \widehat{\varphi}(\lambda) e^{i\lambda x} d\lambda.$$

Suppose that $1 < p, \theta < \infty$. If for some weight function $w : \mathbb{R}^n \to \mathbb{R}^1$ and arbitrary $\varphi \in S(\mathbb{R}^n)$ the quantity

$$\left|\varphi, \Lambda_m^{p,\theta}\right| = \left(\int\limits_{\mathbb{R}^n} \left(\sum_m |\varphi_m(x)|^\theta \right)^{p/\theta} w(x) dx \right)^{1/p}$$

is finite, then the completion of $S(\mathbb{R}^n)$ with respect to this norm will be called a weighted Triebel–Lizorkin space and denoted by $\Lambda_w^{p,\theta}(\mathbb{R}^n)$.

Let X and Y be two function spaces on $\mathbb{R}^n$ with the norms $\|\cdot\|_X$ and $\|\cdot\|_Y$ respectively.

Definition 5.6.1. A distribution $m \in S'$ is called (X, Y) multiplier if for the operator K defined by the Fourier transform equation

$$\widehat{K\varphi} = m\widehat{\varphi}, \quad \varphi \in S(\mathbb{R}^n),$$

there exists a positive constant c such that

$$\|K\varphi\|_Y \le c\|\varphi\|_X$$

for all $\varphi \in S(\mathbb{R}^n)$.

In this case we write that $m \in \mathcal{M}(X, Y)$.

Theorem 5.6.1. *Let $1 < p < q < \infty$ and let $1 < \theta < \infty$. Let $\{\mu_m\}$ ($m = (m_1, \dots, m_n)$, $m_i \in \mathbb{Z}$, $i = 1, \dots, n$) be a set of measures on $S(\mathbb{R}^n)$ such that*

$$\int\limits_{\mathbb{R}^n} |d\mu_m| \le c$$

for some positive constant c independent of m.

Suppose that a measurable function $m(\lambda_1,\dots,\lambda_n)$ is represented on arbitrary Q_m by the relation

$$m(\lambda_1,\dots,\lambda_n)=\int_0^{\lambda_1}\dots\int_0^{\lambda_n}\prod_{j=1}^{n}(\lambda_j-t_j)^{\alpha_j-1}d\mu_m,$$

when $0<\alpha_j<1$, $j=1,2,\dots,n$.

(i) *If $1<p<q<\infty$, $1<\theta<\infty$ and the pairs (v,w) of weights satisfy the conditions supposed in Theorem 4.5.1′, then $m\in\mathcal{M}(\Lambda_w^{p,\theta},\Lambda_v^{q,\theta})$*

(ii) *If the weight function v satisfies the conditions of Theorem 4.5.2′, then $m\in\mathcal{M}(\Lambda^{p,\theta},\Lambda_v^{q,\theta})$.*

Theorem 5.6.2. *Let $1<p,\theta<\infty$. Let $\{\mu_m\}$ be a set of measures which total variations are uniformly bounded.*

Assume that a measurable function $m(\lambda_1,\dots,\lambda_n)$ is represented on arbitrary Q_m by the formula

$$m(\lambda_1,\dots,\lambda_n)=\int_0^{\lambda_1}\dots\int_0^{\lambda_n}d\mu_m.$$

Then for the pairs of weights (v,w) satisfying the conditions assumed in Theorem 5.4.1′ (Theorem 5.4.2′), we have that $m\in\mathcal{M}(\Lambda_w^{p,\theta},\Lambda_v^{q,\theta})$.

Corollary 5.6.1. *Let m be continuous outside coordinate planes and have there continuous derivatives*

$$\frac{\partial^k m}{\partial\lambda_1\cdots\partial\lambda_n^{k_n}},\qquad 0\le k_1+k_2+\dots+k_n=k\le n,\quad k_j=0,1.$$

Moreover, assume that

$$\left|\lambda_1^{k_1+\alpha_1}\lambda_2^{k_2+\alpha_2}\cdots\lambda_n^{k_n+\alpha_n}\frac{\partial^k m}{\partial\lambda_1\cdots\partial\lambda_n^{k_n}}\right|\le M.$$

(i) *If $1<p<q<\infty$, $1<\theta<\infty$, $0<\alpha_j<1$ and (v,w) satisfies the conditions of Theorem 4.5.1′, then $m\in\mathcal{M}(\Lambda_w^{p,\theta},\Lambda_v^{q,\theta})$; if the weight v satisfies the conditions of Theorem 4.5.2′, then $m\in\mathcal{M}(\Lambda^{p,\theta},\Lambda_v^{q,\theta})$;*

(ii) *When $1<p<\infty$, $1<\theta<\infty$, $\alpha_j=1$, $j=1,\dots,n$ and (v,w) satisfies the conditions of Theorem 5.4.1′ (or Theorem 5.4.2′), we have that $m\in\mathcal{M}(\Lambda_w^{p,\theta},\Lambda_w^{p,\theta})$.*

The essential point of the proof of these theorems is the representation of the operator under consideration by certain "elementary" transformations. Two–weighted boundedness criteria for integral transforms with product kernels are also essential for the arguments.

It is worth mentioning the papers [107] and [108] devoted to the unweighted case as the source of the ideas developed here.

Now we establish two–weighted estimate for "elementary" multipliers. The proof for general multipliers is similar to that of the one-dimensional case, exposed in ([39], Chapter 9); therefore we omit the proofs.

For $j = 1, 2, \dots, n$ we set

$$(\lambda_j)_+^{-\beta} = \begin{cases} \lambda_j^{\beta}, & \text{when } \ \lambda_j > 0, \\ 0, & \text{when } \ \lambda_j < 0 \end{cases}$$

and

$$(\lambda_j)^{-\beta} = \begin{cases} 0, & \text{when } \ \lambda_j > 0, \\ \frac{1}{|\lambda_j|^{\beta}}, & \text{when } \ \lambda_j < 0 \end{cases}.$$

Let us introduce the "elementary" multiplier transforms $B_t^{\alpha} : \varphi \to \psi_t$, where $\widehat{\psi}_t(\lambda) = \prod_{j=1}^{n} (\lambda_j - t_j) - \alpha_j \widehat{\varphi}(\lambda)$, $t = (t_1, \dots, t_n) \in \mathbb{R}^n$, $\varphi \in S(\mathbb{R}^n)$.

Lemma 5.6.1. *Let $1 < p < q < \infty$ and (v, w) be a pair of weight functions satisfying the conditions of Theorem 4.5.1′. Then*

$$\|B_t^{\varphi}\|_{q,v} \le c\|\varphi\|_{p,w}, \quad \varphi \in S(\mathbb{R}^n),$$

with a positive constant c not depending of φ and t.

Proof. First we prove that the transform $\varphi \to \psi_0$ is bounded from $L_w^p(\mathbb{R}^n)$ to $L_v^q(\mathbb{R}^n)$. In fact, the Fourier preimage for functions $(\lambda_j)_+^{-\alpha_j}$, $0 < \alpha_j < 1$, $j = 1, \dots, n$ are the functions of the type

$$a(\alpha_j)(x_j)_+^{-(1-\alpha_j)} + b(\alpha_j)(x_j)_-^{-(1-\alpha_j)},$$

where

$$a(\alpha_j) = \frac{i}{\sqrt{2\pi}} e^{-i\alpha_j \frac{\pi}{2}} \Gamma(1 - \alpha_j),$$

$$b(\alpha_j) = -\frac{i}{\sqrt{2\pi}} e^{i\alpha_j \frac{\pi}{2}} \Gamma(1 - \alpha_j), \quad j = 1, \dots, n.$$

Therefore, the Fourier preimage for $(\lambda_j)_+^{-\alpha_j}$ can be represented as a linear combination of the direct product

$$(x_1)_+^{-(1-\alpha_1)} \times \cdots \times (x_n)_+^{-(1-\alpha_n)}$$

or

$$(x_1)_-^{-(1-\alpha_1)} \times \cdots \times (x_n)_-^{-(1-\alpha_n)}.$$

According to Theorem 4.5.1′, convolution of the function φ with the above mentioned function is bounded from $L^p_w(\mathbb{R}^n)$ to $L^q_v(\mathbb{R}^n)$.

Finally, since the shift of t in the Fourier image corresponds to the multiplication by e^{itx} of the Fourier image, the norms of B^α_t coincide with those of B^α.

The proof is complete. □

Theorem 5.6.3. *Suppose that the function m is defined by the formula*

$$m(\lambda) = \int\limits_{-\infty}^{\lambda_1} \cdots \int\limits_{-\infty}^{\lambda_n} \frac{d\mu(t)}{(\lambda_1 - t_1)^{1-\alpha_1} \cdots (\lambda_n - t_n)^{1-\alpha_n}},$$

where μ is a finite measure on $\mathbb{R}^n$.

Let K be the operator defined by the Fourier transform equation

$$\widehat{K\varphi} = m(\lambda)\widehat{\varphi}(\lambda), \quad \varphi \in S.$$

Then under the conditions of Theorem 4.5.1′ on the pair (v, w) of weight function, the operator K is bounded from $L^p_w(\mathbb{R}^n)$ to $L^q_v(\mathbb{R}^n)$.

Proof. We have that

$$K\varphi(x) = \frac{1}{(2\pi)^n} \int\limits_{\mathbb{R}^n} m(\lambda)\widehat{\varphi}(\lambda) e^{i\lambda x} d\lambda$$

$$= \frac{1}{(2\pi)^n} \int\limits_{\mathbb{R}^n} \left(\int\limits_{\mathbb{R}^n} \prod_{j=1}^{n} (\lambda_j - t_j)_+^{-(1-\alpha_j)} d\mu(t) \right) \widehat{\varphi}(\lambda) e^{i\lambda x} d\lambda.$$

Changing the order of integration we get that

$$K\varphi(x) = \frac{1}{(2\pi)^n} \int\limits_{\mathbb{R}^n} \left(\int\limits_{\mathbb{R}^n} \prod_{j=1}^{n} (\lambda_j - t_j)_+^{-(1-\alpha_j)} \widehat{\varphi}(\lambda) e^{i\lambda x} d\lambda \right) d\mu(t)$$

$$= \frac{1}{(2\pi)^n} \int\limits_{\mathbb{R}^n} B^\alpha_t \varphi(\lambda) d\mu(t).$$

Further, Lemma 5.6.1 yields that

$$\|K\varphi\|_{L^q_v(\mathbb{R}^n)} \le c \int\limits_{\mathbb{R}^n} \|B^\alpha_t \varphi\|_{L^q_v(\mathbb{R}^n)} |d\mu(t)| \le c \int\limits_{\mathbb{R}^n} |d\mu| \cdot \|\varphi\|_{L^p_w(\mathbb{R}^n)}.$$

Therefore,

$$\|K\varphi\|_{L^q_v(\mathbb{R}^n)} \le cM \|\varphi\|_{L^p_w(\mathbb{R}^n)},$$

where M is the total variation of μ.

The proof is complete. □

For further details and complementary results we refer the reader to [40], [39, Chapter 9].

5.7 Notes and Comments on Chapter 5

The main results in this chapter were announced in [97], [95].

The one–weight problem for the operator $\mathcal{H}$ was solved in [86], [85]. In the paper [155] the author established the two–weight inequality for the strong Hardy–Littlewood maximal functions under the condition involving the operator itself provided that the weight on the right-hand side satisfies some additional conditions, for instance, belongs to the Muckenhoupt A_p class separately in each variable, or is a product of one–dimensional weights (see also [43] for the related topics).

The Fefferman–Stein weak type inequality

$$w\big(M_S f > \lambda\big) \le c \iint \frac{|f|}{\lambda}\left(1 + \log^+ \frac{|f|}{\lambda}\right) M_S w, \quad \lambda > 0,$$

for the strong maximal function M_S defined on the plane, under the assumption that the two–dimensional weight w satisfies a strong Muckenhoupt condition A_r for some fixed $r > 1$, was proved in [122].

For the product theory of singular integrals we refer to the papers [44], [85], [43], [128].

Criteria guaranteeing the one–weight inequality for the Hilbert transform with single kernel $\mathbb{H}$ were derived in [72]. Moreover, two–weight necessary and sufficient conditions for the operator M under the conditions involving M itself were established in [154].

It should be pronounced that the two–weight problem for the Hilbert transform $\mathbb{H}$ still remains open. Some conditions governing the boundedness of $\mathbb{H}$ between two weighted Lebesgue spaces with different weights were given in [127], [37], [130], [148] and [35] (see also the monographs [53], [39] and [173] and references therein).

The well-known Marcinkiewicz, Mikhlin and Hörmander multiplier theorems have been extended by several authors. General multiplier theorems for Fourier integrals in the unweighted case and their applications to embedding theory for spaces of differentiable functions of several variables were given in [107] and [108]. One-weighted estimates for Fourier multipliers when a weight belongs to Muckenhoupt's A_p class have been established in [104] and [86]. For two–weighted estimates we refer the reader to [40], [38] and [39, Chapter 9].

Chapter 6

Two–weight Estimates for Fourier Operators and Bernstein Inequalities

In this chapter we prove weighted estimates for Fourier operators with different weights and exponents of spaces on both sides of the derived inequalities. Besides this, the Cesàro and Abel means norm convergence criteria in the two–weighted setting are derived. In the one–weighted case the estimates of the rate of summabilities by generalized modulus of smoothness are established. Applying the Fourier operator's norm estimates we extend the well-known Bernstein inequalities for the derivatives of trigonometric polynomials and integral functions of finite degree of one and several variables in the two–weighted setting.

The main tools of the proofs are two–weight norm estimates for one-dimensional and multidimensional Steklov–type means.

6.1 Two–weight Inequalities for Cesàro and Abel–Poisson Means of Fourier Series

Let $\mathbb{T}$ be the interval $[-\pi, \pi]$. Let w be a $2\pi-$ periodic non–negative integrable function. We denote by $L^p_w(\mathbb{T})$, $1 \leq p < \infty$ the Banach function space of all measurable $2\pi-$periodic functions f, for which the norm

$$\|f\|_{p,w} := \left(\int_{\mathbb{T}} |f(x)|^p \, w(x) \, dx \right)^{1/p} < \infty.$$

The uniform boundedness problem of Cesàro and Abel-Poisson means of functions $f \in L^p_w(\mathbb{T})$ $(1 < p < \infty)$ was studied by M. Rosenblum [149] and B. Muckenhoupt [124]. In [149] a characterization of the weights w for which the Cesàro $(C.1)$

and Abel-Poisson means are bounded as operators from $L^p_w(\mathbb{T})$ to $L^p_w(\mathbb{T})$ has been done. Later on B. Muckenhoupt [124] showed that the condition referred in [149] is equivalent to the condition $w \in \mathcal{A}_p(\mathbb{T})$, that is

$$\sup_I \frac{1}{|I|}\int_I w(x)\,dx\left(\frac{1}{|I|}\int_I w^{1-p'}(x)\,dx\right)^{p-1} < \infty, \qquad (6.1.1)$$

where $p' = p/(p-1)$ and the supremum is taken over all intervals whose lengths are not greater than 2π.

In the two–weighted setting, B. Muckenhoupt [125] has shown that the necessary and sufficient condition for the uniform boundedness of the Abel-Poisson means as a sequence of operators from $L^p_w(\mathbb{T})$ to $L^p_v(\mathbb{T})$ is

$$\sup_I \frac{1}{|I|}\int_I v(x)\,dx\left(\frac{1}{|I|}\int_I w^{1-p'}(x)\,dx\right)^{p-1} < \infty. \qquad (6.1.2)$$

Let

$$f(x) \sim \frac{a_0}{2} + \sum_{k=1}^{\infty}(a_k \cos kx + b_k \sin kx) \qquad (6.1.3)$$

be the Fourier series of the function $f \in L^1(\mathbb{T})$. Let $\sigma_n^\alpha(\cdot, f)$ $(\alpha > 0)$ be the Cesàro mean of the series $(6.1.3)$, that is

$$\sigma_n^\alpha(x,f) = \frac{1}{\pi}\int_{\mathbb{T}} f(x+t)\,K_n^\alpha(t)\,dt,$$

where

$$K_n^\alpha(t) = \sum_{k=0}^{n} \frac{A_{n-k}^{\alpha-1} D_k(t)}{A_n^\alpha}$$

is the Fejer kernel and

$$D_k(t) = \frac{\sin\left(k+\frac{1}{2}\right)t}{2\sin\frac{t}{2}}$$

is the Dirichlet kernel, with

$$A_n^\alpha = \begin{pmatrix} n+\alpha \\ \alpha \end{pmatrix} \approx \frac{n^\alpha}{\Gamma(\alpha+1)}.$$

Suppose that $U_r(\cdot, f)$ $(0 \le r < 1)$ is the Abel-Poisson mean of the function f, that is

$$U_r(x,f) = \frac{1}{2\pi}\int_{-\pi}^{\pi} P_r(x-t)\,f(t)\,dt,$$

where

$$P_r(t) = \frac{1 - r^2}{1 - 2r\cos t + r^2}$$

is the Poisson kernel.

In the present section, we investigate the problem regarding the estimates of norms of the operators $\sigma_n^\alpha(\cdot, f)$ $(\alpha > 0)$ and $U_r(\cdot, f)$ $(0 \le r < 1)$ from $L_w^p(\mathbb{T})$ to $L_v^q(\mathbb{T})$, where $1 < p \le q < \infty$.

Let us generalize the Muckenhoupt $A_p(\mathbb{T})$ class as follows:

Definition 6.1.1. Let $1 < p \le q < \infty$. A pair of weight functions (v, w) is said to be of class $\mathcal{A}_{p,q}(\mathbb{T})$ if the condition

$$\sup_I \left(\frac{1}{|I|}\int_I v(x)\,dx\right)^{1/q} \left(\frac{1}{|I|}\int_I w^{1-p'}(x)\,dx\right)^{1/p'} < \infty \tag{6.1.4}$$

holds, where the supremum is taken over all intervals with lengths not more than 2π.

In the case $p = q$ the class $\mathcal{A}_{p,q}(\mathbb{T})$ coincides with the Muckenhoupt $A_p(\mathbb{T})$ class. The following statements are true.

Theorem 6.1.1. *Let $1 < p \le q < \infty$. The the inequality*

$$\|\sigma_n^\alpha(\cdot, f)\|_{q,v} \le c\, n^{\frac{1}{p}-\frac{1}{q}} \|f\|_{p,w}, \quad \alpha > 0 \tag{6.1.5}$$

holds for arbitrary $f \in L_w^p(\mathbb{T})$, where the constant c does not depend on n and f, if and only if $(v, w) \in \mathcal{A}_{p,q}(\mathbb{T})$.

Theorem 6.1.2. *Let $1 < p \le q < \infty$. Then the necessary and sufficient condition for the validity of the inequality*

$$\|U_r(\cdot, f)\|_{q,v} \le c\, (1 - r)^{\frac{1}{q}-\frac{1}{p}} \|f\|_{p,w} \tag{6.1.6}$$

for arbitrary $f \in L_w^p(\mathbb{T})$, where the constant c does not depend on r and f, is that $(v, w) \in \mathcal{A}_{p,q}(\mathbb{T})$.

The proofs of these theorems are based on the next result concerning the following extension of Steklov means

$$f_h^\gamma(x) = \frac{1}{h^{1-\gamma}} \int_{x-h}^{x+h} |f(t)|\,dt, \quad h > 0, \;\; 0 \le \gamma < 1.$$

Theorem 6.1.3. *Let $1 < p \le q < \infty$, $(v, w) \in \mathcal{A}_{p,q}(\mathbb{T})$ and $\gamma = 1/p - 1/q$. Then the inequality*

$$\|f_h^\gamma\|_{q,v} \le c\,\|f\|_{p,w} \tag{6.1.7}$$

holds for every $f \in L_w^p(\mathbb{T})$.

Proof. Let $h \le \pi$ and let N be the least natural number for which $Nh \ge \pi$. Then we have that

$$\int_{\mathbb{T}} [f_h^\gamma(x)]^q v(x)dx \le \sum_{k=-N}^{N-1} \int_{kh}^{(k+1)h} h^{-q(1-\gamma)} \left[\int_{x-h}^{x+h} |f(t)|dt \right]^q v(x)dx$$

$$\le \sum_{k=-N}^{N-1} \int_{kh}^{(k+1)h} h^{-q(1-\gamma)} \left[\int_{(k-1)h}^{(k+2)h} |f(t)|dt \right]^q v(x)dx$$

$$\le \sum_{k=-N}^{N-1} \int_{kh}^{(k+1)h} h^{-q(1-\gamma)} \left[\int_{(k-1)h}^{(k+2)h} |f(t)|^p w(t)dt \right]^{q/p} \left[\int_{(k-1)h}^{(k+2)h} w^{1-p'}(t)dt \right]^{q/p'} v(x)dx$$

$$= \sum_{k=-N}^{N-1} \left(\int_{kh}^{(k+1)h} v(x)dx \right) \left(\int_{(k-1)h}^{(k+2)h} w^{1-p'}(t)dt \right)^{q/p'} h^{-q(1-\gamma)}$$

$$\times \left(\int_{(k-1)h}^{(k+2)h} |f(t)|^p w(t)dt \right)^{q/p}$$

$$= \sum_{k=-N}^{N-1} \left(\frac{1}{h} \int_{kh}^{(k+1)h} v(x)dx \right) \left(\frac{1}{h} \int_{(k-1)h}^{(k+2)h} w^{1-p'}(t)dt \right)^{q/p'} \left(\int_{(k-1)h}^{(k+2)h} |f(t)|^p w(t)dt \right)^{q/p}.$$

Applying the condition $(v, w) \in \mathcal{A}_{p,q}(\mathbb{T})$ we conclude that

$$\int_{-\pi}^{\pi} [f_h^\gamma(x)]^q v(x)dx \le c \sum_{k=-N}^{N-1} \left(\int_{(k-1)h}^{(k+2)h} |f(t)|^p w(t)dt \right)^{q/p}.$$

Hence, by [88, Proposition 5.1.3] we obtain that

$$\int_{-\pi}^{\pi} [f_h^\gamma(x)]^q v(x)dx \le c_1 \|f\|_{p,w}^q.$$

The proof is complete. □

Proof of Theorem 6.1.1. We first prove that

$$|\sigma_n^\alpha(x,f)| \le c_0 \int\limits_{1/n}^{2\pi} \frac{1}{n^\alpha} h^{-1-\alpha} \left(\frac{1}{h} \int\limits_{x-h}^{x+h} |f(t)|dt \right) dh,$$

where the constant c_0 does not depend on f and h. By reversing the order of integration on the right–hand side integral of the previous inequality, we observe that it is more than or equal to

$$I := \int\limits_{x-\pi}^{x+\pi} |f(t)| \left[\int\limits_{\max(|x-t|,1/n)}^{2\pi} \frac{1}{n^\alpha} h^{-2-\alpha} dh \right] dt$$

$$\ge c \int\limits_{x-\pi}^{x+\pi} |f(t)| \frac{1}{n^\alpha} \left[\max\left(|x-t|, \frac{1}{n} \right) \right]^{-1-\alpha} dt$$

since $|x-t| \le \pi$.

Indeed, let us show that for $|x-t| \le \pi$, the inequality

$$\int\limits_{\max(|x-t|,1/n)}^{2\pi} h^{-2-\alpha} dh > c(\max\{|x-t|, 1/n\})^{-\alpha-1}$$

holds, where c does not depend on x, t and n.

In fact, it is obvious that

$$I_1 := \int\limits_{\max(|x-t|,1/n)}^{2\pi} h^{-2-\alpha} dh = \frac{1}{1+\alpha} \left(\frac{1}{(\max\{|x-t|, 1/n\})^{1+\alpha}} \quad \frac{1}{(2\pi)^{1+\alpha}} \right).$$

To prove the latter inequality we consider two cases:

(*a*) Let $|x-t| < 1/n$. Then

$$I_1 = \frac{1}{1+\alpha} \left(n^{1+\alpha} - \frac{1}{(2\pi)^{1+\alpha}} \right) > \frac{1}{1+\alpha} (1 - (2\pi)^{-1-\alpha}) n^{1+\alpha}.$$

(*b*) Let now $|x-t| \ge 1/n$. Then, since $|x-t| \le \pi$, we conclude that

$$I_1 = \frac{1}{1+\alpha} \left(\frac{1}{|x-t|^{1+\alpha}} - \frac{1}{(2\pi)^{1+\alpha}} \right)$$

$$= \frac{1}{2(1+\alpha)} \left(\frac{1}{|x-t|^{1+\alpha}} + \frac{1}{|x-t|^{1+\alpha}} - \frac{2}{(2\pi)^{1+\alpha}} \right)$$

$$> \frac{1}{2(1+\alpha)}\left(\frac{1}{|x-t|^{1+\alpha}}+\frac{1}{\pi^{1+\alpha}}-\frac{2}{(2\pi)^{1+\alpha}}\right)$$
$$\geq \frac{1}{2(1+\alpha)}\left(\frac{1}{|x-t|^{1+\alpha}}+\frac{1}{\pi^{1+\alpha}}-\frac{1}{2^{\alpha}\pi^{1+\alpha}}\right)$$
$$> \frac{1}{2(1+\alpha)}\frac{1}{|x-t|^{1+\alpha}},$$

which implies the desired result.

Therefore,

$$|\sigma_n^\alpha(x,f)| \leq c\int_{1/n}^{2\pi}\frac{1}{n^\alpha}h^{-1-\alpha}h^{-\gamma}f_h^\gamma(x)\,dh.$$

Let now $(v,w) \in \mathcal{A}_{p,q}(\mathbb{T})$ and let $\gamma = 1/p - 1/q$.

Hence, by using the preceding inequality, Minkowski's inequality and (6.1.7) we obtain that

$$\|\sigma_n^\alpha(\cdot,f)\|_{q,v} = \left(\int_{\mathbb{T}}|\sigma_n^\alpha(x,f)|^q v(x)\,dx\right)^{1/q}$$
$$\leq c\left(\int_{\mathbb{T}}\left|\int_{1/n}^{2\pi}\frac{1}{n^\alpha}h^{-1-\alpha}h^{-\gamma}f_h^\gamma(x)\,dh\right|^q v(x)\,dx\right)^{1/q}$$
$$\leq c\int_{1/n}^{2\pi}\left(\int_{\mathbb{T}}|f_h^\gamma(x)|^q v(x)\,dx\right)^{1/q}\frac{1}{n^\alpha}h^{-1-\alpha}h^{-\gamma}dh \leq cn^{\frac{1}{p}-\frac{1}{q}}\|f\|_{p,w}.$$

To prove necessity, let us assume that (6.1.5) holds for every $f \in L_w^p(\mathbb{T})$ and n. Let I_0 be a subinterval of $\mathbb{T}$. If the length I_0 is more than $\pi/4$, then the validity of condition (6.1.4) is clear.

Let $|I_0| \leq \pi/4$ and m be the greatest integer such that

$$m \leq \frac{\pi}{2|I_0|} - 1.$$

Then we have that

$$\left|\left(k+\frac{1}{2}\right)(x-t)\right| \leq (m+1)|x-t| \leq \frac{\pi}{2}, \quad k=0,\ldots,m.$$

By applying Abel's transformation we get, for $x,t \in I_0$, the following estimates:

$$\begin{aligned} K_m^\alpha(x-t) &\geq \sum_{k=0}^m \frac{A_{m-k}^{\alpha-1}}{A_m^\alpha}(2k+1) \geq c(m+2)\frac{1}{(m+1)A_m^\alpha}\sum_{k=0}^m A_{m-k}^{\alpha-1}(k+1) \\ &\geq \frac{c}{|I_0|}\frac{1}{(m+1)A_m^\alpha}\sum_{k=0}^m A_{m-k}^\alpha = \frac{c}{|I_0|}\frac{A_m^{\alpha+1}}{(m+1)A_m^\alpha} \geq \frac{c}{|I_0|}. \end{aligned}$$

Let us take the function

$$f_0(x) = w^{1-p'}(x)\,\chi_{I_0}(x)$$

which belongs to $L^p_w(\mathbb{T})$. If we substitute this function in (6.1.5) for the m indicated above, we find that

$$\left(\frac{1}{|I_0|}\int_{I_0} v(x)\,dx\right)^{1/q}\left(\frac{1}{|I_0|}\int_{I_0} w^{1-p'}(x)\,dx\right)^{1/p'} \le c.$$

Hence (6.1.5) follows from (6.1.4).

The proof is complete. □

From Theorem 6.1.1 we can deduce the following statement which has an independent interest.

Proposition 6.1.1. *Let $1 < p < \infty$ and $(v,w) \in \mathcal{A}_p(\mathbb{T})$. Then $L^p_w(\mathbb{T}) \subset L^p_v(\mathbb{T})$. Moreover*

$$\|f\|_{p,v} \le c\,\|f\|_{p,w} \tag{6.1.8}$$

for arbitrary $f \in L^p_w(\mathbb{T})$.

Proof. Let $f \in L^p_w(\mathbb{T})$. The set of all $2\pi-$periodic continuous functions is dense in $L^p_w(\mathbb{T})$. It is known that the Cesàro means of a continuous function converge uniformly to the initial function. Hence, since $v \in L^1(\mathbb{T})$, it follows that they converge in $L^p_v(\mathbb{T})$ to $f \in L^p_v(\mathbb{T})$, too. By the inequality

$$\|\sigma^\alpha_n(\cdot,f)\|_{p,v} \le c\,\|f\|_{p,w}, \quad \alpha > 0$$

and by Banach–Steinhaus theorem we conclude that the sequence $\{\sigma^\alpha_n(\cdot,f)\}$ converges in $L^p_v(\mathbb{T})$ to some function $\varphi \in L^p_v(\mathbb{T})$. Since $w^{1-p'} \in L^1(\mathbb{T})$ we have that $f \in L^1(\mathbb{T})$. Then $\{\sigma^\alpha_n(\cdot,f)\}$ converges almost everywhere to f. On the other hand, the convergence in norm implies the convergence of a subsequence $\{\sigma^\alpha_{n_k}(\cdot,f)\}$ to φ almost everywhere. Thus $\varphi = f$ almost everywhere and, hence, $f \in L^p_v(\mathbb{T})$.

Further, the identity operator

$$I : L^p_w(\mathbb{T}) \to L^p_v(\mathbb{T}), \quad I(f) = f$$

is closed. By the closed graph theorem (see, for example [150], p. 114) it is continuous and, hence, (6.1.8) holds. □

Proof of Theorem 6.1.2. Let $(v,w) \in \mathcal{A}_{p,q}(\mathbb{T})$ and $\gamma = 1/p - 1/q$. From the estimate (see [125])

$$|U_r(x,f)| \le c\int_{1-r}^{2\pi} (1-r)\,h^{-2} f_h(x)\,dh$$

it follows clearly that

$$|U_r(x,f)| \le c \int_{1-r}^{2\pi} (1-r)\, h^{-2} h^{-\gamma} f_h^{\gamma}(x)\, dh.$$

This estimate, together with Minkowski's inequality and (6.1.7) yields that

$$\|U_r(\cdot,f)\|_{q,v} = \left(\int_{\mathbb{T}} |U_r(x,f)|^q v(x)\, dx \right)^{1/q}$$

$$\le c \left(\int_{\mathbb{T}} \left| \int_{1-r}^{2\pi} (1-r)\, h^{-2} h^{-\gamma} f_h^{\gamma}(x)\, dh \right|^q v(x)\, dx \right)^{1/q}$$

$$\le c \int_{1-r}^{2\pi} \left(\int_{\mathbb{T}} |f_h^{\gamma}(x)|^q v(x)\, dx \right)^{1/q} (1-r)\, h^{-2} h^{-\gamma} dh \le c\,(1-r)^{\frac{1}{q}-\frac{1}{p}} \|f\|_{p,w}\,.$$

Conversely, let (6.1.6) hold for every $f \in L^p_w(\mathbb{T})$ and $0 \le r < 1$. If we take an interval $I_0 \subset \mathbb{T}$ and set

$$r_0 = \max\{1 - |I_0|, 0\}, \quad f_0(x) = w^{1-p'}(x)\, \chi_{I_0}(x),$$

then the estimate

$$U_{r_0}(x, f_0) \ge \frac{c}{|I_0|} \int_I w^{1-p'}(t)\, dt$$

is valid for $x \in I$. If we apply (6.1.6) to f_0 and r_0, and consider the last inequality we get that $(v,w) \in \mathcal{A}_{p,q}(\mathbb{T})$.

The proof is complete. □

In the case $p = q$, Theorem 6.1.1 yields the convergence

$$\|\sigma_n^{\alpha}(\cdot,f) - f\|_{p,v} \to 0, \quad n \to \infty,$$

for $f \in L^p_w(\mathbb{T})$. Moreover, if also $w = v$ we can estimate the rate of this convergence. Indeed, let $w \in \mathcal{A}_p(\mathbb{T})$ and $f \in L^p_w(\mathbb{T})$. It is well known that for the Steklov mean

$$f_h(x) = \frac{1}{h} \int_{x-h}^{x+h} f(t)\, dt, \quad h > 0,$$

the inequality

$$\|f_h\|_{p,w} \le c\, \|f\|_{p,w}$$

holds, where the constant c is independent of h and f. Starting from this we define the modulus of continuity for the function $f \in L^p_w(\mathbb{T})$ as

$$\Omega(\delta, f)_{p,w} := \sup_{h \le \delta} \|f - f_h\|_{p,w}, \quad \delta \ge 0.$$

Theorem 6.1.4. *Let $1 < p < \infty$, $w \in \mathcal{A}_p(\mathbb{T})$ and $f \in L^p_w(\mathbb{T})$. Then there exist a constant c, which does not depend on n and f, such that the estimate*

$$\|\sigma_n(\cdot, f) - f\|_{p,w} \le c\, n\, \Omega\left(\frac{1}{n}, f\right)_{p,w} \tag{6.1.9}$$

holds.

Proof. Let the function $f \in L^p_w(\mathbb{T})$ have the Fourier series

$$f(x) \sim \frac{a_0}{2} + \sum_{k=1}^{\infty} (a_k \cos kx + b_k \sin kx) = \frac{a_0}{2} + \sum_{k=1}^{\infty} A_k(x, f). \tag{6.1.10}$$

Then

$$f(x) - f_h(x) \sim \sum_{k=1}^{\infty} \left(1 - \frac{\sin kh}{kh}\right) A_k(x, f).$$

If $S_n(x, f)$, $n = 0, 1, \dots,$ denote the n-th partial sum of the series (6.1.10), then we have that

$$\|f - S_n(\cdot, f)\|_{p,w} \le c\, \Omega\left(\frac{1}{n}, f\right)_{p,w} \tag{6.1.11}$$

with the constant c independent of n and f (see [65] and [74]). Let us consider the sequence

$$\lambda_k = \begin{cases} \dfrac{k/(n+1)}{n\left(1 - \frac{\sin(k/n)}{k/n}\right)}, & k \le n \\ 0 \qquad\qquad\qquad, & k > n. \end{cases}$$

It is clear that the sequence $\{\lambda_k\}$ satisfies the conditions

$$|\lambda_k| \le c, \qquad \sum_{k=2^{m-1}}^{2^m - 1} |\lambda_{k+1} - \lambda_k| \le c, \quad k, m = 1, 2, \dots. \tag{6.1.12}$$

Taking into account (6.1.12) and a theorem by D. S. Kurtz (see [103] and [104]), we obtain that

$$\|S_n(\cdot, f) - \sigma_n(\cdot, f)\|_{p,w} = \left\| \sum_{k=1}^{n} \frac{k}{n+1} A_k(\cdot, f) \right\|_{p,w}$$

$$= \left\| \sum_{k=1}^{n} \lambda_k n \left(1 - \frac{\sin(k/n)}{k/n}\right) A_k(\cdot, f) \right\|_{p,w}$$

$$= n \left\| \sum_{k=1}^{\infty} \lambda_k \left(1 - \frac{\sin(k/n)}{k/n}\right) A_k(\cdot, f) \right\|_{p,w}$$

$$\leq c\, n \left\| \sum_{k=1}^{\infty} \left(1 - \frac{\sin(k/n)}{k/n}\right) A_k(\cdot, f) \right\|_{p,w} = c\, n \left\| f - f_{1/n} \right\|_{p,w}$$

$$\leq c\, n\, \Omega\left(\frac{1}{n}, f\right)_{p,w}.$$

This estimate together with (6.1.11) and Minkowski's inequality yields (6.1.9).

The set of all 2π−periodic continuous functions is dense in $L^p_w(\mathbb{T})$. Since the Abel–Poisson means of a continuous function converges uniformly to the initial function (see, for example, [177], Theorem 6.6) and since $v \in L^1(\mathbb{T})$, they converge in $L^p_v(\mathbb{T})$, too. Hence, in the case $p = q$, by Banach-Steinhaus theorem we obtain the convergence

$$\|U_r(\cdot, f) - f\|_{p,v} \to 0, \quad r \to 1$$

for $f \in L^p_w(\mathbb{T})$. If also $w = v$, then we can estimate the rate of this convergence as follows: □

Theorem 6.1.5. *Let $1 < p < \infty$, $w \in \mathcal{A}_p(\mathbb{T})$ and $f \in L^p_w(\mathbb{T})$. Then there exists a constant c, which does not depend on r and f, such that the estimate*

$$\|U_r(\cdot, f) - f\|_{p,w} \leq c\, \frac{1}{1-r}\, \Omega(1-r, f)_{p,w}$$

holds.

Proof. If the function $L^p_w(\mathbb{T})$ has the Fourier series (6.1.10), then the Fourier series of the means $U_r(x, f)$ is

$$U_r(x, f) \sim \frac{a_0}{2} + \sum_{k=1}^{\infty} r^k A_k(x, f).$$

Hence, we have that

$$f(x) - U_r(x, f) \sim \sum_{k=1}^{\infty} (1 - r^k) A_k(x, f).$$

Further, the sequence

$$\lambda_k = \frac{(1-r)\left(1 - r^k\right)}{1 - \frac{\sin k(1-r)}{k(1-r)}}, \quad k = 1, 2, \ldots,$$

satisfies the condition (6.1.12). Therefore, by taking into account a theorem by D. S. Kurtz (see [103], [104]) we find that,

$$\|U_r(\cdot,f)-f\|_{p,w}=\left\|\sum_{k=1}^{\infty}(1-r^k)A_k(\cdot,f)\right\|_{p,w}$$

$$=\frac{1}{1-r}\left\|\sum_{k=1}^{\infty}\lambda_k\left(1-\frac{\sin k(1-r)}{k(1-r)}\right)A_k(\cdot,f)\right\|_{p,w}$$

$$\leq c\,\frac{1}{1-r}\left\|\sum_{k=1}^{\infty}\left(1-\frac{\sin k(1-r)}{k(1-r)}\right)A_k(\cdot,f)\right\|_{p,w}=c\,\frac{1}{1-r}\|f-f_{1-r}\|_{p,w}$$

$$\leq c\,\frac{1}{1-r}\Omega(1-r,f)_{p,w}.$$

The proof is complete. □

Let us now discuss the two-dimensional case.

Let $\mathbb{T}^2=\mathbb{T}\times\mathbb{T}$ and w be a weight function on $\mathbb{T}^2$. We denote by $L^p_w(\mathbb{T}^2)$, $1\leq p<\infty$, the space of functions $f(x,y)$ which are 2π−periodic with respect to each variable, such that

$$\|f\|_{p,w}=\left(\int_{\mathbb{T}^2}|f(x,y)|^p\,w(x,y)\,dxdy\right)^{1/p}<\infty.$$

Let the function $f\in L^1(\mathbb{T}^2)$ have the Fourier series

$$f(x,y)\sim\sum_{m,n=0}^{\infty}\lambda_{mn}\left(a_{mn}\cos mx\cos ny+b_{mn}\sin mx\sin my\right.\tag{6.1.13}$$
$$\left.+c_{mn}\cos mx\sin ny+d_{mn}\sin mx\sin ny\right),$$

where

$$\lambda_{mn}=\begin{cases}\frac{1}{4}, & m=n=0,\\ \frac{1}{2}, & m=0,\ n>0\ \text{ or }\ m>0,\ n=0,\\ 1, & m>0,\ n>0.\end{cases}$$

Let also

$$\sigma^{(\alpha,\beta)}_{mn}(x,y,f)=\frac{\sum\limits_{i=0}^{m}\sum\limits_{j=0}^{n}A^{\alpha-1}_{m-i}A^{\beta-1}_{n-j}S_{ij}(x,y,f)}{A^{\alpha}_{m}A^{\beta}_{n}}\qquad(\alpha,\beta>0)$$

be the Cesàro means of the function f, where $S_{ij}(\cdot,\cdot,f)$ are the partial sums of the series (6.1.13), and

$$U_{r\rho}(x,y,f)=\frac{1}{(2\pi)^2}\int_{\mathbb{T}^2}f(t,s)\,P_r(x-t)\,P_\rho(y-s)\,dtds,\quad 0\leq r,\rho<1$$

be the Abel–Poisson means of f (see, for example [177, Vol. II, pp. 300–305]).

Definition 6.1.2. The pair (v,w) is said to belongs to the class $\mathcal{A}_{p,q}\left(\mathbb{T}^2,\mathbb{J}\right)$ if the condition

$$\sup_{J\in\mathbb{J}}\left(\frac{1}{|J|}\int\limits_J v\left(x,y\right)dxdy\right)^{1/q}\left(\frac{1}{|J|}\int\limits_J w^{1-p'}\left(x,y\right)dxdy\right)^{1/p'}<\infty \tag{6.1.14}$$

holds, where $\mathbb{J}$ denote the set of all rectangles with the sides parallel to the coordinate axes.

Now we give the two-dimensional analogues of Theorems 6.1.1 and 6.1.2.

Theorem 6.1.6. *Let $1<p\le q<\infty$. Then the condition $(v,w)\in\mathcal{A}_{p,q}\left(\mathbb{T}^2,\mathbb{J}\right)$ is necessary and sufficient for the validity of the inequality*

$$\left\|\sigma_{mn}^{(\alpha,\beta)}\left(\cdot,\cdot,f\right)\right\|_{q,v}\le c\;(mn)^{\frac{1}{p}-\frac{1}{q}}\left\|f\right\|_{p,w},\quad\alpha>0,\;\beta>0 \tag{6.1.15}$$

for every $f\in L^p_w\left(\mathbb{T}^2\right)$, where the constant c is independent of m, n and f.

Theorem 6.1.7. *Let $1<p\le q<\infty$. The inequality*

$$\left\|U_{r\rho}\left(\cdot,\cdot,f\right)\right\|_{q,v}\le c\left(1-r\right)^{\frac{1}{q}-\frac{1}{p}}\left(1-\rho\right)^{\frac{1}{q}-\frac{1}{p}}\left\|f\right\|_{p,w} \tag{6.1.16}$$

holds for arbitrary $f\in L^p_w\left(\mathbb{T}^2\right)$, where the constant c does not depend on r, ρ and f, if and only if $(v,w)\in\ \mathcal{A}_{p,q}\left(\mathbb{T}^2,\mathbb{J}\right)$.

To prove these theorems we need the boundedness of the generalized multiple Steklov means

$$f_{hk}^{\gamma}\left(x,y\right):=\frac{1}{(hk)^{1-\gamma}}\int\limits_{x-h}^{x+h}\int\limits_{y-k}^{y+k}\left|f\left(t,s\right)\right|dtds,\quad 0\le\gamma<1,\quad h,k>0.$$

We set $f_{hk}:=f_{hk}^{0}$.

Theorem 6.1.8. *Let $1<p\le q<\infty$, $(v,w)\in\mathcal{A}_{p,q}\left(\mathbb{T}^2,\mathbb{J}\right)$ and $\gamma=1/p-1/q$. Then the inequality*

$$\left\|f_{hk}^{\gamma}\right\|_{q,v}\le c\left\|f\right\|_{p,w} \tag{6.1.17}$$

holds for every $f\in L^p_w\left(\mathbb{T}^2\right)$.

Proof. Let $h\le\pi$ and $k\le\pi$. Let M and N be the least natural numbers for which $Mh\ge\pi$ and $Nk\ge\pi$. Then

$$\int\limits_{\mathbb{T}^2}[f_{hk}^{\gamma}(x,y)]^q v(x,y)dxdy\le\sum_{i=-M}^{M}\sum_{j=-N}^{N}\int\limits_{ih}^{(i+1)h}\int\limits_{jk}^{(j+1)k}(hk)^{-q(1-\gamma)}$$

$$\times \left[\int_{x-h}^{x+h} \int_{y-k}^{y+k} |f(t,\tau)| dt d\tau \right]^q v(x,y) dx dy$$

$$\leq \sum_{i=-M}^{M} \sum_{j=-N}^{N} \int_{ih}^{(i+1)h} \int_{jk}^{(j+1)k} (hk)^{-q(1-\gamma)}$$

$$\times \left[\int_{(i-1)h}^{(i+2)h} \int_{(j-1)k}^{(j+1)k} |f(t,\tau)| dt d\tau \right]^q v(x,y) dx dy.$$

By using Hölder's inequality we get that

$$\int_{\mathbb{T}^2} [f_{hk}^{\gamma}(x,y)]^q v(x,y) dx dy \leq \sum_{i=-M}^{M} \sum_{j=-N}^{N} \int_{ih}^{(i+1)h} \int_{jk}^{(j+1)k} (hk)^{-q(1-\gamma)}$$

$$\times \left[\int_{(i-1)h}^{(i+2)h} \int_{(j-1)k}^{(j+1)k} |f(t,\tau)|^p w(t,\tau) dt d\tau \right]^{q/p}$$

$$\times \left[\int_{(i-1)h}^{(i+2)h} \int_{(j-1)k}^{(j+1)k} w^{1-p'}(x,y) dx dy \right]^{q/p'} v(x,y) dx dy.$$

By using the condition $\mathcal{A}_p(\mathbb{T}^2, \mathbb{J})$ we find that

$$\int_{\mathbb{T}^2} [f_{hk}^{\gamma}(x,y)]^q v(x,y) dx dy \leq c \sum_{i=-M}^{M} \sum_{j=-N}^{N} \left(\int_{(i-1)h}^{(i+2)h} \int_{(j-1)k}^{(j+1)k} |f(t,\tau)|^p w(t,\tau) dt d\tau \right)^{q/p}.$$

Consequently,

$$\int_{\mathbb{T}^2} [f_{hk}^{\gamma}(x,y)]^q v(x,y) dx dy \leq c \|f\|_{p,w}^q.$$

The proof is complete. □

Proof of Theorem 6.1.6. We begin by proving the following inequality

$$\left| \sigma_{mn}^{(\alpha,\beta)}(x,y,f) \right| \leq c \int_{1/n}^{2\pi} \int_{1/m}^{2\pi} \frac{1}{m^{\alpha} n^{\beta}} h^{-1-\alpha} k^{-1-\beta} f_{hk}(x,y) \, dh dk,$$

where the constant does not depend on f, x, y, m and n.

If we reverse the order of integration on the right–hand side of this inequality, then by using the arguments similar to those used in the one-dimensional case we obtain that

$$I := \int_{x-\pi}^{x+\pi}\int_{y-\pi}^{y+\pi} |f(t,s)| \left[\int_{\max(|x-t|,1/m)}^{2\pi} \int_{\max(|y-s|,1/n)}^{2\pi} \frac{1}{m^\alpha n^\beta} h^{-2-\alpha} k^{-2-\beta} dh dk \right] dt ds$$

$$\geq c \int_{x+\pi}^{x-\pi}\int_{y-\pi}^{y+\pi} |f(t,s)| \frac{1}{m^\alpha n^\beta} \left[\max\left(|x-t|, \frac{1}{m}\right)\right]^{-1-\alpha} \left[\max\left(|y-s|, \frac{1}{n}\right)\right]^{-1-\beta} dt ds.$$

Applying the well–known estimates for the Cesàro kernel, the last inequality yields

$$I \geq c \int_{\mathbb{T}^2} |f(t,s)| K_m^\alpha (x-t) K_n^\beta (y-s) dt ds \geq c \left| \sigma_{mn}^{(\alpha,\beta)}(x,y,f) \right|.$$

Hence we have proved the desired estimate. Further, the obtained inequality yields

$$\left| \sigma_{mn}^{(\alpha,\beta)}(x,y,f) \right| \leq c \int_{1/n}^{2\pi}\int_{1/m}^{2\pi} \frac{1}{m^\alpha n^\beta} h^{-1-\alpha} k^{-1-\beta} (hk)^{-\gamma} f_{hk}^\gamma (x,y)\, dh dk.$$

By using this inequality, Minkowski's inequality and (6.1.17) we have that

$$\left\| \sigma_{mn}^{(\alpha,\beta)}(\cdot,\cdot,f) \right\|_{q,v} = \left(\int_{\mathbb{T}^2} \left| \sigma_{mn}^{(\alpha,\beta)}(x,y,f) \right|^q v(x,y)\, dx dy \right)^{1/q}$$

$$\leq c \left(\int_{\mathbb{T}^2} \left| \int_{1/n}^{2\pi}\int_{1/m}^{2\pi} \frac{1}{m^\alpha n^\beta} h^{-1-\alpha} k^{-1-\beta} (hk)^{-\gamma} f_{hk}^\gamma (x,y)\, dh dk \right|^q v(x,y)\, dx dy \right)^{1/q}$$

$$\leq c \frac{1}{m^\alpha n^\beta} \int_{1/n}^{2\pi}\int_{1/m}^{2\pi} \left(\int_{\mathbb{T}^2} \left| f_{hk}^\gamma (x,y) \right|^q v(x,y)\, dx dy \right)^{1/q} h^{-1-\alpha} k^{-1-\beta} (hk)^{-\gamma} dh dk$$

$$\begin{aligned} &\leq c \frac{1}{m^\alpha n^\beta} \left(\int_{1/n}^{2\pi}\int_{1/m}^{2\pi} h^{-1-\alpha} k^{-1-\beta} (hk)^{-\gamma} dh dk \right) \|f\|_{p,w} \\ &\leq c \frac{1}{m^\alpha n^\beta} \frac{1}{\alpha+\gamma} \frac{1}{\beta+\gamma} m^{\alpha+\gamma} n^{\beta+\gamma} \|f\|_{p,w} = c(\alpha,\beta,\gamma)(mn)^\gamma \|f\|_{p,w}. \end{aligned}$$

Now we prove that the inequality

$$\left\|\sigma_{mn}^{(\alpha,\beta)}\left(\cdot,\cdot,f\right)\right\|_{q,v}\leq c\left(mn\right)^{\frac{1}{p}-\frac{1}{q}}\left\|f\right\|_{p,w},\quad f\in L_{w}^{p}\left(\mathbb{T}^{2}\right),\ \alpha>0,\ \beta>0,$$

implies that $(v,w)\in\mathcal{A}_{p,q}\left(\mathbb{T}^{2},\mathbb{J}\right)$.

Let $J_0=I_1\times I_2$ with $|I_i|\leq\pi/4$, $i=1,2$, and let m, n be the greatest natural numbers with

$$\frac{\pi}{2\left(m+2\right)}\leq|I_1|\leq\frac{\pi}{2\left(m+1\right)},\quad\frac{\pi}{2\left(n+2\right)}\leq|I_2|\leq\frac{\pi}{2\left(n+1\right)}.$$

For $(x,y)\in J_0$ and $(t,s)\in J_0$, by using Abel transform we find that

$$K_m^{\alpha}\left(x-t\right)\geq\frac{c}{|I_1|}\quad\text{and}\quad K_n^{\beta}\left(y-s\right)\geq\frac{c}{|I_2|}.$$

Now, for m, n indicated above, substituting the function

$$f_0\left(x,y\right)=w^{1-p'}\left(x,y\right)\chi_{J_0}\left(x,y\right)$$

in (6.1.15), we obtain that

$$\left(\int\limits_{J_0}\left|\frac{1}{|J_0|}\int\limits_{J_0}w^{1-p'}\left(t,s\right)dtds\right|^{q}v\left(x,y\right)dxdy\right)^{1/q}$$

$$\leq c\left(mn\right)^{1/p-1/q}\left(\int\limits_{J_0}w^{1-p'}\left(x,y\right)dxdy\right)^{1/p}$$

$$\leq c\left(\frac{1}{|I_1|}\frac{1}{|I_2|}\right)^{1/p-1/q}\left(\int\limits_{J_0}w^{1-p'}\left(x,y\right)dxdy\right)^{1/p},$$

which implies that

$$\left(\frac{1}{|J_0|}\int\limits_{J_0}v\left(x,y\right)dxdy\right)^{1/q}\left(\frac{1}{|J_0|}\int\limits_{J_0}w^{1-p'}\left(x,y\right)dxdy\right)^{1/p'}\leq c.$$

Taking the supremum over all rectangles $J_0=I_1\times I_2$ with $|I_i|\leq\pi/4$, $i=1,2$, we get $(v,w)\in\mathcal{A}_{p,q}\left(\mathbb{T}^{2},\mathbb{J}\right)$.

The proof is complete. □

From Theorem 6.1.6 it follows that when $(v,w) \in \mathcal{A}_p(\mathbb{T}^2,\mathbb{J})$, we have that

$$\lim_{\substack{m\to\infty\\ n\to\infty}} \|\sigma_{mn}^{(\alpha,\beta)}(\cdot,\cdot,f) - f\|_{q,v} = 0$$

for arbitrary $f \in L^p_w(\mathbb{T}^2)$.

Indeed, let us consider the sequence of operators

$$U_{mn} : f \to \sigma_{mn}^{(\alpha,\beta)}(\cdot,\cdot,f).$$

It is evident that U_{mn} is linear and bounded for each (m,n) because

$$\int_{\mathbb{T}^2} v(x,y)dxdy < \infty, \qquad \int_{\mathbb{T}^2} w^{1-p'}(x,y)dxdy < \infty.$$

Then, since $(v,w) \in \mathcal{A}_p(\mathbb{T}^2,\mathbb{J})$ by Theorem 6.1.6 the sequence of operators norms

$$\{\|U_{mn}\|_{L^p_w \to L^p_v}\}_{m,n=1}^{\infty}$$

is bounded. On the other hand, the set of 2π-periodic functions which are continuous in the plane is dense in $L^p_w(\mathbb{T}^2)$. Then it is known that the Cesàro means of Lipschitz functions of two variables converges uniformly. Since $v \in L^1(\mathbb{T}^2)$ the last convergence we have by means of L^p_v norms as well. Applying the Banach–Steinhaus theorem we conclude that the norm convergence holds for arbitrary $f \in L^p_w(\mathbb{T}^2)$.

Proof of Theorem 6.1.7. Let $(v,w) \in \mathcal{A}_{p,q}(\mathbb{T})$ and $\gamma = 1/p - 1/q$. Let us prove that

$$|U_{r\rho}(x,y,f)| \le c \int_{1-\rho}^{2\pi}\int_{1-r}^{2\pi} (1-r)(1-\rho)\, h^{-2}k^{-2}\, (hk)^{-\gamma}\, f_{hk}^{\gamma}(x,y)\, dhdk. \qquad (6.1.18)$$

Indeed, we have that

$$\int_{1-\rho}^{2\pi}\int_{1-r}^{2\pi} (1-r)(1-\rho)\, h^{-2}k^{-2}\, (hk)^{-\gamma}\, f_{hk}^{\gamma}(x,y)\, dhdk$$

$$= \int_{1-\rho}^{2\pi}\int_{1-r}^{2\pi} (1-r)(1-\rho)\, h^{-2}k^{-2} f_{hk}(x,y)\, dhdk$$

$$= \int_{1-\rho}^{2\pi}\int_{1-r}^{2\pi} (1-r)(1-\rho)\, h^{-2}k^{-2} \left(\frac{1}{hk} \int_{x-h}^{x+h}\int_{y-k}^{y+k} |f(t,s)|\, dtds \right) dhdk$$

$$= \int_{x-\pi}^{x+\pi}\int_{y-\pi}^{y+\pi} |f(t,s)|(1-r)(1-\rho)\left(\int_{\max\{|x-t|,1-r\}}^{2\pi}\int_{\max\{|y-s|,1-\rho\}}^{2\pi} h^{-3}k^{-3}dhdk\right)dtds$$

$$= \int_{x-\pi}^{x+\pi}\int_{y-\pi}^{y+\pi} |f(t,s)|\left(\int_{\max\{|x-t|,1-r\}}^{2\pi}(1-r)h^{-3}dh\right)$$

$$\times\left(\int_{\max\{|y-s|,1-\rho\}}^{2\pi}(1-\rho)k^{-3}dk\right)dtds$$

$$\geq c\frac{1}{(2\pi)^2}\int_{x-\pi}^{x+\pi}\int_{y-\pi}^{y+\pi}|f(t,s)|P_r(x-t)P_\rho(y-s)\,dtds$$

$$\geq c\left|\frac{1}{(2\pi)^2}\int_{x-\pi}^{x+\pi}\int_{y-\pi}^{y+\pi}f(t,s)P_r(x-t)P_\rho(y-s)\,dtds\right| = c\,|U_{r\rho}(x,y,f)|\,.$$

Hence (6.1.18) holds.

By using the estimate (6.1.18), Minkowski's inequality and Theorem 6.1.8, we get that

$$\|U_{r\rho}(\cdot,\cdot,f)\|_{q,v} = \left(\int_{\mathbb{T}^2}|U_{r\rho}(x,y,f)|^q v(x,y)\,dxdy\right)^{1/q}$$

$$\leq c\left(\int_{\mathbb{T}^2}\left|\int_{1-\rho}^{2\pi}\int_{1-r}^{2\pi}(1-r)(1-\rho)h^{-2}k^{-2}(hk)^{-\gamma}f^{\gamma}_{hk}(x,y)\,dhdk\right|^q v(x,y)\,dxdy\right)^{1/q}$$

$$\leq c\int_{1-\rho}^{2\pi}\int_{1-r}^{2\pi}(1-r)(1-\rho)h^{-2-\gamma}k^{-2-\gamma}\left(\int_{\mathbb{T}^2}|f^{\gamma}_{hk}(x,y)|^q v(x,y)\,dxdy\right)^{1/q}dhdk$$

$$\leq c\left(\int_{1-\rho}^{2\pi}\int_{1-r}^{2\pi}(1-r)(1-\rho)h^{-2-\gamma}k^{-2-\gamma}dhdk\right)\|f\|_{p,w}$$

$$\leq c\,(1-r)^{-\gamma}(1-\rho)^{-\gamma}\,\|f\|_{p,w}\,.$$

Let us now prove that the estimate (6.1.16) implies the condition (6.1.14). For the rectangle $J_0 = I_1 \times I_2$, if we set

$$r_0 := \max\{1-|I_1|,0\},\quad \rho_0 := \max\{1-|I_2|,0\}$$

and

$$f_0(x,y) := w^{1-p'}(x,y)\,\chi_{J_0}(x,y),$$

we obtain that, for $(x,y) \in J_0$,

$$U_{r_0\rho_0}(x,y,f_0) \geq \frac{c}{|J_0|}\int_{J_0} w^{1-p'}(t,s)\,dtds,$$

where c is a constant independent of x, y, w, p and J_0. If we apply the inequality (6.1.16) to f_0, r_0 and ρ_0, and use the last inequality we find that

$$\left(\int_{J_0}\left|\frac{1}{|J_0|}\int_{J_0} w^{1-p'}(t,s)\,dtds\right|^q v(x,y)\,dxdy\right)^{1/q}$$

$$\leq c\left[(1-r_0)(1-\rho_0)\right]^{\frac{1}{q}-\frac{1}{p}}\left(\int_{J_0} w^{1-p'}(x,y)\,dxdy\right)^{1/p}$$

$$\leq c\left(|J_0|\right)^{\frac{1}{q}-\frac{1}{p}}\left(\int_{J_0} w^{1-p'}(x,y)\,dxdy\right)^{1/p},$$

and from this it follows that $(v,w) \in \mathcal{A}_{p,q}\left(\mathbb{T}^2,\mathbb{J}\right)$.

The proof is complete. □

Remark 6.1.1. We note that the analogous results are valid for n-dimensional Fourier series.

6.2 On the Means of Fourier Integrals

The Cesàro mean of a function $f \in L^1(\mathbb{R})$ can be represented as

$$S_R(x,f) = \frac{1}{\pi}\int_{-\infty}^{\infty} f(t)\,F_R(x-t)\,dt, \quad -\infty < x < \infty,\ R > 0,$$

where

$$F_R(x) = \frac{1-\cos Rx}{Rx^2}$$

is the Fejer kernel (see [61], p. 15). Let also

$$U_t(x,f) = \frac{1}{\pi}\int_{-\infty}^{\infty} f(y)\,P_t(x-y)\,dy, \quad t > 0,$$

be the Abel-Poisson mean of the function $f \in L^1(\mathbb{R})$, where

$$P_t(x) = \frac{t}{t^2 + x^2}$$

is the Poisson kernel.

For the kernel $F_R(x)$ it is easy to show that the inequalities

$$|F_R(x)| \leq \frac{R}{2}, \quad |F_R(x)| \leq \frac{2}{Rx^2}, \; x \neq 0, \tag{6.2.1}$$

hold.

Definition 6.2.1. Let $1 < p \leq q < \infty$. A pair of weight functions (v, w) is said to be of the class $\mathcal{A}_{p,q}(\mathbb{R})$ if the condition

$$\sup_I \left(\frac{1}{|I|} \int_I v(x)\, dx \right)^{1/q} \left(\frac{1}{|I|} \int_I w^{1-p'}(x)\, dx \right)^{1/p'} < \infty \tag{6.2.2}$$

holds, where $p' = p/(p-1)$ and the supremum is taken over all intervals with $I \subset \mathbb{R}$. In the case $p = q$ the class $\mathcal{A}_{p,q}(\mathbb{R})$ is denoted by $\mathcal{A}_p(\mathbb{R})$.

The following statements hold:

Theorem 6.2.1. *Let $1 < p \leq q < \infty$. Then the inequality*

$$\|S_R(\cdot, f)\|_{q,v} \leq c\, R^{\frac{1}{p} - \frac{1}{q}} \|f\|_{p,w} \tag{6.2.3}$$

holds for arbitrary $f \in L^p_w(\mathbb{R})$, where the constant c does not depend on R and f, if and only if $(v, w) \in \mathcal{A}_{p,q}(\mathbb{R})$.

Theorem 6.2.2. *Let $1 < p \leq q < \infty$. Then necessary and sufficient condition for the validity of the inequality*

$$\|U_t(\cdot, f)\|_{q,v} \leq c\, t^{\frac{1}{q} - \frac{1}{p}} \|f\|_{p,w} \tag{6.2.4}$$

for arbitrary $f \in L^p_w(\mathbb{R})$, where the constant c does not depend on t and f, is that $(v, w) \in \mathcal{A}_{p,q}(\mathbb{R})$.

In the case $p = q$ and $v = w$ we have the necessary and sufficient condition for the convergence of Abel–Poisson means in the norm of $L^p_w(\mathbb{R})$.

Theorem 6.2.3. *Let $1 < p < \infty$. Then the convergence*

$$\lim_{t \to 0} \|U_t(\cdot, f) - f\|_{p,w} = 0 \tag{6.2.5}$$

holds for every $f \in L^p_w(\mathbb{R})$ if and only if $w \in \mathcal{A}_p(\mathbb{R})$.

To prove these theorems we need the boundedness of the Steklov–type means

$$f_h^\gamma(x) = \frac{1}{h^{1-\gamma}} \int_{x-h}^{x+h} |f(t)|\, dt, \quad h > 0,\ 0 \le \gamma < 1.$$

In particular, we have the following statement

Theorem 6.2.4. *Let* $1 < p \le q < \infty$, $(v, w) \in \mathcal{A}_{p,q}(\mathbb{R})$ *and* $\gamma = 1/p - 1/q$. *Then the inequality*

$$\left\| f_h^\gamma \right\|_{q,v} \le c \left\| f \right\|_{p,w} \tag{6.2.6}$$

holds for every $f \in L_w^p(\mathbb{R})$.

The proof is similar to that of Theorem 6.1.3.

Proof of Theorem 6.2.1. Let $(v, w) \in A_{p,q}(\mathbb{R})$ *and* $\gamma = 1/p - 1/q$. Let us show that

$$|S_R(x, f)| \le c \int_{1/R}^{\infty} \frac{1}{R} h^{-2} h^{-\gamma} f_h^\gamma(x)\, dh. \tag{6.2.7}$$

By reversing the order of integration and taking into account (6.2.1), we have that

$$\int_{1/R}^{\infty} \frac{1}{R} h^{-2-\gamma} f_h^\gamma(x)\, dh = \int_{1/R}^{\infty} \frac{1}{R} h^{-2} \left(\frac{1}{h} \int_{x-h}^{x+h} |f(t)|\, dt \right) dh$$

$$= \int_{-\infty}^{\infty} |f(t)| \left(\frac{1}{R} \int_{\max\{|x-t|, 1/R\}}^{\infty} h^{-3} dh \right) dt = \int_{-\infty}^{\infty} |f(t)| \left(\frac{1}{2R} \left[\max\{|x-t|, 1/R\}\right]^{-2} \right) dt$$

$$\ge c \int_{-\infty}^{\infty} |f(t)|\, F_R(x-t)\, dt \ge c \left| \int_{-\infty}^{\infty} f(t)\, F_R(x-t)\, dt \right| = c\, |S_R(x, f)|,$$

and (6.2.7) follows from this. Hence, by (6.2.7), Minkowski's inequality and (6.2.6) we obtain that

$$\|S_R(\cdot, f)\|_{q,v} = \left(\int_{-\infty}^{\infty} |S_R(x, f)|^q\, v(x)\, dx \right)^{1/q}$$

$$\le c \left(\int_{-\infty}^{\infty} \left| \int_{1/R}^{\infty} \frac{1}{R} h^{-2} h^{-\gamma} f_h^\gamma(x)\, dh \right|^q v(x)\, dx \right)^{1/q}$$

$$\leq \int_{1/R}^{\infty}\left(\int_{-\infty}^{\infty}\left|f_h^{\gamma}(x)\right|^q v(x)\,dx\right)^{1/q}\frac{1}{R}h^{-2-\gamma}dh \leq c\left(\frac{1}{R}\int_{1/R}^{\infty}h^{-2-\gamma}dh\right)\|f\|_{p,w}$$

$$= cR^{\gamma}\|f\|_{p,w}\,.$$

Let us now prove that (6.2.3) implies the condition $(v,w)\in\mathcal{A}_{p,q}(\mathbb{R})$.

Let I_0 be an interval of $\mathbb{R}$. If we choose R_0 so that

$$\frac{2}{|I_0|}\leq R_0\leq\frac{\pi}{|I_0|},$$

then we get that

$$F_{R_0}(x-t)\geq\frac{2R_0}{\pi^2}\geq\frac{4}{\pi^2|I_0|}$$

for $x,t\in I_0$. Consider the function

$$f_0(x)=w^{1-p'}(x)\,\chi_{I_0}(x)\,.$$

If $f_0\notin L^p_w(\mathbb{R})$, then by the Muckenhoupt method used in [125] one can show that $(v,w)\in\ \mathcal{A}_{p,q}(\mathbb{R})$. Let $f_0\in L^p_w(\mathbb{R})$. If we put f_0 in (6.2.3) for R_0, then we obtain the inequality

$$\left(\int_{I_0}\left|\int_{I_0}f_0(t)\,F_{R_0}(x-t)\,dt\right|^q v(x)\,dx\right)^{1/q}\leq cR_0^{\frac{1}{p}-\frac{1}{q}}\left(\int_{I_0}w^{1-p'}(x)\,dx\right)^{1/p},$$

which yields

$$\left(\frac{1}{|I_0|}\int_{I_0}v(x)\,dx\right)^{1/q}\left(\frac{1}{|I_0|}\int_{I_0}w^{1-p'}(x)\,dx\right)^{1/p'}\leq c.$$

Hence $(v,w)\in\mathcal{A}_{p,q}(\mathbb{R})$. The proof is complete. □

Proof of Theorem 6.2.2. Let $(v,w)\in\mathcal{A}_{p,q}(\mathbb{R})$ *and* $\gamma=1/p-1/q$. It is known that (see [125])

$$|U_t(x,f)|\leq c\int_t^{\infty}th^{-2}f_h(x)\,dh.$$

This gives that

$$|U_t(x,f)|\leq c\int_t^{\infty}th^{-2}h^{-\gamma}f_h^{\gamma}(x)\,dh,$$

and from this inequality, Minkowski's inequality and Theorem 6.2.4 we obtain that

$$\|U_t(\cdot,f)\|_{q,v}=\left(\int_{-\infty}^{\infty}|f(t,x)|^q v(x)\,dx\right)^{1/q}$$

$$\leq c\left(\int_{-\infty}^{\infty}\left|\int_t^{\infty} th^{-2}h^{-\gamma}f_h^{\gamma}(x)\,dh\right|^q v(x)\,dx\right)^{1/q}$$

$$\leq c\int_t^{\infty} th^{-2-\gamma}\left(\int_{-\infty}^{\infty}|f_h^{\gamma}(x)|^q v(x)\,dx\right)^{1/q}dh\leq c\|f\|_{p,w}\int_t^{\infty} th^{-2-\gamma}dh$$

$$=ct^{-\gamma}\|f\|_{p,w}\,.$$

To prove the converse inequality let us assume that (6.2.4) holds, and take any interval I_0 of $\mathbb{R}$. Let us take $t_0=|I_0|$ and set

$$f_0(x):=w^{1-p'}(x)\chi_{I_0}(x)\,.$$

In the case $f_0\notin L^p_w(\mathbb{R})$, the necessity of the condition $(v,w)\in\mathcal{A}_{p,q}(\mathbb{R})$ can be proved as in [125]. Let $f_0\in L^p_w(\mathbb{R})$. For any $x\in I_0$, it yields

$$|U_{t_0}(x,f_0)|=\frac{1}{\pi}\left|\int_{-\infty}^{\infty}\frac{t_0f_0(y)\,dy}{t_0^2+(x-y)^2}\right|=\frac{1}{\pi}\int_{I_0}\frac{|I_0|\,w^{1-p'}(y)\,dy}{|I_0|^2+(x-y)^2}\geq\frac{1}{2\pi|I_0|}\int_{I_0}w^{1-p'}(y)\,dy.$$

If we apply (6.2.4) to the function f_0 and consider the last inequality, then it follows that

$$\left(\int_{I_0}\left|\frac{1}{|I_0|}\int_{I_0}w^{1-p'}(y)\,dy\right|^q v(x)\,dx\right)^{1/q}\leq ct_0^{\frac{1}{q}-\frac{1}{p}}\left(\int_{I_0}w^{1-p'}(x)\,dx\right)^{1/p}.$$

This yields that

$$\left(\frac{1}{|I|}\int_I v(x)\,dx\right)^{1/q}\left(\frac{1}{|I|}\int_I w^{1-p'}(x)\,dx\right)^{1/p'}\leq c,$$

which implies that $(v,w)\in\mathcal{A}_{p,q}(\mathbb{R})$.

The proof is complete. □

Proof of Theorem 6.2.3. Let $w \in \mathcal{A}_p(\mathbb{R})$. It is known that w satisfies

$$\int_{-\infty}^{\infty} \frac{w^{1-p'}(x)}{(1+|x|)^{p'}} dx < \infty,$$

and, hence,

$$\int_{-\infty}^{\infty} \frac{|f(x)|}{1+|x|} dx < \infty$$

for $f \in L^p_w(\mathbb{R})$ (see [124]). Consequently,

$$\int_{-\infty}^{\infty} \frac{|f(x)|}{1+x^2} dx < \infty, \tag{6.2.8}$$

and for such functions the relation

$$\lim_{t\to\infty} U_t(x,f) = f(x)$$

holds for almost all $x \in \mathbb{R}$ (see, for example [98], pp. 109–110]). Let $\mathcal{M}$ denote the Hardy-Littlewood maximal operator. Since the condition (6.2.8) is satisfied, it follows from Theorem 4.2 in [51] (see p. 22) that

$$\sup_{t>0} |U_t(x,f)| \le c\mathcal{M}f(x), \quad x \in \mathbb{R}, \tag{6.2.9}$$

where the constant c does not depend on f. On the other hand, by considering (6.2.9) we obtain that

$$|U_t(x,f) - f(x)|^p \le 2^{p-1}\left(|U_t(x,f)|^p + |f(x)|^p\right)$$

$$\le 2^{p-1}\left(\left(\sup_{t>0}|U_t(x,f)|\right)^p + |f(x)|^p\right) \le c2^{p-1}\left((\mathcal{M}f(x))^p + |f(x)|^p\right).$$

Since $w \in \mathcal{A}_p(\mathbb{R})$ it follows that $\mathcal{M}$ is bounded in $L^p_w(\mathbb{R})$ (see [124]). Hence the right–hand side of the preceding inequality belongs to $L^p_w(\mathbb{R})$. Now, by Lebesgue's dominated convergence theorem we conclude that

$$\lim_{t\to\infty} \int_{-\infty}^{\infty} |U_t(x,f) - f(x)|^p w(x)\, dx = 0.$$

Now let $\lim_{t\to 0} \|U_t(\cdot,f) - f\|_{p,w} = 0$ for every $f \in L^p_w(\mathbb{R})$. From this and Banach-Steinhauss theorem it follows that the family of operator norms $\{\|U_t\|\}$ is bounded. Thus, by Theorem 6.2.2 in the special case $p = q$ and $v = w$ we conclude that $w \in \mathcal{A}_p(\mathbb{R})$.

The proof is complete. □

In the sequel, the similar problems for multidimensional Fourier integrals will be discussed. Let $R = (R_1, \dots, R_n)$ and $\tau = (\tau_1, \dots, \tau_n)$, $R_j, \tau_j > 0$ $(j = 1, \dots, n)$. Further, let $x = (x_1, \dots, x_n)$ and $y = (y_1, \dots, y_n)$ be points in $\mathbb{R}^n$.

We consider n-dimensional Cesàro and iterated Abel-Poisson means of a function $f \in L^1(\mathbb{R}^n)$ given by the following formulae:

$$S_R(x, f) = \frac{1}{\pi^n} \int\limits_{\mathbb{R}^n} f(y) \prod_{j=1}^{n} F_{R_j}(x_j - y_j) dy$$

and

$$U_\tau(x, f) = \frac{1}{\pi^n} \int\limits_{\mathbb{R}^n} f(y) \prod_{j=1}^{n} P_{\tau_j}(x_j - y_j) dy,$$

respectively.

Some properties of the iterated Abel-Poisson means can be found in [162] (see pp. 69–70).

We denote by $\mathbb{J}$ the set of all rectangles of $\mathbb{R}^n$ with sides parallel to the coordinate axis.

Definition 6.2.2. Let $1 < p \leq q < \infty$. A pair of weights (v, w) is said to be of class $\mathcal{A}_{p,q}(\mathbb{R}^n, \mathbb{J})$ if the condition

$$\sup_{J \in \mathbb{J}} \left(\frac{1}{|J|} \int\limits_J v(y) dy \right)^{1/q} \left(\frac{1}{|J|} \int\limits_J w^{1-p'}(y) dy \right)^{1/p'} < \infty$$

holds.

We have the following results:

Theorem 6.2.5. *Let $1 < p \leq q < \infty$. Then the inequality*

$$\|S_R(\cdot, f)\|_{q,v} \leq c \left(\prod_{j=1}^{n} R_j \right)^{\frac{1}{p} - \frac{1}{q}} \|f\|_{p,w}, \quad f \in L^p_w(\mathbb{R}^n), \tag{6.2.10}$$

holds if and only if $(v, w) \in \mathcal{A}_{p,q}(\mathbb{R}^n, \mathbb{J})$.

Theorem 6.2.6. *Let $1 < p \leq q < \infty$. The condition $(v, w) \in \mathcal{A}_{p,q}(\mathbb{R}^n, \mathbb{J})$ is necessary and sufficient for the validity of the inequality*

$$\|U_\tau(\cdot, f)\|_{q,v} \leq c \left(\prod_{j=1}^{n} \tau_j \right)^{\frac{1}{q} - \frac{1}{p}} \|f\|_{p,w}. \tag{6.2.11}$$

The proofs of these theorems rely on the uniform norm boundedness of n-dimensional generalized Steklov means

$$f_h^\gamma(x) = \left(\prod_{j=1}^n h_j\right)^{\gamma-1} \int\limits_{x_j-h_j}^{x_j+h_j} \cdots \int\limits_{x_n-h_n}^{x_n+h_n} |f(y)|dy, \ \ 0 \le \gamma < 1,$$

where $h = (h_1, \dots, h_n)$, $h_j > 0$ (j=1,...,n).

Theorem 6.2.7. *Let $1 < p \le q < \infty$ and $(v, w) \in \mathcal{A}_{p,q}(\mathbb{R}^n, \mathbb{J})$. Then the inequality*

$$\|f_h^\gamma\|_{q,v} \le c\|f\|_{p,w}, \quad f \in L_w^p(\mathbb{R}^n), \tag{6.2.12}$$

holds for $\gamma = \frac{1}{p} - \frac{1}{q}$ and where the positive constant c does not depend on f and h.

Proof. From the definition of f_h^γ and Hölder's inequality we get that

$$\|f_h^\gamma\|_{q,v}^q = \int\limits_{\mathbb{R}^n} \left(\prod_{j=1}^n h_j\right)^{q(\gamma-1)} \left(\int\limits_{x_1-h_1}^{x_1+h_1} \cdots \int\limits_{x_n-h_n}^{x_n+h_n} |f(y)|dy\right)^q v(x)dx$$

$$= \sum_{i_1,\dots,i_n=-\infty}^{\infty} \int\limits_{i_1h_1}^{(i_1+1)h_1} \cdots \int\limits_{i_nh_n}^{(i_n+1)h_n} \left(\prod_{j=1}^n h_j\right)^{q(\gamma-1)} \left(\int\limits_{x_1-h_1}^{x_1+h_1} \cdots \int\limits_{x_n-h_n}^{x_n+h_n} |f(y)|dy\right)^q v(x)dx$$

$$\le \sum_{i_1,\dots,i_n} \int\limits_{i_1h_1}^{(i_1+1)h_1} \cdots \int\limits_{i_nh_n}^{(i_n+1)h_n} \left(\prod_{j=1}^n h_j\right)^{q(\gamma-1)} \left(\int\limits_{(i_1-1)h_1}^{(i_1+2)h_1} \cdots \int\limits_{(i_n-1)h_n}^{(i_n+2)h_n} |f(y)|dy\right)^q v(x)dx$$

$$\le \sum_{i_1,\dots,i_n} \int\limits_{i_1h_1}^{(i_1+1)h_1} \cdots \int\limits_{i_nh_n}^{(i_n+1)h_n} \left(\prod_{j=1}^n h_j\right)^{q(\gamma-1)} \left(\int\limits_{(i_1-1)h_1}^{(i_1+2)h_1} \cdots \int\limits_{(i_n-1)h_n}^{(i_n+2)h_n} |f(y)|^p w(y)dy\right)^{q/p}$$

$$\times \left(\int\limits_{(i_1-1)h_1}^{(i_1+2)h_1} \cdots \int\limits_{(i_n-h_n)}^{(i_n+2)h_n} w^{1-p'}(y)dy\right)^{q/p'} v(x)dx$$

$$= \sum_{i_1,\dots,i_n} \left(\left(\prod_{j=1}^n h_j\right)^{-1} \int\limits_{i_1h_1}^{(i_1+1)h_1} \cdots \int\limits_{i_nh_n}^{(i_n+h_n)h_n} v(y)dy\right)$$

$$\times\left(\left(\prod_{j=1}^{n}h_j\right)^{-1}\int\limits_{(i_1-1)h_1}^{(i_1+2)h_1}\cdots\int\limits_{(i_n-1)h_n}^{(i_n+2)h_n}w^{1-p'}(y)dy\right)^{q/p'}$$

$$\times\left(\int\limits_{(i_1-1)h_1}^{(i_1+2)h_1}\cdots\int\limits_{(i_n-1)h_n}^{(i_n+2)h_n}|f(y)|^p w(y)dy\right)^{q/p}.$$

Considering the condition $(v,w)\in\mathcal{A}_{p,q}(\mathbb{R}^n,\mathbb{J})$ and using Proposition 5.1.3 of [88] we conclude that

$$\|f_h^\gamma\|_{q,v}^q\le c\sum_{i_1,\dots,i_n}\left(\int\limits_{(i_1-1)h_1}^{(i_1+2)h_1}\cdots\int\limits_{(i_n-1)h_n}^{(i_n+2)h_n}|f(y)|^p dy\right)^{q/p}\le c\|f\|_{p,w}^q.$$

The proof is complete. □

Proof of Theorem 6.2.1. Let $\gamma=1/p-1/q$ and let $(v,w)\in\mathcal{A}_{p,q}(\mathbb{R}^n,\mathbb{J})$.

We have that

$$\int\limits_{1/R_1}^{\infty}\cdots\int\limits_{1/R_n}^{\infty}\left(\prod_{j=1}^{n}R_j\right)^{-1}\prod_{j=1}^{n}h_j^{-2-\gamma}f_h^\gamma(x)dh$$

$$=\int\limits_{1/R_1}^{\infty}\cdots\int\limits_{1/R_n}^{\infty}\left(\prod_{j=1}^{n}R_j\right)^{-1}\prod_{j=1}^{n}h_j^{-2}\left(\prod_{j=1}^{n}h_j^{-1}\int\limits_{x_1-h_1}^{x_1+h_1}\cdots\int\limits_{x_n-h_n}^{x_n+h_n}|f(y)|dy\right)dh$$

$$=\int\limits_{\mathbb{R}^n}|f(y)|\left(\int\limits_{\max\{|x_1-y_1|,1/R_1\}}\cdots\int\limits_{\max\{|x_n-y_n|,1/R_n\}}\left(\prod_{j=1}^{n}R_j\right)^{-1}\left(\prod_{j=1}^{n}h_j^{-3}\right)dh\right)$$

$$=\int\limits_{\mathbb{R}^n}|f(y)|\frac{1}{2R_1}\left(\max\left\{|x_1-y_1|,\frac{1}{R_1}\right\}\right)^{-2}\cdots\frac{1}{2R_n}\left(\max\left\{|x_n-y_n|,\frac{1}{R_n}\right\}\right)^{-2}dy$$

$$\ge c\int\limits_{\mathbb{R}^n}|f(y)|\prod_{j=1}^{n}F_{R_j}(x_j-y_j)dy\ge c|S_R(x,f)|.$$

This inequality gives that

$$|S_R(x,f)|\le c\int\limits_{1/R_1}^{\infty}\cdots\int\limits_{1/R_n}^{\infty}\left(\prod_{j=1}^{n}R_j\right)^{-1}\left(\prod_{j=1}^{n}h_j\right)^{-2-\gamma}f_h^\gamma(x)dx \qquad (6.2.13)$$

By (6.2.13), Minkowski's inequality and (6.2.12) we get that

$$\|S_R(\cdot, f)\|_{q,v}$$

$$\leq c\left(\int_{\mathbb{R}^n}\left|\int_{1/R_1}\cdots\int_{1/R_n}\left(\prod_{j=1}^n R_j\right)^{-1}\left(\prod_{j=1}^n h_j\right)^{-2-\gamma} f_h(x)dh\right|^q v(x)dx\right)^{1/q}$$

$$\leq c\int_{1/R_1}^{\infty}\cdots\int_{1/R_n}^{\infty}\left(\prod_{j=1}^n R_j\right)^{-1}\left(\prod_{j=1}^n h_j\right)^{-2-\gamma}\left(\int_{\mathbb{R}^n}|f_h^{\gamma}(x)|^q v(x)dx\right)^{1/q} dh$$

$$\leq c\|f\|_{p,w}\left(\int_{1/R_1}^{\infty}\cdots\int_{1/R_n}^{\infty}\left(\prod_{j=1}^n R_j\right)^{-1}\left(\prod_{j=1}^n h_j\right)^{-2-\gamma} dh\right) = c\left(\prod_{j=1}^n R_j\right)^{\gamma}\|f\|_{p,w}.$$

To prove the converse inequality let us take $J_0 = I_1 \times \cdots \times I_n \in \mathbb{J}$.

Choosing R_j $(j = 1, \ldots, n)$ so that

$$\frac{2}{|I_j|} \leq R_j \leq \frac{\pi}{|I_j|},$$

we have that

$$F_{R_j}(x_j - y_j) \geq \frac{2R_j}{\pi^2} \geq \frac{4}{\pi^2|I_j|}$$

for $x_j, y_j \in I_j$, $j = 1, \ldots, n$. If we put the function

$$f_0(x) := w^{1-p'}(x)\chi_{J_0}(x)$$

in (6.2.10), then we obtain that

$$\|S_R(\cdot, f_0)\|_{q,v} \leq c\left(\prod_{j=1}^n R_j\right)^{1/p-1/v}\|f_0\|_{p,w}$$

for R_j $(j = 1, \ldots, n)$. Taking into account the definition of f_0 and the above–derived inequalities for $F_{R_j}(x_j - y_j)$ we find that

$$\left(\frac{1}{|J_0|}\right)^{1-1/p+1/q}\left(\int_{J_0} w^{1-p'}(y)dy\right)^{1-1/p}\left(\int_{J_0} v(y)dy\right)^{1/q} \leq c$$

with a positive constant c independent of J_0, which means that $(v, w) \in \mathcal{A}_{p,q}(\mathbb{R}^n, \mathbb{J})$.

The proof is complete. □

Theorem 6.2.6 can be proved in a similar way as Theorem 6.2.5 was proved; we therefore omit the proof.

6.3 Bernstein Inequalities in the Two–weighted Setting

The Bernstein inequality which is very important in approximation theory, states that for any trigonometric polynomial $T_n(x)$ of degree $\le n$, it yields that

$$\left\|T_n'\right\|_p \le c\, n \left\|T_n\right\|_p, \quad 1 \le p \le \infty.$$

By virtue of the inequality (6.2.3) we can prove the following generalization of the Bernstein's inequality:

Theorem 6.3.1. *Let $1 < p \le q < \infty$, $(v, w) \in \mathcal{A}_{p,q}(\mathbb{T})$ and $T_n(x)$ be a trigonometric polynomial of degree at most n. Then the inequality*

$$\left\|T_n'\right\|_{q,v} \le c\, n^{1+\frac{1}{p}-\frac{1}{q}} \left\|T_n\right\|_{p,w} \tag{6.3.1}$$

holds, where the constant c does not depend on n and T_n. Moreover, for the conjugate trigonometric polynomial $\widetilde{T}_n$, we have that

$$\left\|\left(\widetilde{T}_n\right)'\right\|_{q,v} \le c\, n^{1+\frac{1}{p}-\frac{1}{q}} \left\|T_n\right\|_{p,w}. \tag{6.3.2}$$

Proof. We have the representation

$$T_n(x) = \frac{1}{\pi}\int_{\mathbb{T}} T_n(t)\, D_n(t-x)\, dt, \quad x \in \mathbb{T},$$

where

$$D_n(t) = \frac{1}{2} + \sum_{k=1}^{n} \cos kt$$

is the Dirichlet kernel of order n. By differentiation, we obtain that

$$\begin{aligned}
T_n'(x) &= -\frac{1}{\pi}\int_{\mathbb{T}} T_n(t)\, D_n'(t-x)\, dt = -\frac{1}{\pi}\int_{\mathbb{T}} T_n(x+t)\, D_n'(t)\, dt \\
&= \frac{1}{\pi}\int_{\mathbb{T}} T_n(x+t) \left\{\sum_{k=1}^{n} k \sin kt\right\} dt \\
&= \frac{1}{\pi}\int_{\mathbb{T}} T_n(x+t) \left\{\sum_{k=1}^{n} k \sin kt + \sum_{k=1}^{n-1} k \sin(2n-k)\, t\right\} dt \\
&= \frac{1}{\pi}\int_{\mathbb{T}} T_n(x+t)\, 2n \sin nt \left\{\frac{1}{2} + \sum_{k=1}^{n-1} \frac{n-k}{n} \cos kt\right\} dt \\
&= 2n\frac{1}{\pi}\int_{\mathbb{T}} T_n(x+t) \sin nt K_{n-1}(t)\, dt,
\end{aligned}$$

where K_{n-1} is the Fejer's kernel of order $n-1$. By taking absolute values, we get that (see [177], Vol I, p. 85)

$$\left|T_n'(x)\right| \le 2n\frac{1}{\pi}\int_{\mathbb{T}} |T_n(x+t)|\, K_{n-1}(t)\, dt = 2n\sigma_{n-1}(x, |T_n|)\,.$$

Hence, by (6.1.5), we find that

$$\begin{aligned}\left\|T_n'\right\|_{q,v} &= \left(\int_{\mathbb{T}} \left|T_n'(x)\right|^q v(x)\, dx\right)^{1/q} \le 2\, n\left(\int_{\mathbb{T}} \left|\sigma_{n-1}(x, |T_n|)\right|^q v(x)\, dx\right)^{1/q} \\ &= 2\, n\left\|\sigma_{n-1}(\cdot, |T_n|)\right\|_{q,v} \le c\, n\,(n-1)^{\frac{1}{p}-\frac{1}{q}}\left\|T_n\right\|_{p,w},\end{aligned}$$

and inequality (6.3.1) follows.

For the conjugate of T_n we have that

$$\widetilde{T}_n(x) = \frac{1}{\pi}\int_{\mathbb{T}} T_n(t)\, \widetilde{D}_n(t-x)\, dt,$$

where

$$\widetilde{D}_n(t) = \sum_{k=1}^{n} \sin kt$$

is the conjugate Dirichlet kernel. By differentiation we get that

$$\left(\widetilde{T}_n\right)'(x) = 2\, n\frac{1}{\pi}\int_{\mathbb{T}} T_n(x+t)\cos nt K_{n-1}(t)\, dt.$$

Hence

$$\left|\left(\widetilde{T}_n\right)'(x)\right| \le 2n\sigma_{n-1}(x, |T_n|)\,.$$

Thus estimate (6.1.5) yields (6.3.2).

The proof is complete. □

The following extended Bernstein inequality for trigonometric polynomials of two variables is also valid:

Theorem 6.3.2. *Let $1 < p \le q < \infty$ and $(v, w) \in \mathcal{A}_{p,q}\left(\mathbb{T}^2, \mathbb{J}\right)$. Then for every trigonometric polynomial $T_{mn}(x, y)$ of degree $\le m$ with respect to x and of degree $\le n$ with respect to y, the inequality*

$$\left\|\frac{\partial^2 T_{mn}}{\partial x \partial y}\right\|_{q,v} \le c\,(mn)^{1+\frac{1}{p}-\frac{1}{q}}\left\|T_{mn}\right\|_{p,w} \tag{6.3.3}$$

holds, where the constant c does not depend on m,n and T_{mn}.

Proof. It is known that (see [177, Vol II, pp. 302–303]) the representations

$$\sigma_{mn}(x,y,f) = \frac{1}{\pi^2}\int\limits_{\mathbb{T}^2} f(x+t,y+s)\,K_m(t)\,K_n(s)\,dtds, \quad f\in L^1\left(\mathbb{T}^2\right),$$

and

$$T_{mn}(x,y) = \frac{1}{\pi^2}\int\limits_{\mathbb{T}^2} T_{mn}(t,s)\,D_m(t-x)\,D_n(s-y)\,dtds$$

are valid. If we take the partial derivatives of $T_{mn}(x,y)$ with respect to x and y, from the last relation, we get that

$$\frac{\partial^2 T_{mn}(x,y)}{\partial x\partial y} = \frac{1}{\pi^2}\int\limits_{\mathbb{T}^2} T_{mn}(t,s)\,D'_m(t-x)\,D'_n(s-y)\,dtds.$$

By using the process used in the proof of Theorem 6.3.1 we have that

$$\frac{\partial^2 T_{mn}(x,y)}{\partial x\partial y} = \frac{2m2n}{\pi^2}\int\limits_{\mathbb{T}^2} T_{mn}(x+t,y+s)\sin mt\sin ns K_{m-1}(t)\,K_{n-1}(s)\,dtds.$$

Consequently,

$$\left|\frac{\partial^2 T_{mn}(x,y)}{\partial x\partial y}\right| \le 4mn\sigma_{(m-1)(n-1)}\left(x,y,|T_{mn}|\right).$$

Taking the norms in the last inequality and using (6.1.15) we obtain that

$$\left\|\frac{\partial^2 T_{mn}}{\partial x\partial y}\right\|_{q,v} \le cmn\,(m-1)^{\frac{1}{p}-\frac{1}{q}}\,(n-1)^{\frac{1}{p}-\frac{1}{q}}\,\|T_{mn}\|_{p,w},$$

which completes the proof of the theorem. □

If $G_\sigma(x)$ is an integral function of degree $\le\sigma$ which belongs to $L^p(\mathbb{R})$, $1\le p\le\infty$, then the Bernstein inequality

$$\left\|G'_\sigma\right\|_p \le \sigma\left\|G_\sigma\right\|_p$$

holds (see [170, p. 217]). A similar inequality is valid for integral functions of several variables (see [170, p. 253]). In this section we give also some generalizations of Bernstein's inequality for integral functions of one and two variables in the two–weighted setting.

Theorem 6.3.3. *Let $1<p\le q<\infty$, $(v,w)\in\mathcal{A}_{p,q}(\mathbb{R})$, and let $G_\sigma(x)$ be an integral function of degree $\le\sigma$, bounded on the whole real axis which belongs to $L^p_w(\mathbb{R})$. Then the inequality*

$$\left\|G'_\sigma\right\|_{q,v} \le c\,\sigma^{1+\frac{1}{p}-\frac{1}{q}}\left\|G_\sigma\right\|_{p,w} \tag{6.3.4}$$

holds.

Proof. It is known that for the function $G'_\sigma(x)$ the representation

$$G'_\sigma(x) = c\sigma^2 \int_{-\infty}^{\infty} G_\sigma(x+t)\, h_\sigma(t) \left(\frac{\sin \sigma t}{\sigma t}\right)^2 dt \tag{6.3.5}$$

holds, where $|h_\sigma(t)| \le 1$ (see [170, p. 205]). Since

$$\left(\frac{\sin \sigma t}{\sigma t}\right)^2 = \frac{1-\cos 2\sigma t}{2(\sigma t)^2}$$

we have that

$$G'_\sigma(x) = c\sigma \int_{-\infty}^{\infty} G_\sigma(x+t)\, h_\sigma(t)\, F_{2\sigma}(t)\, dt.$$

If we take the absolute values and consider the fact $|h_\sigma(t)| \le 1$ we obtain that

$$|G'_\sigma(x)| \le c\sigma \int_{-\infty}^{\infty} |G_\sigma(x+t)|\, F_{2\sigma}(t)\, dt = c\,\sigma S_{2\sigma}(x, |G_\sigma|).$$

Hence, by inequality (6.2.3) we get that

$$\begin{aligned} \|G'_\sigma\|_{q,v} &= \left(\int_{-\infty}^{\infty} |G'_\sigma(x)|^q v(x)\, dx\right)^{1/q} \le c\sigma \left(\int_{-\infty}^{\infty} |S_{2\sigma}(x, |G_\sigma|)|^q v(x)\, dx\right)^{1/q} \\ &= c\sigma \|S_{2\sigma}(\cdot, |G_\sigma|)\|_{q,v} \le c\,\sigma\,(2\sigma)^{\frac{1}{p}-\frac{1}{q}} \|G_\sigma\|_{p,w} = c\,\sigma^{1+\frac{1}{p}-\frac{1}{q}} \|G_\sigma\|_{p,w}. \end{aligned}$$

The proof is complete. □

Theorem 6.3.4. *Let $1 < p \le q < \infty$, $(v,w) \in \mathcal{A}_{p,q}(\mathbb{R}^n, \mathbb{J})$, and let $G_{\sigma_1,\dots,\sigma_n}$ be a bounded integral function of degree $\le \sigma_i$ with respect to x_j for arbitrary $j = 1,\dots,n$. Then the inequality*

$$\left\|\frac{\partial^n G_{\sigma_1,\dots,\sigma_n}}{\partial x_1 \dots \partial x_n}\right\|_{q,v} \le c \left(\prod_{j=1}^{n} \sigma_j\right)^{1+1/p-1/q} \|G_{\sigma_1,\dots,\sigma_n}\|_{p,w} \tag{6.3.6}$$

holds with a positive constant c independent of $G_{\sigma_1,\dots,\sigma_n}$.

Proof. It is clear that the function $\frac{\partial^k G_{\sigma_1,\dots,\sigma_n}(x_1,\dots,x_n)}{\partial x_1,\dots,\partial x_{k-1}}$ is an integral function of degree $\le \sigma_k$ $(k = 1,\dots,n)$ and, thus, according to (6.3.5), we obtain that

$$\frac{\partial^n G_{\sigma_1,\dots,\sigma_n}(x_1,\dots,x_n)}{\partial x_1,\dots,\partial x_n}$$

$$= c\left(\prod_{j=1}^{n}\sigma_j\right)^n \int\limits_{\mathbb{R}^n} G_{\sigma_1,\dots,\sigma_n}(x_1+y_1,\dots,x_n+y_n)\prod_{j=1}^{n}h_{\sigma_j}(y_j)\prod_{j=1}^{n}\left(\frac{\sin\sigma_j y_j}{\sigma_j y_j}\right)^2 dy,$$

where $\left|h_{\sigma_j}(y_j)\right| \le 1$. Hence,

$$\frac{\partial^n G_{\sigma_1,\dots,\sigma_n}(x_1,\dots,x_n)}{\partial x_1,\dots,\partial x_n}$$

$$= c\prod_{j=1}^{n}\sigma_j \int\limits_{\mathbb{R}^n} G_{\sigma_1,\dots,\sigma_n}(x_1+y_1,\dots,x_n+y_n)\prod_{j=1}^{n}h_{\sigma_j}(y_j)\prod_{j=1}^{n}F_{2\sigma_j}(y_j).$$

From this equality we get that

$$\left|\frac{G_{\sigma_1,\dots,\sigma_n}}{\partial x_1,\dots,\partial x_n}\right| \le c\prod_{j=1}^{n}\sigma_j \int\limits_{\mathbb{R}^n} \left|G_{\sigma_1,\dots,\sigma_n}(x_1+y_1,\dots,x_n+y_n)\right|\prod_{j=1}^{n}F_{2\sigma_j}(y_j)dy$$

$$= c\prod_{j=1}^{n}\sigma_j S_{2\sigma_1\dots 2\sigma_n}\left(x,\left|G_{\sigma_1\dots\sigma_n}\right|\right).$$

This inequality and (6.2.10) yield (6.3.6) and, consequently, the proof is complete. □

6.4 Notes and Comments on Chapter 6

One-weighted boundedness criteria for the Abel-Poisson means of univariable trigonometric Fourier series were established in [149] and [124]. One-weighted estimates for norms of some linear summability operators generated by multiple trigonometric Fourier series have been obtained in [129].

In the two–weighted setting, boundedness criteria in $L^p_w(\mathbb{T})$ $(1 < p < \infty)$ for the Abel–Poisson means were established in [125], and Cesàro and Abel-Poisson means norm estimates and convergence were given in [62]. Moreover, in [63] and [64] the same problem was studied in the more general case when we have different weights and exponents of spaces on both sides of the norm inequalities for the above mentioned Fourier operators. In these papers some applications to the extension of Bernstein's inequality for derivatives of trigonometric polynomials and integral functions of finite order were also given in the two–weighted setting. We notice that in [83] the similar problem was studied from the point of view of rate of summability in unweighted Lebesgue spaces $L^p(1 < p < \infty)$ for various methods of the linear summability.

Appendix: Multidimensional Lorentz Spaces

The approach presented here was initiated in the PhD thesis [9] by Sorina Barza. Some further developments can be found in the papers [8], [12] and [18]. For another (equivalent definition) of the two-dimensional rearrangement (see Remark below) we refer also to the paper [24] by A.P. Blozinski. For simplicity we are going to reduce our definitions to the two-dimensional case, although there are natural extensions to higher dimensions too (see the references given below). Our approach is to give a geometric definition of the rearrangement of a measurable set of $\mathbb{R}^2$ and extend also it to rearrange functions by looking at the level sets and the use of the "layer-cake formula".

Definition 1. Let $E \subset \mathbb{R}^2$ and $\varphi_E(x) = |\{y \in \mathbb{R} : (x,y) \in E\}|, x \in \mathbb{R}$. Let the function φ_E^*, defined by

$$\varphi_E^*(s) := \inf\{l : |\{x \in \mathbb{R} : \varphi_E(x) > l\}| \leq s \quad (s \geq 0)$$

be the usual decreasing rearrangement of φ_E. Then, the two-dimensional decreasing rearrangement of E is the set

$$E^* := \{(s,t) \in \mathbb{R}_+^2 : 0 < t < \varphi_E^*(s)\}.$$

Definition 2. The two-dimensional decreasing rearrangement f_2^* of a function f on $\mathbb{R}^2$ is given by

$$f_2^*(s,t) = \int_0^\infty \chi_{\{|f|>\tau\}^*}(s,t)d\tau.$$

Remark: Observe that the function f_2^* is decreasing in each variable and is equimeasurable with the function itself. This definition is equivalent with the definition considered in [24] of the "iterated rearrangement" $R_{1,2}f(s,t)$, which is the two-dimensional rearrangement of f, when we first rearrange with respect to y and then with respect to x.

Definition 3. We say that a measurable function f on $\mathbb{R}^2$ belongs to the (multidimensional) Lorentz space $\Lambda^p_{1,2}(w)$, $0 < p < \infty$, provided $\|f\|_{\Lambda^p_{1,2}(w)}$, defined by

$$\|f\|_{\Lambda^p_{1,2}(w)} := \left(\int_{\mathbb{R}^2_+} (f_2^*(s,t))^p \, w(s,t) ds dt \right)^{1/p}$$

is finite. Here w is a nonnegative, locally integrable function on $\mathbb{R}^2_+$, not identically equal to 0.

Some of the properties of the two-dimensional Lorentz spaces are presented in the following theorem:

Theorem 1.

1. *If* $0 < p < \infty$, *then* $\Lambda^p_{1,2}(1) = L^p(\mathbb{R}^2)$.

2. *Let* $0 < p < \infty$. *Then* $\| \quad \|_{\Lambda^p_{1,2}(w)}$ *is a quasinorm if and only if there exists a constant* $C > 0$ *such that*

$$\int_D w(2x, 2y) dx dy \le C \int_D w(x,y) dx dy$$

for all decreasing sets $D \in R^2_+$. *Moreover, with this quasinorm,* $\Lambda^p_{1,2}(w)$ *becomes a complete quasinormed space.*

3. *If* $\Lambda^p_{1,2}(w)$ *is a Banach space, then* $p \ge 1$.

4. *Let* $1 \le p < \infty$. *Then* $\| \quad \|_{\Lambda^p_{1,2}(w)}$ *is a norm if and only* $w((A \bigcap B)^*) + w((A \bigcup B)^*) \le w(A^*) + w(B^*)$, *which is also equivalent to the fact that* $w(s,t) = v(t)$, *with* v *decreasing.*

In the classical case it is known that the weighted Lorentz spaces are normable when $p > 1$ (i.e., the defining functional is equivalent with a norm) for a larger class of weights than only decreasing ones, namely the weights which are in the class B_p. This is the class of weights which is equivalent with the boundedness of the Hardy operator on the cone of decreasing functions. Some progresses have been made in the multidimensional case, when the weight is of product type. The following theorem was proved in [12], Theorem 2.4:

Theorem 2. *Let* $w(s,t) = u(s)v(t)$ *and* $p > 1$. *The following conditions are equivalent:*

1. $\Lambda^p_{1,2}(uv)$ *is normable.*

2. u *and* $v \in B_p$.

3. *If* $\|f\|^*_{\Lambda^p_{1,2}(uv)} = \left(\int_0^\infty \int_0^\infty \left(\frac{1}{s}\int_0^s \left(\frac{1}{t} f_y^*(\cdot,\tau)d\tau\right)_x^* (\sigma)d\sigma\right)^p u(s)v(t)dsdt\right)^{1/p}$, *then* $\| \ \|^*_{\Lambda^p_{1,2}(uv)}$ *is a norm equivalent to* $\| \ \|_{\Lambda^p_{1,2}(uv)}$.

4. *The Hardy operator* $Hf(s,t) = \frac{1}{st}\int_0^s \int_0^t f(\sigma,\tau)d\sigma d\tau$ *is bounded from* $L^p_{dec}(\mathbb{R}^2_+, uv)$ *to* $L^p(\mathbb{R}^2_+, uv)$.

5. *The operator* $f \to \frac{1}{s}\int_0^s \left(\frac{1}{t} f_y^*(\cdot,\tau)d\tau\right)_x^* (\sigma)d\sigma$ *is bounded from* $L^p_{dec}(\mathbb{R}^2_+, uv)$ *to* $L^p(\mathbb{R}^2_+, uv)$.

6. *The quasinorm* $\|f\|^{**}_{\Lambda^p_{1,2}(uv)} = \| \int_0^\infty \int_0^\infty (Hf_2^*(s,t))^p u(s)v(t)dsdt)^p\|$ *is equivalent with* $\|f\|_{\Lambda^p_{1,2}(uv)}$.

Moreover, in [8] (see Theorem 3.1), the complete characterization of the boundedness of the Hardy operator $Hf(x,y) = \frac{1}{xy}\int_0^x \int_0^y f(s,t)dsdt$ on the cone of decreasing functions of two variables was given. Suppose that $f(x,y)$ is a decreasing function in each variable and w is a weight function. We first introduce some notation:

$$H_1 f(x,y) = \frac{1}{x}\int_0^x f(s,y)ds, \quad H_2 f(x,y) = \frac{1}{y}\int_0^y f(x,t)dt,$$

and $D_x^y = D\bigcap([0,x] \times [0,y])$, where D is a decreasing set. $L^p_{dec}(w)$ is the usual cone of functions in $L^p(w)$, which are decreasing in each variable. Then we have the following result:

Theorem 3. *Let* $0 < p < \infty$. *The following are equivalent conditions:*

1. $H : L^p_{dec}(w) \to L^p(w)$.

2. $H_1, H_2 : L^p_{dec}(w) \to L^p(w)$.

3. *There exists a constant* $C > 0$ *such that, for every decreasing set D*

$$\int_{\mathbb{R}^2_+ \setminus D} \frac{|D_x^y|^p}{(xy)^p} u(x,y)dxdy \leq C \int_D u(x,y)dxdy.$$

Open Problems

Now we give some problems which seem to have resisted solution up to now.

(1) Characterize the two variables weights v and w ensuring that (1.1.1) holds for any of the following three remaining cases:

(a) $1 \leq q < p < \infty$;

(b) $1 = p \leq q < \infty$;

(c) $0 < q \leq 1 \leq p < \infty$.

Remark 1: The special case with product weights is of interest also in these cases.

Remark 2: The corresponding questions as those above and in Theorem 1.1.1 are of interest and still open also for any dimension $N = 3, 4,$.

Remark 3: For the one–dimensional case Hardy inequality for $1 < p \leq q < \infty$ can be characterized by many different conditions (see, e.g., Theorem 1.1.2), even some scales of conditions (see [60]). Moreover, as seen in Chapter 1 for the case when one of the weights in (1.1.1) is of product type then (1.1.1) can be characterized by just one condition but this condition is not unique. These considerations lead to the following:

(2) For the case $1 < p \leq q < \infty$ and when one of the weights v and w in (1.1.1) is of product type: does there exist some scales of conditions (of the type as those in [60] for the one–dimensional case) which, in particular, implies all corresponding two–dimensional results in Chapter 1.

Remark 4: This question is of interest also for the cases (a)-(c) considered in Open Question (1) (concerning case (a) in the one dimensional case see [136]) .

(3) The characterization of the Hardy–type inequality in Theorem 2.6.1, part 2, is open for all parameters except those covered here. The same yields for the reversed Hardy–type inequality in Theorem 2.6.1, part 2, and the mixed case (with operators on both sides)in Theorem 2.6.1, part 1. Most of these cases are not known even for the one-dimensional case in this generality.

Remark 5: Concerning the embedding case handled in Theorem 2.6.1, part 4, the complementary information given in Theorem 2.7.1 certify that this case is more or less solved at least for absolutely continuous measures.

(4) Find necessary and sufficient conditions to characterize (2.5.15) for the case $1 < p \le q < \infty$ at least for so general cases so we as special cases obtain the one–dimensional result of this type.

(5) Establish necessary and sufficient conditions on a weight v governing the inequality

$$\|I_{\alpha,\beta} f\|_{L^q_v(\mathbb{R}^2)} \le c\|f\|_{L^p(\mathbb{R}^2)}, \quad 1 < p \le q < \infty$$

where both α and β are between 0 and $1/p$ and $I_{\alpha,\beta}$ is the Riesz potential with product kernels.

(6) Give a complete characterization of a class of weights v guaranteeing the trace inequality

$$\|\mathcal{R}_{\alpha,\beta} f\|_{L^q_v(\mathbb{R}^2_+)} \le c\|f\|_{L^p(\mathbb{R}^2_+)},$$

for the Riemann–Liouville transform $\mathcal{R}_{\alpha,\beta}$ on $\mathbb{R}^2_+$ with product kernels in the remaining cases:

(a) $0 < \alpha, \beta \le 1/p, \quad 1 < p \le q < \infty;$

(b) $\alpha = 1/p, \quad 1/p < \beta < 1, \ 1 < p \le q < \infty;$

(c) $0 < \alpha, \beta < 1, \quad 0 < q < p < \infty, \ p > 1.$

(7) Characterize completely a class of weight pairs (v, w) guaranteeing the two–weight inequality for multiple Hardy transform in Lebesgue spaces with mixed norms, for example, provided that v or w is a product of one–dimensional weights.

Remark 6. For one–weight criteria with weights of product type for maximal operators and singular integrals in mixed–normed spaces we refer to [105].

(8) To prove or disprove the Fefferman–Stein inequality

$$\iint\limits_{\mathbb{R}^2} ((M_S f)(x,y))^p v(x,y)dxdy \le c \iint\limits_{\mathbb{R}^2} |f(x,y)|^p (M_S v)(x,y)dxdy, \ 1 < p < \infty,$$

where M_S is the strong maximal operator defined on the plane. Similar inequality for the strong fractional maximal operators was proved in Chapter 4 (see Theorem 4.2.2). For the Fefferman–Stein weak type inequality we refer the reader to [122] (see also Notes and Comments on Chapter 5).

Remark 7. By using the iteration and Fubini's theorem it is easy to prove the inequality

$$\iint_{\mathbb{R}^2} ((M_1M_2)f(x,y))^p v(x,y)dxdy \le c \iint_{\mathbb{R}^2} |f(x,y)|^p((M_2M_1)v)(x,y)dxdy,$$

where $1 < p < \infty$, the positive constant c does not depend on f and v, M_1M_2 is the iterated maximal operator on the plane defined at the beginning of Chapter 5 and M_2M_1 is defined as follows:

$$(M_2M_1)f(x,y) = \sup_{y\neq y_1} \frac{1}{y-y_1}\int_{y_1}^{y}\left(\sup_{x\neq x_1}\frac{1}{x-x_1}\int_{x_1}^{x}|f(s,\tau)|ds\right)d\tau,$$

where $x_1, x, y_1, y \in \mathbb{R}$.

(9) Solve the two–weight problem for the following two–dimensional Hardy–Steklov operator

$$T_2f(x,y) := \int_{\varphi_1(x)}^{\varphi_2(x)}\int_{\psi_1(y)}^{\psi_2(y)} f(t,\tau)dtd\tau$$

under the appropriate conditions on φ_1, φ_2, ψ_1 and ψ_2. As regards the Hardy–Steklov operator $T_2f(x) = \int_{a(x)}^{b(x)} f(t)dt$, its one–dimensional boundedness was studied in [70].

(10) Find necessary and sufficient condition on a weight v defined on $\mathbb{R}^{1+2}_+ := \mathbb{R}_+ \times \mathbb{R}^2$ for which the trace inequality

$$\left(\int_{\mathbb{R}^{1+2}_+} |T_{\alpha,\beta}f(x,y,t)|^q v(x,y,t)dxdydt\right)^{1/q}$$

$$\le c\left(\int_{\mathbb{R}^2}|f(x,y)|^p dxdy\right)^{1/p}, \quad 0<\alpha,\beta<1, \quad 1<p<q<\infty,$$

holds, where

$$T_{\alpha,\beta}f(x,y,t) := \iint_{\mathbb{R}^2} \frac{f(s,\tau)}{(|s-x|+t)^{1-\alpha}(|\varepsilon-y|+t)^{1-\beta}}dsd\varepsilon.$$

Solve the same problem for the related maximal functions:

$$\mathcal{M}_{\alpha,\beta}f(x,y,t) := \sup_{\substack{I\ni x, J\ni y\\ |I|,|J|>t}} \frac{1}{|I|^{1-\alpha}|J|^{1-\beta}}\iint_{I\times J}|f(s,\varepsilon)|dsd\varepsilon, \quad 0\le\alpha,\beta<1.$$

For the solution of the two–weight problem for the appropriate operators with single kernel we refer to [53, Chapter 3] and references therein.

Bibliography

[1] S. Abramovich, S. Banić and M. Matić, Superquadratic functions in several variables, *J. Math. Anal. Appl.* **327** (2007), No. 2, 1444–1460.

[2] S. Abramovich, G. Jameson and G. Sinnamon, Refining of Jensen's inequality, *Bull. Math. Soc. Sci. Math. Roumanie* (N.S.) **47**(95) (2004), 3–14.

[3] S. Abramovich, G. Jameson and G. Sinnamon, Inequalities for averages of convex and superquadratic functions, *J. Inequal. Pure Appl. Math.* **5** (2004), No. 4, Art. 91 [ONLINE: http//jipam.vu.edu.au/].

[4] D. R. Adams, A trace inequality for generalized potentials, *Studia Math.* **48** (1973), 99–105.

[5] D. R. Adams, *Lectures on L^p-potential theory*, Umeå Univ. Reports, No. 2, 1981.

[6] D. R. Adams and L. I. Hedberg, *Function spaces and potential theory*, Springer–Verlag, Berlin, 1996.

[7] K. Andersen and E. Sawyer, Weighted norm inequalities for the Riemann–Liouville and Weyl fractional integral operators, *Trans. Amer. Math. Soc.* **308** (1988), No. 2, 547–558.

[8] N. Arcozzi, S. Barza, J. L. Garcia-Domingo and J. Soria, Hardy's inequalities for monotone functions on partially ordered measure spaces, *Proc. Roy. Soc. Edinburgh Sect.* A **136** (2006), No. 5, 909–919.

[9] S. Barza, *Weighted multidimensional integral inequalities and applications*, PhD Thesis, Department of Mathematics, Luleå University of Technology, 1999.

[10] S. Barza, Multidimensional integral inequalities with homogeneous weights, *Gene. Math.* **13** (2005), No. 1, 73–82.

[11] S. Barza, P. H. Heinig and L.–E. Persson, Duality theorem over the cone of monotone functions and sequences in higher dimensions, *J. Inequal. Appl.* **7** (2002), No. 1, 79–108.

[12] S. Barza, A. Kaminska, L.–E. Persson and J. Soria, Mixed norm and multidimensional Lorentz spaces, *Positivity* **10** (2006), No. 3, 539–554.

[13] S. Barza, M. Johansson and L.–E. Persson, A Sawyer duality principle for radially monotone functions in $\mathbb{R}^n$, *J. Inequal. Pure Appl. Math.* **6** (2005), No. 2, Article 44, 13 pp.

[14] S. Barza, and L.–E. Persson, Weighted multidimensional inequalities for monotone functions, *Math. Bohemica* **124** (1999), 329–335.

[15] S. Barza, J. Pecaric and L.–E. Persson, Reversed H ölder type inequalities for monotone functions of several variables, *Math. Nachr.* **186** (1997), 67–80.

[16] S. Barza, L.–E. Persson and J. Soria, Sharp weighted multidimensional integral inequalities for monotone functions, *Math. Nachr.* **210** (2000), 43–58.

[17] S. Barza, L.–E. Persson and J. Soria, Sharp weighted multidimensional integral inequalities of Chebyshev type, *J. Math. Anal. Appl.* **236** (1999), No. 2, 243–253.

[18] S. Barza, L.–E. Persson and J. Soria, Multidimensional rearrangements and Lorentz spaces, *Acta Math. Hungar.* **104** (2004), No. 3, 203–224.

[19] S. Barza, L.–E. Persson and V. D. Stepanov, On weighted multidimensional embeddings for monotone functions, *Math. Scand.* **88** (2001), No. 2, 303–319.

[20] G. Bennett, Some elementary inequalities, *Quart. J. Math. Oxford* **(2) 38** (1987), No. 152, 401–425.

[21] G. Bennett, Some elementary inequalities, *Quart. J. Math. Oxford* **(2) 39** (1988), No. 156, 401–425.

[22] J. Bergh, V. I. Burenkov and L.–E. Persson, Best constants in Reversed Hardy's Inequalities for Quasimonotone functions, *Acta Sci.Math.* (Szeged) **59** (1994), No. 1–2, 221–239.

[23] S. Bloom and R. Kerman, Weighted norm inequalities for operators of Hardy type, *Proc. Amer. Math. Soc.* **113** (1991), No. 1, 135–141.

[24] A.P. Blozinski, Multivariate rearrangements and Banach function spaces with mixed norms, *Trans. Amer. Math. Soc.,* **263** (1981), 149–167.

[25] L. Carleson, An interpolation problem for bounded analytic functions, *Amer. J. Math.* **80** (1958), 921–930.

[26] M. J. Carro and J. Soria, Weighted Lorentz spaces and the Hardy operator, *J. Funct. Anal.* **112** (1993), No. 2, 480–494.

[27] S. Chanillo, D. K. Watson and R. L. Wheeden, Some integral and maximal operators related to starlike sets, *Studia Math.* **107** (1993), No. 3, 223–255.

[28] A. Čižmešija, J. Oguntuase and L. E. Persson, Multidimensional Hardy-type inequalities via convexity, *Bull. Austral. Math.* Soc. **77** (2008), 245–260.

[29] A. Čižmešija and J. Pečarić, Some new generalisations of inequalities of Hardy and Levin–Cochran–Lee, *Bull. Austral. Math. Soc.* **63** (2001), No. 1, 105–113.

[30] A. Čižmešija and J. Pečarić, Multivariable mixed means and inequalities of Hardy and Levin–Cochran–Lee type, *Math. Inequal. Appl.* **5(3)** (2002), 397–415.

[31] A. Čižmešija, J. Pečarić and L.–E. Persson, On strengthened Hardy and Pólya–Knopp's inequalities, *J. Approx. Theory* **125** (2003), No. 1, 74–84.

[32] A. Čižmešija, L.E. Persson and A. Wedestig, Weighted integral inequalities for Hardy and geometric mean operators with kernals over cones in R^n, *Italian J. Pure Appl. Math.* **18** (2005), 89–118.

[33] J. A. Cochran and C.-S. Lee, Inequalities related to Hardy's and Heinig's, *Math. Proc. Cambridge Philos. Soc.* **96** (1984), No. 1, 1–7.

[34] R. R. Coifman and C. Fefferman, Weighted norm inequalities for maximal functions and singular integrals, *Studia Math.* **51** (1974), 241–249.

[35] D. Cruz-Uribe, SFO, J. M. Martell and C. Perez, Sharp two–weight inequalities for singular integrals, with applications to the Hilbert transform and the Sarason conjecture, *Adv. Math.* **216** (2007), No. 2, 647–676.

[36] E. M. Dynkin and B. P. Osilenker, Weighted inequalities for singular integrals and their applications (Russian), *Math Surveys Ser. Math. Anal. Vol.* **21**, *VINITI, Moscow,* 1983.

[37] D. E. Edmunds and V. Kokilashvili, Two–weight inequalities for singular integrals, *Canadian Math. Bull.* **38** (1995), 119–125.

[38] D. E. Edmunds and V. Kokilashvili, On Fourier multipliers, *J. Anal. Appl.* **1** (2003), No. 3, 143–155.

[39] D. E. Edmunds, V. Kokilashvili and A. Meskhi, *Bounded and compact integral operators*, Kluwer Academic Publishers, Dordrecht, Boston, London, 2002.

[40] D. E. Edmunds, V. Kokilashvili and A. Meskhi, On Fourier multipliers in weighted Triebel–Lizorkin spaces, *J. Inequal. Appl.* **7** (2002), No. 4, 555–591.

[41] C. Fefferman and E. Stein, Some maximal inequalities, *Amer. J. Math.* **93** (1971), 107–115.

[42] R. Fefferman, Strong differentiation with respect to measures, *Amer. J. math.* **103** (1981), 33–40.

[43] R. Fefferman, Multiparameter Fourier analysis, In : *Beijing Lectures in Harmonic Analysis, Beijing,* (Ed. E. M. Stein), pp. 47–130, 1984, *Ann. Math. Stud.*, Vol. **112**, *Princeton Univ. Press, Princeton–NJ,* 1986.

[44] R. Fefferman and E. M. Stein, Singular integrals on product spaces, *Adv. in Math.* **45** (1982), 117–143.

[45] M. Gabidzashvili, Weighted inequalities for anisotropic potentials, (Russian) *Trudy Tbiliss. Mat. Inst. Razmadze Akad. Nauk Gruzin. SSR* **82** (1986), 25–36.

[46] M. Gabidzashvili and V. Kokilashvili, Two weight weak type inequalities for fractional type integrals, *Preprint, No.* **45**, *Mathematical Institute Czech Acad. Sci., Prague,* 1989.

[47] M. Gabidzashvili, I. Genebashvili and V. Kokilashvili, Two–weight inequalities for generalized potentials, (Russian) *Trudy Mat. Inst. Steklov* **194** (1992), 89–96. English transl. in: *Proc. Steklov Inst. Mat.* **4** (1993), 91–99.

[48] I. Gabisonija and A. Meskhi, Two–weight inequalities for a discrete Hilbert transform, *Proc. A. Razmadze Math. Inst* **116** (1998), 107–122.

[49] J. García–Cuerva and J. L. Rubio de Francia, *Weighted norm inequalities and related topics*, North-Holland Mathematics Studies, **116**, Mathematical Notes, 104, North-Holland Publishing Co., Amsterdam, 1985.

[50] J. L. Garcia-Domingo, Embeddings for monotone functions and Bp weights, between discrete and continuous, *Math. Scand.* **97** (2005), No. 2, 281–297.

[51] J. B. Garnett, *Bounded analytic functions*, Academic Press, 1981.

[52] I. Genebashvili, A. Gogatishvili and V. Kokilashvili, Solution of two–weight problems for integral transforms with positive kernels, *Georgian Math. J.* **3** (1996), No. 1, 319–342.

[53] I. Genebashvili, A. Gogatishvili, V. Kokilashvili and M. Krbec, *Weight theory for integral transforms on spaces of homogeneous type*, Pitman Monographs and Surveys in Pure and Applied Mathematics, **92**, Longman, Harlow, 1998.

[54] V. Georgiev, Weighted estimates for the wave equation. *Nonlinear waves, Proceedings of the Fourth MSJ International Research Institute,* Vol. 1, 71–80. *Hokkaido Univ.*, 1996.

[55] V. Georgiev, H. Lindblad and C. Sogge, Weighted Strichartz estimates and global existence for semilinear wave equations, *American J. Math.* **119** (1997), 1291–1319.

[56] V. Georgiev and A. Meskhi, On a weighted Strichartz estimate for inhomogeneous wave equations, *Proc. A. Razmadze Math. Inst.* **139** (2005), 96–99.

[57] E. K. Godunova, Inequalities based on convex functions, (Russian)*Izv. Vysh. Uchebn. Zaved. Matematika,* **47(4)** (1965), 45–53; *English transl. in Amer. Math. Soc., Transl., II Ser.* **88** (1970), 57–66.

[58] E. K. Godunova, Integral inequalities with convex functions, (Russian) *Izv. Vysš. Učebn. Zaved. Matematika* **1968**, No. 1, (68), 47–49.

[59] A. Gogatishvili and V. Kokilashvili, Criteria of two–weight strong type inequalities for fractional maximal functions, *Georgian Math. J.* **3** (1996), No. 5, 423–446.

[60] A. Gogatishvili, A. Kufner, L.–E. Persson and A. Wedestig, An equivalence theorem for integral conditions related to Hardy's inequality, *Real Anal. Exchange* **29** (2003/04), No. 2, 867–880.

[61] R. R. Goldberg, *Fourier Transforms*, Cambridge University Press, 1961.

[62] A. Guven and V. Kokilashvili, On the mean summability by Cesaro method of Fourier trigonometric series in two–weighted setting, *J. Inequal. Appl.* Article ID 41837, 15 pages, 2006.

[63] A. Guven, V. Kokilashvili, Two–weight estimates for Fourier Operators and Bernstein inequality, *Studia Sci. Math. Hungarica,* DOI: 10.1556/SScMath. 2009,1109, 1-23.

[64] A. Guven, V. Kokilashvili, On the means of Fourier integrals and Bernstein inequality in two–weighted setting, *Positivity* (to appear).

[65] E. A. Haciyeva, *Investigation of the properties of functions with quasimnotone Fourier coefficients in generalized Nikolskii-Besov spaces,* (Russian) Authors Summary of Candidates Dissertation, Tbilisi, 1986.

[66] G. H. Hardy, Note on a theorem of Hilbert, *Math. Z.* **6** (1920), No. 3-4, 314–317.

[67] G. H. Hardy, Notes on some points in the integral calculus, LX. *An inequality between integrals, Messenger of Math.* **54** (1925), 150–156.

[68] G. H. Hardy, J. E. Littlewood and G. Pólya, *Inequalities*, Cambridge University Press, Cambridge, 1959.

[69] H. P. Heinig, R. Kerman and M. Krbec, Weighted exponential inequalities, *Georgian Math. J.* **8** (2001), No. 1, 69–86.

[70] H. P. Heinig and G. Sinnamon, Mapping properties of integral averaging operators, *Studia Math.* **129** (1998), 157–177.

[71] L. Hörmander, L^p estimates for (pluri-) subharmonic functions, *Math. Scand.* **20** (1967), 65–78.

[72] D. A. Hunt, B. Muckenhoupt and R. L. Wheeden, Weighted norm inequalities for the conjugate function and Hilbert transform, *Trans. Amer. Math. Soc.* **176** (1973), 227–251.

[73] C. O. Imoru, On some integral inequalites related to Hardy's, *Canad. Math. Bull.* **20** (1977), No. 3, 307–312.

[74] D. M. Israfilov, A. Guven, Approximation by trigonometric polynomials in weighted Orlicz spaces, *Studia Math.* **174** (2006), 147–168.

[75] P. Jain and R. Hassija, Some remarks on two dimensional Knopp type inequalities, *Appl. Math. Lett.* **16** (2003), No. 4, 459–464.

[76] P. Jain, L.E. Persson and A. Wedestig, From Hardy to Carleman and general mean-type inequalities, In: *Function Spaces and Applications, Narosa Publishing House, New Delhi*, 2000, 117–130.

[77] M. Johansson, *Carleman type inequalities and Hardy type inequalities for monotone functions*, PhD. Thesis, Department of Mathematics, Luleå University of Technology, 2007.

[78] F. John, Blow-up of solutions of nonlinear wave equations in three space dimensions, *Manuscripta Math.* **28** (1979), No. 1-3, 235–268.

[79] S. Kaijser, L. Nikolova, L.–E. Persson and A. Wedestig, Hardy-type inequalities via convexity, *Math. Inequal. Appl.* **8** (2005), No. 3, 403–417.

[80] S. Kaijser, L.–E. Persson and A. Öberg, On Carleman and Knopp's inequalities, *J. Approx. Theory* **117** (2002), No. 1, 140–151.

[81] J. Kato and T. Ozawa, Weighted Strichartz estimates for the wave equation in even space dimensions. *Math. Z.* **247** (2004), No. 4, 747–764.

[82] K. Knopp, Uber Reihen mit positiven Gliedern, *J. London Math. Soc.* **3** (1928), 205–211.

[83] V. Kokilashvili, On Approximation of periodic functions, (Russian) *Trudy Tbilisi Mat. Inst. im Razmadze Akad. Nauk. Gruz. SSR* **34** (1968), 51–81.

[84] V. M. Kokilashvili, On Hardy's inequalities in weighted spaces, (Russian) *Soobsch. Akad. Nauk Gruz. SSR* **96** (1979), 37–40.

[85] V. M. Kokilashvili, Bisingular integral operators in weighted spaces, (Russian) *Soobshch. Akad. Nauk Gruzin. SSR* **101** (1981), No. 2, 289–292.

[86] V. M. Kokilashvili, Weighted Lizorkin–Triebel spaces. Singular integrals, multipliers, imbedding theorems, (Russian) in: *Studies in the Theory of Differentiable Functions of Several Variables and Its Applications* IX, *Trudy Mat. Inst. Steklov* **161** (1983), 125–149. English transl. In: *Proc. Steklov Inst. Math.* **3** (1984), 135–162.

[87] V. Kokilashvili, New aspects in weight theory. In: *Function Spaces, Differential Operators and Nonlinear Analysis. Prometeus Publishing House, Prague,* 1996, 51–70.

[88] V. Kokilashvili and M. Krbec, *Weighted inequalities in Lorentz and Orlicz spaces*, World Scientific, Singapore, New Jersey, London, Hong Kong, 1991.

[89] V. Kokilashvili and A. Meskhi, On a trace inequality for one–sided potentials and applications to the solvability of nonlinear integral equations, *Georgian Math. J.* **8** (2001), No. 3, 521–536.

[90] V. Kokilashvili and A. Meskhi, Boundedness and compactness criteria for the generalized truncated potentials, (Russian) *Proc. Steklov Math. Inst. RAN* **232** (2001), 164–178.

[91] V. Kokilashvili and A. Meskhi, On a trace inequality for one–sided potentials with multiple kernels, *Frac. Calc. Appl. Anal.* **6** (2003), No. 4, 461–472.

[92] V. Kokilashvili and A. Meskhi, On one–sided potentials with multiple kernels, *Integr. Transf. Spec. Funct.* **16** (2005), No. 8, 669–683.

[93] V. Kokilashvili and A. Meskhi, A note on two–weight estimates for strong fractional maximal functions and potentials with product kernels, *Proc. A. Razmadze Math. Inst.* **137** (2005), 135–140.

[94] V. Kokilashvili and A. Meskhi, Two–weighted criteria for integral transforms with multiple kernels, In : *Proceedings of Conference on the Occasion of the* 70 *Anniversary of Professor Zbigniew Ciesielski, Bedlewo,* 20–24 *September,* 2004, *Institute of Mathematics, Polish Academy of Sciences, Banach Center publications* **72** (2006), 119–140.

[95] V. Kokilashvili and A. Meskhi, Two–weight Estimates for Strong Maximal Functions, Potentials and Singular Integrals with Multiple Kernels, *Proceedings of Conference on the Occasion of the 70 Anniversary of Professor Ryskul*

Oinarov, Astana, Kazakhstan, September 3–5, 2007. Published in: *Eurazian Mathematical Journal* No. 2, 2007, 31–46.

[96] V. Kokilashvili and A. Meskhi, Two–weight estimates for strong fractional maximal functions and potentials with multiple kernels, *J. Korean Math. Soc.* **46**(2009), No.3, 523–550. .

[97] V. Kokilashvili, A. Meskhi and Ts. Tsanava, Strong and iterated maximal functions, and applications to the mean summability of the double trigonometric Fourier series, *Proc. A. Razmadze Math. Inst.* **139** (2005), 132–136.

[98] P. Koosis, *Introduction to H_p Spaces*, Cambridge University Press, 1998.

[99] K. Krulić, J. Pečarić and L.–E. Persson, Some new Hardy type inequalities with general kernels, *Math. Inequal. Appl.* (to appear).

[100] A. Kufner, L. Maligranda and L.–E. Persson, *The Hardy inequality–about its history and some related results*, Vydavatelsky Servis Publishing House, Pilsen, 2007.

[101] A. Kufner, L. Maligranda and L.–E. Persson, The prehistory of the Hardy inequality, *Amer. Math. Monthly* **113** (2006), No. 8, 715–732.

[102] A. Kufner and L.–E. Persson, *Weighted inequalities of Hardy type,* World Scientific Publishing Co, Singapore, New Jersey, London, Hong Kong, 2003.

[103] D. S. Kurtz, Littlewood–Paley and multiplier theorems on weighted L^p wpaces, *PhD. Dissertation, Rutgers University*, 1978.

[104] D. S. Kurtz, Littlewood–Paley and multiplier theorems in weighted L^p spaces, *Trans. Amer. Math. Soc.***259** (1980), 235–263.

[105] D. S. Kurtz, Classical operators on mixed–normed spaces with product weights, *Rocky Mountain J. Math.* **37** (2007), No. 1, 2007.

[106] H. Lindblad and C. Sogge, Long-time existence for small amplitude semilinear wave equations, *Amer. J. Math.* **118** (1996), 1047–1135.

[107] P. I. Lizorkin, (L_p, L_q)–multipliers of Fourier integrals, (Russian) *Dokl. Akad. Nauk SSSR* **152** (1963), 808–811, Engl. transl. in *Soviet Math. Dokl.* **4** (1963), 1420–1424.

[108] P. Lizorkin, On multipliers of Fourier integrals in the spaces $L_{p,\theta}$, (Russian) *Tr. Mat. Inst. Steklova* **89** (1967), 231–248. English Transl. in *Proc. Steklov Inst. Math.* **89** (1967), 269–290.

[109] M. Lorente, A characterization of two–weight norm inequalities for one–sided operators of fractional type, *Canad. J. Math.* **49** (1997), No. 5, 1010–1033.

[110] E. R. Love, Inequalities related to those of Hardy and of Cochran and Lee, *Math. Proc. Cambridge Philos. Soc.* **99** (1986), No. 3, 395–408.

[111] F. J. Martin–Reyes, Weights, one–sided operators, singular integrals, and ergodic theorems, *In: Nonlinear Analysis, Function Spaces and Applications,* Vol **5**, *Proceedings, M. Krbec et. al. (eds), Prometheus Publishing House, Prague,* 1994, 103–138.

[112] F. J. Martin-Reyes, L. Pick and A. de la Torre, A_∞^+– condition, *Canad. J. Math.* **45** (1993), No. 6, 1231–1244.

[113] F. J. Martin–Reyes and A. de la Torre, Two–weight norm inequalities for fractional one–sided maximal operators, *Proc. Amer. Math. Soc.* **117** (1993), 483–489.

[114] F. J. Martin–Reyes and E. Sawyer, Weighted inequalities for Riemann–Liouville fractional integrals of order one and greater, *Proc. Amer. Math. Soc.* **106** (1989), 727–733.

[115] V. G. Maz'ya, Sobolev spaces, *Springer, Berlin*, 1985.

[116] V. G. Maz'ya and I. Verbitsky, Capacitary inequalities for fractional integrals, with applications to partial differential equations and Sobolev multipliers, *Ark. Mat.* **33** (1995), 81–115.

[117] A. Meskhi, Solution of some weight problems for the Riemann–Liouville and Weyl operators, *Georgian Math. J.* **5** (1998), No. 6, 565–574.

[118] A. Meskhi, On the boundedness and compactness of ball fractional integral operators, *Fract. Calc. Appl. Anal.* **3** (2000), No. 1, 13–30.

[119] A. Meskhi, Criteria for the boundedness and compactness of integral transforms with positive kernels, *Proc. Edinburgh Math. Soc.* **44** (2001), No. 2, 267–284.

[120] A. Meskhi, A note on two–weight inequalities for multiple Hardy-type operators, *J. Funct. Spaces Appl.* **3** (2005), 223–237.

[121] A. Meskhi, Two–weight inequalities for multiple Hardy-type operators, *Proc. A. Razmadze math. Inst.* **136** (2004), 149–153.

[122] T. Mitsis, The weighted weak type inequality for the strong maximal function, *J. Fourier Anal. Appl.* **12** (2006), No. 6, 645–652.

[123] B. Muckenhoupt, Hardy's inequality with weights, *Studia Math.* **4** (1972), 31–38.

[124] B. Muckenhoupt, Weighted norm inequalities for the Hardy maximal function, *Trans. Amer. Math. Soc.* **165** (1972), 207–226.

[125] B. Muckenhoupt, Two weight function norm inequalities for the Poisson integral, *Trans. Amer. Math. Soc.* **210** (1975), 225–231.

[126] B. Muckenhoupt and R. L. Wheeden, Weighted norm inequalities for fractional integrals, *Trans. Amer. Math. Soc.* **192** (1974), 261–276.

[127] B. Muckenhout and R. L. Wheeden, Two–weight function norm inequalities for the Hardy–Littlewood maximal function and the Hilbert transform, *Studia Math.* **55** (1976), No. 3, 279–294.

[128] A. Nagel and E. M. Stein, On the product theory of singular integrals, *Rev. Mat. Iberoamericana* **20** (2004), 531–561.

[129] A. D. Nakhman, B. P. Osilenker, Estimates of weighted norms of some operators generated by multiple trigonometric Fourier series, (Russian) *Izvestiya Vysshikh Uchebnykh Zavedeniĭ. Matematika* **239** (1982), No. 4, 39–50.

[130] F. Nazarov, S.Treil and A. Volberg, Two–weight estimate for the Hilbert transform and corona decomposition for non doubling measures, *Preprint,* 2004, pp. 1–38.

[131] C. Niculescu and L.–E. Persson, *Convex functions and their applications. A contemporary approach,* CMS Books in Mathematics/Ouvrages de Mathématiques de la SMC **23**. Springer, New York, 2006.

[132] J. Oguntuase, C. Okpoti, L.E. Persson and F. Allotey, Weighted multidimensional Hardy type inequalities via Jensen's inequality, *Proc. A. Razmadze Inst.* **144** (2007), 91–105.

[133] J. Oguntuase, C. Okpoti and L.–E. Persson Multidimensional Hardy type inequalities for $p < 0$ and $0 < p < 1$, *J. Math. Inequal.* **1** (2007), No. 1, 1–11.

[134] J. A. Oguntuase and L.–E. Persson, *Refinement of Hardy's inequalities via superquadratic and subquadratic functions, J. Math. Anal. Appl.* **339** (2008), 1305–1312.

[135] J. A. Oguntuase, L.–E. Persson and E. K. Essel, Multidimensional Hardy–type inequalities with general kernels, *J. Math. Anal. Appl.* **348**(2008), 411–418.

[136] J. A. Oguntuase, L.–E. Persson, E. K. Essel and B. A. Popoola, Refined multidimensional Hardy–type inequalities via superquadracity, *Banach J. Math. Anal.* **2** (2008), No. 2, 129–139.

[137] C. Okpoti, L.E. Persson and G. Sinnamon, An equivalence theorem for some integral conditions with general measures related to Hardy's inequality, *J. Math. Anal. Appl.* **326** (2007), No. 1, 398–413.

[138] C. Okpoti, L.E. Persson and G. Sinnamon, An equivalence theorem for some integral conditions with general measures related to Hardy's inequality II, *J. Math. Anal. Appl.* **337** (2008), No. 1, 219–230.

[139] B. Opic and P. Gurka, Weighted inequalities for geometric means, *Proc. Amer. Math. Soc.* **120** (1994), No. 3, 771–779.

[140] B. Opic and A. Kufner, *Hardy-type inequalities*, Pitman series, **219**, Harlow, 1990.

[141] J. Pečarić, I. Perić, and L.–E. Persson, A multidimensional integral inequality for monotone functions of several variables, *Acta Sci.Math.* (Szeged) **62** (1996), No. 3–4, 407–412.

[142] L.–E. Persson and V.D. Stepanov, Weighted integral inequalities with the geometric mean operator, *J. Inequal. Appl.* **7** (2002), 727–746.

[143] L.–E. Persson, V. Stepanov and P. Wall, Some scales of equivalent weight characterizations of Hardy's inequality: the case $q < p$, *Math. Inequal. Appl.* **10** (2007), No. 2, 267–279.

[144] L.–E. Persson, V. Stepanov and E. Ushakova, Equivalence of Hardy-type inequalities with general measures on the cones of non-negative respective non-increasing functions, *Proc. Amer. Math. Soc.* **134** (2006), No. 8, 2363–2372.

[145] L.–E. Persson and E. Ushakova, Some multi-dimensional Hardy type integral inequalities, *Math. Inequal. Appl.* Vol. 1(2007), No. 3, 301–319.

[146] D. V. Prokhorov, On the boundedness of a class of integral operators, *J. London Math. Soc.*, **61** (2000), No. 2, 617–628.

[147] D. V. Prokhorov and V. Stepanov, Weighted estimates for the Riemann–Liouville operators and applications, *Trudy Mat.Inst. Steklov.* **243** (2003), 289–312.

[148] Y. Rakotontratsimba, Two–weight norm inequality for Calderón–Zygmund operators, *Acta Math. Hungar.* **80** (1998), 39–54.

[149] M. Rosenblum, Summability of Fourier series in $L^p(d\mu)$, *Trans. Amer. Math. Soc.* **165** (1962), No. 1, 32–42.

[150] W. Rudin, *Real and Complex Analysis*, 3rd edition, McGraw-Hill, 1987.

[151] S. G. Samko, A. A. Kilbas and O. I. Marichev, *Fractional Integrals and Derivatives. Theory and Applications*, Gordon and Breach Sci. Publishers, New York, 1993.

[152] E. T. Sawyer, Multipliers of Besov and power weighted L^2 spaces, *Indiana Univ. Math. J.* **33(3)** (1984), 353–366.

[153] E. T. Sawyer, Weighted norm inequalities for fractional maximal operators, in: *Seminar on Harmonic Analysis,Montreal, Que.*, 1980, pp. 283–309, *CMS Conf. Proc., 1, Amer. Math. Soc.Providence, R.I.*, 1981.

[154] E. T. Sawyer, A characterization of a two–weight norm inequality for maximal operators, *Studia Math.* **75** (1982), 1–11.

[155] E. T. Sawyer, Two weight norm inequalities for certain maximal and integral operators, in: *Harmonic analysis, Minneapolis, Minn.* 1981, pp. 102–127, *Lecture Notes in Math.* Vol. **908**, *Springer, Berlin, New York*, 1982.

[156] E. T. Sawyer, A two–weight weak type inequality for fractional integrals, *Trans. Amer. Math. Soc.* **281** (1984), 339–345.

[157] E. Sawyer, Weighted inequalities for two-dimensional Hardy operator, *Studia Math.* **82** (1985), No. 1, 1–16.

[158] E. T. Sawyer, A characterization of two weight norm inequalities for fractional and Poisson integrals, *Trans. Amer. Math. Soc.* **308** (1988), 533–545.

[159] E. T. Sawyer and R. L. Wheeden, Carleson conditions for the Poisson integral, *Indiana Univ. Math. J.* **40** (1991), No. 2, 639–676.

[160] E. T. Sawyer and R. L. Wheeden, Weighted inequalities for fractional integrals on Euclidean and homogeneous spaces, *Amer. J. Math.* **114** (1992), 813–874.

[161] E. T. Sawyer, R. L. Wheeden and S. Zhao, Weighted norm inequalities for operators of potential type and fractional maximal functions, *Potential Anal.* **5** (1996), 523–580.

[162] E. M. Stein, G. Weiss, *Introduction to Fourier analysis on Euclidean spaces*, Princeton University Press, 1971.

[163] V. D. Stepanov, Two–weight estimates for Riemann-Liouville integrals, (Russian) *Izv. Akad. Nauk SSSR Ser. Mat.* **54** (1990), No. 3, 645–656; translation in *Math. USSR-Izv.* **36** (1991), No. 3, 669–681.

[164] V. Stepanov, The weighted Hardy's inequality for nonincreasing functions, *Trans. Amer. Math. Soc.***338** (1993), 173–186.

[165] V. D. Stepanov, Weighted norm inequalities for integral operators and related topics. In: *Nonlinear Analysis, Function spaces and Applications,* **5**, 139–176, *Olimpia Press, Prague,*1994.

[166] W. Strauss, Nonlinear scattering theory, *Scattering Theory in Mathematical Physics*, 53–79. *Reidel, Dordrecht,* 1979.

[167] W. Strauss, Nonlinear scattering theory at low energy. *J. Funct. Anal.* **41** (1981), No. 1, 110–133.

[168] J. O. Strömberg and A. Torchinsky, *Weighted Hardy spaces*, Lecture Notes in Math., Vol. **1381**, Springer Verlag, Berlin, 1989.

[169] K. Tachizava, On weighted dyadic Carleson's inequalities, *J. Inequal. Appl.* **6** (2001), No. 4, 415–433.

[170] A. F. Timan, *Theory of approximation of functions of a real variable*, Pergammon Press, 1963.

[171] E. Ushakova. *Norm inequalities of Hardy and Pólya–Knopp types. PhD. Thesis*, Department of Mathematics, Luleå University of Technology, 2006.

[172] I. E. Verbitsky and R. L. Wheeden, Weighted norm inequalities for integral operators, *Trans. Amer. Math. Soc.* **350** (1998), 3371–3391.

[173] A. Volberg, *Calderón–Zygmund capacities and operators on nonhomogeneous spaces*, CBMS Regional Conference Series in Mathematics, Vol. **100**, 2003.

[174] A. Wedestig, *Weighted inequalities of Hardy-type and their limiting inequalities*, PhD Thesis, Department of Mathematics, Luleå University of Technology, 2003.

[175] A. Wedestig, Weighted inequalities for the Sawyer two-dimensional Hardy operator and its limiting geometric mean operator, *J. Inequal. Appl.* **4** (2005), 387–394.

[176] R. L. Wheeden, A characterization of some weighted norm inequalities for the fractional maximal functions, *Studia Math.* **107** (1993), 251–272.

[177] A. Zygmund, *Trigonometric Series*, Vols. **I-II**, Cambridge Univ. Press, 2nd edition, 1959.

Index

A

B

C

D

E

F

G

H

I

K

L

M

N

O

P

R

S

T

U

V

W

Y